Corrosion

Material Performance and Cathodic Protection

Corrosion

Material Performance and Cathodic Protection

Edited by

S. Syngellakis
Wessex Institute, UK

WIT*PRESS* Southampton, Boston

Editor:

S. Syngellakis
Wessex Institute, UK

Published by

WIT Press
Ashurst Lodge, Ashurst, Southampton, SO40 7AA, UK
Tel: 44 (0) 238 029 3223; Fax: 44 (0) 238 029 2853
E-Mail: witpress@witpress.com
http://www.witpress.com

For USA, Canada and Mexico

Computational Mechanics International Inc
25 Bridge Street, Billerica, MA 01821, USA
Tel: 978 667 5841; Fax: 978 667 7582
E-Mail: infousa@witpress.com
http://www.witpress.com

British Library Cataloguing-in-Publication Data

A Catalogue record for this book is available
from the British Library

ISBN: 978-1-78466-249-3
eISBN: 978-1-78466-250-9

Library of Congress Catalog Card Number: 2017938203

Printed by Lightning Source, UK.

Preface

Corrosion is a degrading material process frequently encountered in engineering structures and components, which may lead to costly and catastrophic failures if not properly and timely addressed. The significance of the phenomenon prompted the collection of selected articles dedicated to various areas of corrosion research from the Transactions of the Wessex Institute.

Corrosion investigations are generally concerned with material resistance to the phenomenon. They may either characterise it, or propose methods for improving material performance. The collected articles address issues, which can be categorised into two main areas: the first is concerned with material science approaches to corrosion, that is, visual or instrumental means of assessing existing behaviour or effectiveness of corrective measures and techniques. The second part deals exclusively with numerical simulations, largely based on the boundary element method, of cathodic protection schemes for the purpose of predicting and optimising their performance.

The volume begins with a group of papers using visual inspection combined with advanced detection techniques such as optical microscopy, X-ray fluorescence, X-ray diffraction, ultrasonic scanning, electron scanning, and electron probe microanalysis. Saeed *et al.* examine corrosion in historical military vehicles focusing on elemental identification and material loss while, incidentally, providing an overview of the various modes of corrosion. McIntyre *et al.* detect strain fields arising from stress corrosion cracking in a nickel alloy exposed to hydrothermal, controlled electrochemical conditions. Two subsequent contributions assess corrosion resistance: the first, by Leiro and Ayllón investigates the behaviour of a type of weathering steel containing non-metallic inclusions under long-term atmospheric exposure; the second, by Wilton-Smith *et al.* tests and evaluates turbocharger zinc-coated and stainless steel components subjected to accelerated corrosion induced by exposure to salt solution and dust particles.

Belufa and Caminha *et al.* look into the effects of fretting corrosion, the former investigating contact surface damage of automotive connectors made of high copper alloys, the latter evaluating failure mechanisms on the head-cone interface of hip prostheses made of stainless steel.

A second group of material investigations rely on electrochemical tests and analysis in their assessments of corrosive activity. Leiro and Ayllón provide the basics of electrochemistry related to atmospheric corrosion and subsequently present atmospheric corrosion results for weathering steels with and without non-metallic inclusions for comparison purposes. Abd-elrhman *et al.* and Kimura look into improved corrosion resistance of titanium alloys as biomaterials, the former by introducing novel compositions, the latter by conducting in-air oxidation and reactive sputtering.

Abosrra *et al.* and Toloei *et al.* examine and compare the contrasting corrosion behaviour of two types of metal for various grades of surface roughness. Hussein *et al.* consider protective coatings produced by plasma electrolytic oxidation and investigate their dependence on the processing parameters. Kimura and Hirano correlate corrosion characteristics of coated stainless steel with existing defects in the coating under polymer electrolyte fuel cell environment. Brown and Barnard model numerically the relevant electrochemical process to predict localized anodic dissolution on ZnAl alloy coatings. Bigdeli Karimi *et al.* investigate surface patterning as a means of improving the corrosion performance of nickel.

The two final papers in this part of the book are concerned with reinforced concrete. Koleva *et al.* study corrosion resistance and self-healing of steel reinforcement in the presence of "empty" and "Ca-containing" polymeric vesicles. Zhang and Li use finite element modelling to assess the effect of stirrup corrosion on the seismic integrity of a reinforced concrete column.

In the second part of the book, various boundary element formulations for modelling cathodic protection are applied to a range of engineering problems. Amaya *et al.* monitor the coating condition on a ballast tank wall; Wang *et al.* predict the impressed current cathodic protection of an offshore platform; Mujezinović *et al.* combine analytical with numerical modelling to calculate the protective current density distribution in the case of an electrolyte with discontinuous electrical conductivity; Peratta et al. employ an advanced efficient formulation for pipelines and deep well casings in multi-layer soils; Xing *et al.* and Wu *et al.* compare boundary element predictions with field measurements; the former optimize a ship's cathodic protection system by minimizing its electric and magnetic signature, the latter assess the influence of coating damage on the cathodic protection of a ship's hull. Finally, Irša *et al.* formulate and solve numerically an inverse galvanic corrosion problem for recovering the current density from electric potential data.

In summary, this volume describes a wide spectrum of experimental and analytical studies, which provide a fairly comprehensive account of corrosion manifestations and methodologies for addressing them in structural and industrial design. As such, it is expected to make a valuable reference publication for engineers and scientists interested in the protection of structures and components from harmful and potentially ruinous corrosive action.

Stavros Syngellakis
The New Forest, 2017

Acronyms

ADRS Acceleration Displacement Response Spectrum
AFM Atomic Force Microscopy
BEM Boundary Element Method
BF Breakdown Factor
BVP Boundary Value Problem
CA Cellular Automata
CE Cement Extract
CIS Close Interval Survey
COF Coefficient of Friction
CP Cathodic Protection, Collocation Point
CPE Constant Phase Element
CPET Corrosion and Protection Evaluation Tool
CR Cauchy-Riemann (equations)
CRM Corrosion Related Magnetic (field)
CSM Capacity Spectrum Method
CVA Cyclic Voltammetry
CPCD Critical Passivation Current Density (method)
DCVG Direct Current Voltage Gradient
DLS Dynamic Light Scattering
EBSD Electron Backscattering Diffraction
EDM Electrical Discharge Machining
EDS Energy Dispersive Spectroscopy
EDX Energy Dispersive X-ray (analysis or spectrometry or spectroscopy)
EIS Electrochemical Impedance Spectroscopy
EPMA Electron Probe Microanalysis
FDM Finite Difference Method
FE-SEM Field Emission type Scanning Electron Microscopy
FIB Focused (gallium) Ion Beam
GPIB General Purpose Interface Bus
ICP Inductive Coupled Plasma (spectrometry)

ICCP	Impressed Current Cathodic Protection
IR	Infrared
LNG	Liquefied Natural Gas
LPR	Linear Polarization Resistance
NAB	Nickel-Aluminium-Bronze (alloy)
OCP	Open-Circuit Potential
PA	Patina-Acceleration (steel)
PDP	Potentio-Dynamic Polarization
PEFC	Polymer Electrolyte Fuel Cell
PEO	Plasma Electrolytic Oxidation; Polyethylene Oxide
PS	Polystyrene
PSM	Physical Scale Model
RBF	Radial Basis Function
RC	Reinforced Concrete
RH	Relative Humidity
SCC	Stress Corrosion Cracking
SCCM	Standard Cubic Centimetres per Minute
SCE	Saturated Calomel Electrode
SEM	Scanning Electron Microscopy
SIM	Scanning Ion Microscopy
SSE	Saturated Sulfate Electrode
SVET	Scanning Vibrating Electrode Technique
TEM	Transmission Electron Microscopy
UEP	Underwater Electrical Potential
WE	Working Electrode
WLI	White Light Interferometry
WS	Weathering Steel
XPS	X-ray Photoelectron Spectroscopy
XRD	X-Ray Diffraction
XRF	X-Ray Fluorescence

Contents

Section 1: Material performance

Material characterisation to understand various modes of corrosion failures in large military vehicles of historical importance
A. Saeed, Z. Khan, N. Garland & R. Smith 3

Studies of microscopic strains on Alloy 600 surfaces arising from stress-corrosion cracking
N. S. McIntyre, T. Simpson, J. Qin, N. Sherry, M. Bauer, J. Ulaganathan, A. G. Carcea, R. C. Newman, M. Kunz & N. Tamura... 15

Long-lasting outdoor atmospheric corrosion tests
M. Leiro & E. S. Ayllón 27

Accelerated corrosion tests on waste-gated turbocharger's adjustable and fixed end-links
K. Wilton-Smith, Z. Khan, A. Saeed & M. Hadfield 39

Investigation on contact surface damage of an automotive connector by fretting corrosion
M. A. Beloufa 47

Evaluation of the fretting corrosion mechanisms on the head–cone interface of hip prostheses
I. Caminha, C. R. M. Roesler, H. Keide, C. Barbosa, I. Abud & J. L. Nascimento 61

Electrochemical analysis of corrosion products from long-lasting atmospheric tests
M. Leiro & E. S. Ayllón 69

Electrochemical and corrosion resistance of new Ti-Mo-Fe alloys for biomedical applications
Y. M. Abd-Elrhman, M. A. Gepreel,
A. Abd El-Moneim & S. Kobayashi 81

Improvement of corrosion resistance and biocompatibility of medical nickel–titanium alloys through forming TiO_2 thin film
Y. Kimura.. 91

Corrosion of mild steel and 316L austenitic stainless steel with different surface roughness in sodium chloride saline solutions
L. Abosrra, A. F. Ashour, S. C. Mitchell & M. Youseffi......................... 107

Simultaneous effect of surface roughness and passivity on corrosion resistance of metals
A. S. Toloei, V. Stoilov & D. O. Northwood 119

Plasma electrolytic oxidation coatings on Mg-alloys for improved wear and corrosion resistance
R. O. Hussein, X. Nie & D. O. Northwood 133

Corrosion resistance of Au/Ni thin film-coated stainless steel used as a polymer electrolyte fuel-cell separator
Y. Kimura & M. Hirano.. 149

Field-based prediction of localized anodic dissolution events taking place on ZnAl alloy coatings in the presence of 5% NaCl solution
S. G. R. Brown & N. C. Barnard 161

Improving corrosion performance by surface patterning
M. Bigdeli Karimi, V. Stoilov & D. O. Northwood................................ 173

Self-healing of steel corrosion in a model alkaline medium: electrochemical response and surface analysis
D. A. Koleva, J. Hu, H. Kolev & K. Van Breugel 183

The corrosion of stirrups and its effect on the seismic fragility of a corroded reinforced concrete column
G. Zhang & B. Li 197

Section 2: Cathodic protection

Numerical-analysis-assisted monitoring method for the coating condition on a ballast tank wall
K. Amaya, A. Nakayama & N. Yamamoto 211

Boundary element modelling of the impressed current cathodic protection of an offshore platform
X. Wang, B. Hou & Z. Lan.. 221

Calculation of the protective current density distribution of a cathodic protection system with galvanic anodes in terms of double-layer electrolyte
A. Mujezinović, Ai. Muharemović, I. Turković & Ad. Muharemović..... 231

Computational modelling of cathodic protection systems for pipelines in multi-layer soil
A. B. Peratta, J. M. W. Baynham & R. A. Adey 245

Numerical modelling of cathodic protection systems for deep-well casings
A. B. Peratta, J. M. W. Baynham & R. A. Adey 259

Optimization of ship's cathodic protection system to minimize electric and magnetic signature by mathematical simulation
S. Xing, J. Wu & Y. Yan... 271

The influence of coating damage on the impressed current cathodic protection effect
J. Wu, S. Xing & F. Yun... 281

Recovering current density from electric potential data
J. Irša, A. N. Galybin & A. Peratta.. 289

Author index 301

Section 1
Material performance

Material characterisation to understand various modes of corrosion failures in large military vehicles of historical importance

A. Saeed[1], Z. Khan[1], N. Garland[1] & R. Smith[2]
[1]*Bournemouth University, UK*
[2]*The Tank Museum, Bovington, UK*

Abstract

Large military vehicles within museum collections are stored in two distinct environments, controlled and uncontrolled, with an intermittent transitional mode where vehicles travel between the two. Variable environmental conditions combined with operational factors pose significant risks to the reliability, durability and longevity of these vehicles. Although there are methods for retarding or decelerating aspects of failure, to maintain the integrity and originality of these vehicles as artefacts, a sustainable methodology for conserving them should be developed. Corrosion is one of the significant contributors to the structural damage and material aging of historical military vehicles; therefore, an experimental study was conducted to understand the prevailing mechanisms of failures due to corrosion with the various types occurring in these vehicles identified. This paper describes the various modes of corrosion in historic vehicles while X-ray fluorescence and ultrasonic scanning corrosion mapping techniques characterise corrosion inhibiting materials and subsequent material loss. Understanding material profiles and their link to environmental exposure during use and non use of these vehicles will lead to a sustainable methodology for their conservation.
Keywords: sustainable, military, vehicle, corrosion, materials, conservation.

1 Introduction

The structural and functional deterioration of the large metal structures due to their environmental exposure and the operating conditions are responsible for their structural aging [1]. Aging mechanisms in large metal structures such as bridges, aircrafts, vehicles etc is recognised as a growing problem [2]. Aging mechanisms such as corrosion, surface and sub surface cracks, wear due to interacting surfaces and undesired stresses are leading to mechanical failures, higher maintenance costs and compromise structural integrity. Corrosion has been identified as one of the major contributor to the aging mechanisms; it affects all metal structures by deteriorating the material's properties [3].

This research focuses upon the corrosion damage within large historic vehicles in the museum environment. Corrosion is indiscriminate in affecting metal structures and museum artefacts are no exception. The military vehicles from World Wars I and II were designed and manufactured according to the automotive technologies in place at those times [4] with only a limited life-expectancy, however their participation in the war and the subsequent historic significance means it is important to devise sustainable methodologies for long-term preservation. Continuous corrosion problems within large historic military vehicles in the museum environment are leading to concerns about their longevity, in particular where failures in protective measures are also identified [5]. For the purpose of sustainable methods, it is important to understand the various corrosion modes, material characterisation and the material loss profile due to corrosion.

For this research, military vehicles were inspected for visual signs of corrosion and novel methods such as X-ray fluorescence for elemental identification and ultrasonic scanning for material loss were conducted. Visual inspection identified different modes of corrosion depending on their exposure to extreme environments during the war, operating conditions at the time of their service life, types of materials used and their current working or non-working environments.

X-Ray fluorescence method found Fe, the main constituent; other elements such as W, Mo and Mn were also identified. Ultrasonic methods found severe material loss due to corrosion in the selected samples.

2 Experimental methodologies

The following methods are used in the research. Only the most common and reoccurring types of corrosion which were found on a larger scale in the vehicles are presented here.

2.1 Visual inspection

Visual inspection was completed on more than 20 large military vehicles from the World War I, II and post war eras. The investigation found one or more forms of corrosion in the same vehicles. Some widespread and most occurring modes of corrosion are described from four military tanks in Table 1.

2.2 X-ray fluorescence

X-ray fluorescence (XRF) was conducted on Sherman M4A1 for elemental identification. The test sample was bombarded by an intense x-ray beam resulting in the emission of fluorescent x-rays from the test object. The emitted fluorescent x-rays are then detected by XRF analyser [6] which identifies the elements by measuring the energies of the emitted fluorescent x-rays and counting the number of rays in each energy spectrum [6, 7].

Table 1: Military tanks presented in this research.

No	Vehicle	Environment	Manufacture	Country of origin	Era
1	Mark II	Controlled	1917	United Kingdom	WW1
2	Sherman*	Uncontrolled	1940–1941	United States	WW2
3	King Tiger	Controlled	1943	Germany	WW2
4	Centurion	Uncontrolled	1945	United Kingdom	Post war
5	Scorpion	Uncontrolled	1973	United Kingdom	Post war

XRF and ultrasonic scanning is conducted on Sherman.

2.3 Ultrasonic scanning

Ultrasonic scanning uses pulse waves from 0.10 to 15.0 MHz frequencies range to measure material loss and other dimensional anomalies in the test objects [8]. This research work presents ultrasonic scanning using the pulse echo method whereby pulses are transmitted and received back on the same side of the test object to detect and characterise defects.

3 Results

3.1 Uniform corrosion

With uniform corrosion, a direct chemical attack affects the surface of the metal and then spreads out evenly across the entire surface; the results are not usually fatal and it is not classified as one of the severe forms [9]. Fig. 1 of the King Tiger's flywheel shows uniform corrosion spreading across the surface. Uniform corrosion was also found in Tog II and FT 17 on considerably larger surface areas. Uniform corrosion, if left to continue, could lead to general thinning of the surface and may transform into another form of corrosion [10].

Figure 1: King tiger flywheel.

3.2 Fretting corrosion

Fretting corrosion occurs at the interface of the interacting surfaces which are subject to relative motions. The King Tiger planetary gears (Fig. 2) are a classic example of this form. Interacting mechanical wear leads to the removal of material particles, followed by the oxidation of the material debris and the newly exposed surface [11]. The oxidised debris would act as an abrasive media leading to further wear and fatigue failures could occur [9, 12].

Figure 2: King tiger planetary gears.

3.3 Stress corrosion cracking

Stress corrosion cracking is considered one of the severe types of corrosion; it could cause a partial or complete failure of a structure especially in the presence of dynamic or static loading and initiation of a stress corrosion crack [13]. The Mark-II Tank glacis plate in Fig. 3 shows stress corrosion cracking where a crack is initiated and penetrating through the surface with no significant material loss on the surface. Stress corrosion is difficult to detect, hard to predict and unexpected failure can occur [13, 14].

3.4 Galvanic corrosion

Galvanic corrosion is the electrochemical reaction between two dissimilar metals in the presence of electron conductive path and the preferential corrosion of one metal more than the other [15, 16]. This mode of corrosion was identified in the Scorpion, see Fig. 4. The presence of moisture among the bolt, washer and the armoured skirt has caused this type of corrosion.

Figure 3: Mark II glacis plate.

Figure 4: Galvanic corrosion in Scorpion.

3.5 Pitting corrosion

Centurion in Fig. 5 shows signs of pitting corrosion. In this case, the corrosion is confined to a small point with resulting cavities in the metal. The localised phenomenon of corrosion creates holes in the metal and then penetrates inwards, deteriorating the metal from the inside. The corrosion products cover the pit and create a small hole with no significant material loss on the surface leading to a catastrophic malfunction; it is one of the severe forms of corrosion, which is very difficult to detect and hard to predict the result [14, 17, 18].

3.6 X-ray Fluorescence of the Sherman M4A1

X-Ray fluoresce (XRF) was conducted on three points A, B, C on corroded, E, F, and G on sand blasted surfaces and point D on the cross section of the sample, see Fig. 6. Table 2 shows the results from points A, B and C. On averages, Fe at 92.07% is the highest, second highest is the Si at 5.39%, Mn at 0.91% was the third highest and the lowest was Cu with a trace of 0.04%.

Figure 5: Pitting in Centurion.

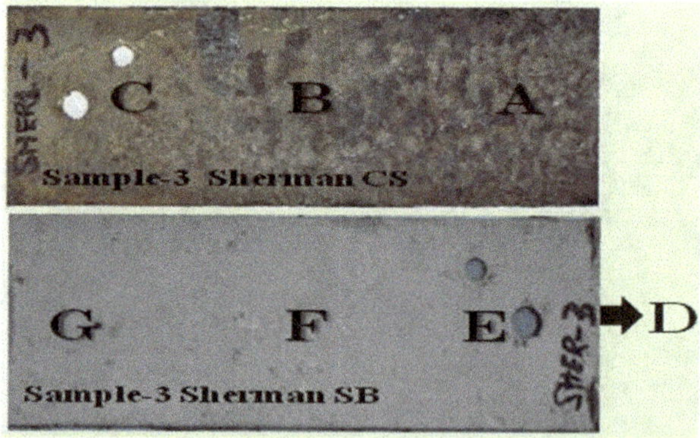

Figure 6: XRF points on Sherman.

Fig. 7 shows the graphical representation of the XRF conducted at point D where Fe at 98.30% is the highest, second highest is Si at 0.65%, Mn at 0.63% the third highest and no traces of Ti, V, Co and Mo were identified.

XRF conducted at points E, F and G on the sandblasted surface shown on averages Fe at 98.12% the highest, Mn at 0.66% second highest, Si at 0.59% the third highest while W at 0.02% the lowest and no traces of Co or V were identified. Fig. 8 shows the graphical representation of the XRF analysis at point G.

3.7 Pulse echo ultrasonic scanning

Pulse Echo scanning was conducted on the Sherman M4A1 with a probe of 4.00 MHz twin crystal and 10.00 mm diameter using 0.25 MHz Hi pass filter and 10.00 MHz low pass filter frequencies with a probe delay of 7.630 μS. Compression waves with a velocity of 5903.00 m/s are used to scan a total area of 174.00 mm length with 6.040 pulses/mm and 68.00 mm width with 3.580 pulses/mm mapping material loss every 2.00 mm on the sample.

Table 2: XRF of the corroded surface of M4A1.

Vehicle Type: Sherman				
Sample 1			Surface: Corroded	
Time duration: 30 seconds			MQ for points: A = 5.70 B = 4.50 C = 4.60	
Material Standard		06 American	06 American	06 American
Constituents %		A-Reading 254	B-Reading 255	C-Reading 256
1	Si	4.89	5.92	5.38
2	P	0.46	0.54	0.85
3	Ti	0.19	0.19	0.2
4	V	0.05	0.05	0.05
5	Cr	0.11	0.08	0.09
6	Mn	0.86	1.11	0.78
7	Fe	92.68	91.48	92.05
8	Co	0.21	0.22	0.12
9	Ni	0.06	0.08	0.1
10	Cu	0.05	0.03	0.04
11	Mo	0.05	0.06	0.06
12	W	0.33	0.2	0.12

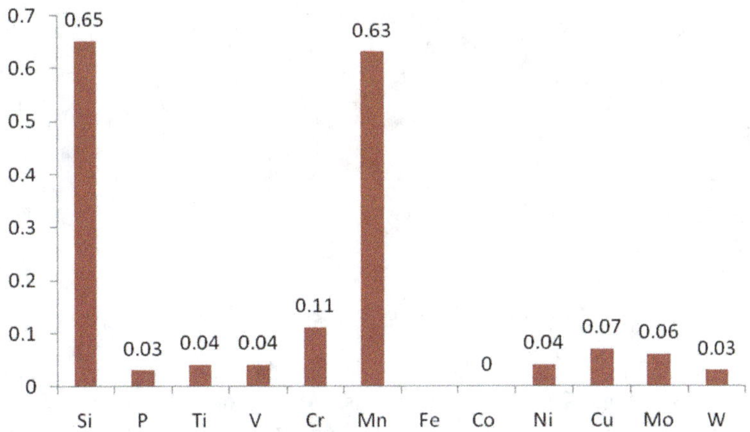

Figure 7: XRF analysis of point D, No Fe (98.30%) included.

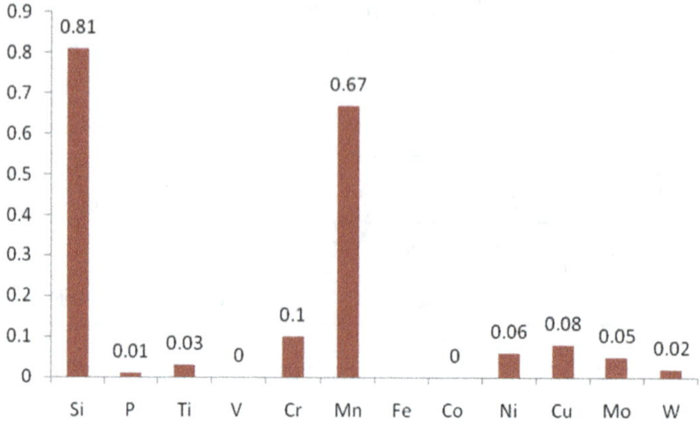

Figure 8: XRF analysis of point G, No Fe (98.09%) included.

Fig. 9 shows the end views and the c-scan of the sample with implementation of colour coding Red for 0.00 mm and Blue for 10.00 mm. Two cursors Red and Blue show the x-axis, y-axis location and the corresponding depths of the particular points on the sample.

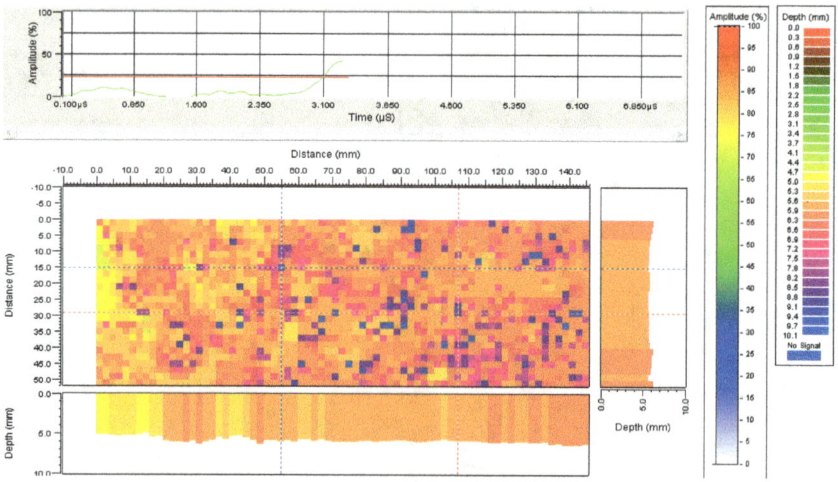

Figure 9: Pulse echo scan of M4A1.

A user defined area is taken for finding the maximum and minimum thicknesses starting at 2.00 mm on the x-axis, 2.00 mm on the y-axis in Fig. 10 and finishing at 172.00 mm on the x-axis and 66.00 mm at the y-axis in Fig. 11 respectively to avoid any distortion in signals at the edges. The scan detected maximum remaining thickness of 9.80 mm at horizontal position of 150.00 mm and vertical position of 26.00 mm with a drastic loss in material to a minimum of

4.60 mm at horizontal position of 28.00 mm and vertical position of 8.00 mm. The maximum thickness of 9.80 mm is recorded at two different points. The second lowest depth is 4.70 mm and is detected at seven different positions; third lowest of 4.80 mm at six various points and the remaining thickness of 4.90 mm is recorded at four different points in the samples. The minimum remaining thickness of 4.60 mm is recorded at three different points.

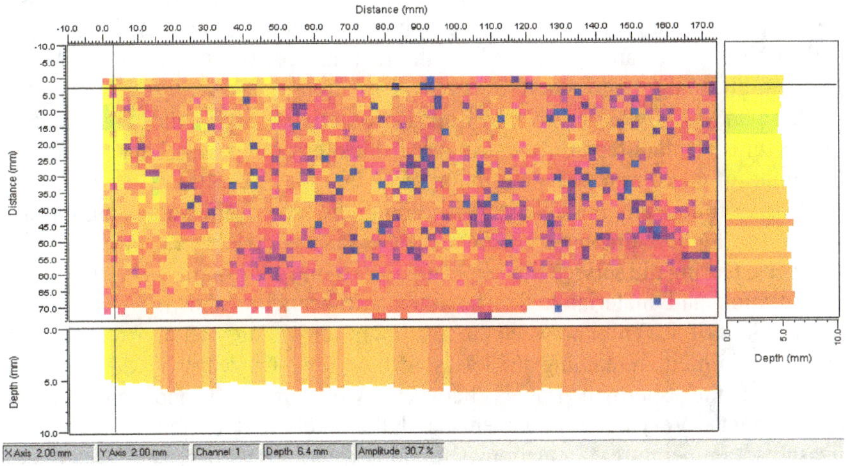

Figure 10: User defined area starting at 2.00 mm on *x*-axis and at 2.00 mm on *y*-axis.

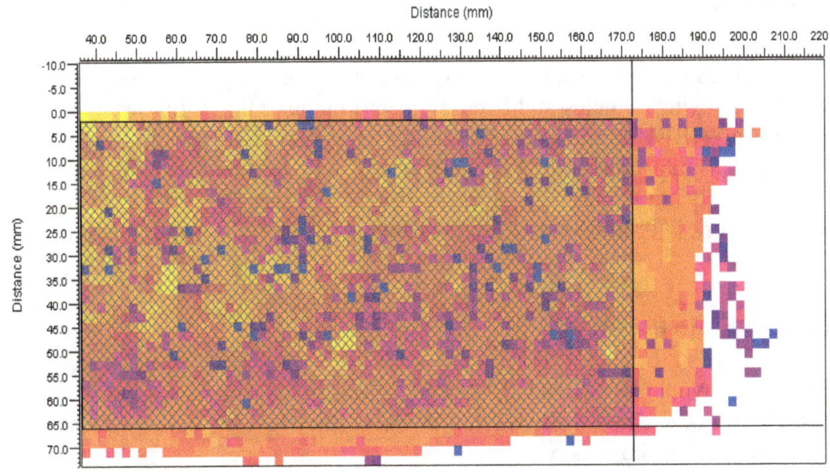

Figure 11: User defined area finishing at 174 mm on *x*-axis and at 66 mm on *y*-axis.

4 Discussion

The modes of corrosion identified above can be retarded by using protective measures such as coatings and paint for uniform corrosion, application of lubrication and exclusion of air in the fretting corrosion, alleviating the stress in stress corrosion cracking, using same metal types in galvanic corrosion or applying insulation between two different metals.

The alloying elements such as Mn, Cr, V, W and Mo identified though XRF are used as hardening agents to improve hardness, ductility and tensile strength [19]. Also Mo, Mn and P are used for the rust and corrosion prevention. Mo is used in high strength alloys and super alloys for strength and wear resistance [20].

Sherman M4A1 was designed by Lima Locomotive Works in the United States. At the time of manufacture, the armoured thickness range was kept between a maximum of 62.00 mm and a minimum of 12.00 mm [21, 22]. However it is not practical to follow any possible modification in the armour thickness and, for this reason, the highest thickness recorded in the scan is considered the maximum thickness. During the service life, exposure to extreme environments has led to the deterioration and material loss. The ultrasonic scan has shown a remarkable material loss of 5.20 mm between the maximum remaining thickness of 9.80 mm and the minimum remaining thickness of 4.60 mm. Currently in the museum collection, the Sherman travels between two controlled and uncontrolled environments. Varying environments and the likelihood of failing protective measures, temperature fluctuations, humidity ratio variations, rain water and the extent of chemical reaction etc will influence the rate of corrosion and, in the controlled environment, corrosion residues may result in indoor atmospheric corrosion [23, 24].

5 Conclusion

The presented research work uses a conventional method of analysing corrosion through surveying vehicles in the museum environment; this is followed by novel methods of XRF for the elemental identification and ultrasonic scanning to measure and map corrosion. Further research will be conducted on corrosion modelling and life expectancy of large historic vehicles based upon the results from this work in order to maximise their longevity, durability and reliability.

Acknowledgements

The authors would like to thank Mark Nel, Mark Clark in AGR, Jon Previtt in West-Dean College and Mike Hayton in The Tank Museum Bovington for their in kind support during this research work.

References

[1] Colavita, M., Service life and aging of military equipment. *Corrosion: Environments and Industries*, ASM Handbook, vol. 13C, eds. S.D. Cramer & B.S. Covino, Jr., pp. 220–228, 2006.

[2] Agarwala, V.S., Corrosion in the military. *Corrosion: Environments and Industries*, ASM Handbook, vol. 13C, eds. S.D. Cramer & B.S. Covino, Jr., pp. 126–135, 2006.

[3] Ahmad, Z., *Principles of Corrosion Engineering and Corrosion Control, Ch. 2: Basic Concepts in Corrosion*, Butterworth-Heinemann: Oxford, pp. 9–56, 2006.

[4] Handsy, I.C. & Repp, J., Ground vehicle corrosion. *Corrosion: Environments and Industries*, ASM Handbook, vol. 13C, eds. S.D. Cramer & B.S. Covino, Jr., pp. 148–150, 2006.

[5] Gudze, M.T. & Melchers, R.E., Operational based corrosion analysis in naval ships. *Corrosion Science*, **50**, pp. 3296–3307, 2008.

[6] Beckhoff, B., *Handbook of Practical X-ray Fluorescence Analysis*, Springer: Berlin, London, 2006.

[7] Grieken, R. v. & Markowicz, A.A., *Handbook of X-ray Spectrometry*, 2nd ed., M. Dekker: New York, NY, 2002.

[8] Drury, J.C., *Ultrasonic Flaw Detection for Technicians*, 3rd ed. Silverwing Ltd.: Swansea, 2004.

[9] Baboian, R., *Corrosion Tests and Standards: Application and Interpretation*, ASTM: Philadelphia, PA, 1995.

[10] Uhlig, H.H. & Revie, R.W., *Corrosion and Corrosion Control: An Introduction to Corrosion Science and Engineering*, 4th ed., Wiley-Interscience: Hoboken, NJ, 2008.

[11] Bill, R.C., The role of oxidation in the fretting wear process. *Proc. of the Int. Conf. on Wear of Materials*, San Francisco, CA, ASCE: New York, pp. 238–250, 1981.

[12] Neyman, A. & Olszewski, O., Research on fretting wear dependence of hardness ratio and friction coefficient of fretted couple. *Wear*, **162-164**, pp. 939–943, 1993.

[13] Procter, R.P.M., Omideyi, M.V.E., Newman, R.C. & Zheng, W., Stress-corrosion cracking of C---Mn steels in methanol-ammonia environments–I. Effects of environmental and mechanical variables. *Corrosion Science*, **33**, pp. 1009–1031, 1992.

[14] Newman, R.C., Stress corrosion cracking. *Shreir's Corrosion*, ed. J.A.R. Tony, Elsevier: Oxford, pp. 864–901, 2010.

[15] Hampel, C.A., *The Encyclopedia of Electrochemistry*, Reinhold; Chapman & Hall, 1964.

[16] Hack, H.P., Galvanic corrosion. *Shreir's Corrosion*, ed. J.A.R. Tony, Elsevier: Oxford, pp. 828–856, 2010.

[17] Szklarska-Smialowska, Z., *Pitting Corrosion of Metals*, National Association of Corrosion Engineers: Houston, TX, 1986.

[18] Baboian, R., *Automotive Corrosion Tests and Standards*, Society of Manufacturing Engineers, 1996.

[19] Ashby, M.F. & Jones, D.R.H. *Engineering Materials 2: An Introduction to Microstructures, Processing and Design*, 3rd ed. Butterworth-Heinemann: Oxford, 2006.

[20] Schmidt, F.F., *The Engineering Properties of Molybdenum and Molybdenum alloys*, Defense Metals Information Center: Columbus, Ohio, 1963.

[21] Chamberlain, P. & Ellis, C., *British and American Tanks of World War II: The Complete Illustrated History of British, American, and Commonwealth Tanks, 1939–1945*, Cassell & Co.: London, New York, 2000.

[22] Jackson, R., *Tanks and Armoured Fighting Vehicles*, Parragon: Bath, 2007.

[23] Mendoza, A.R. & Corvo, F., Outdoor and indoor atmospheric corrosion of non-ferrous metals. *Corrosion Science*, **42**, pp. 1123–1147, 2000.

[24] Odnevall, I. & Leygraf, C., The formation of Zn4Cl2(OH)4SO4 5H2O in an urban and an industrial atmosphere. *Corrosion Science*, **36**, pp. 1551–1559, 1994.

Studies of microscopic strains on Alloy 600 surfaces arising from stress-corrosion cracking

N. S. McIntyre[1], T. Simpson[1], J. Qin[1], N. Sherry[1], M. Bauer[1], J. Ulaganathan[2], A. G. Carcea[2], R. C. Newman[2], M. Kunz[3] & N. Tamura[3]
[1]*Faculty of Science, University of Western Ontario, Canada*
[2]*University of Toronto, Canada*
[3]*Advanced Light Source, Lawrence Berkeley National Laboratory, USA*

Abstract

Microscopic distributions of elastic and plastic strains have been studied in Alloy 600 during accelerated corrosion. Laue diffraction is employed with a sub-micron beam of highly coherent polychromatic (white) radiation. The diffraction patterns are analysed to detect elastic and plastic deformations associated with the crack initiation and propagation processes. Stressed C-ring and unstressed mill annealed samples of Alloy 600 were corroded under hydrothermal, controlled electrochemical conditions. In the C-ring, the filamentous surface cracks produced had compressive strain fields along each crack. In the mill annealed sample, changes in strain fields in the same area were measured as oxidation progressed. Cracking at the metal grain boundaries appeared to be induced by expansive growth of surface oxides. For the mill annealed sample, accumulation of elastic strains in the grain boundaries appeared in advance of crack propagation.
Keywords: strain, stress-corrosion cracking, Laue diffraction.

1 Introduction

Stress-corrosion cracking (SCC) is the single most common mode of degradation for high performance nickel alloys. SCC of nickel and its alloys is generally believed to begin at an interface, possibly resulting from stresses arising from the expansion of corrosion products in a grain boundary that can interact with local strains already present in the material.

These and other stress-related properties are being studied using microscopic beams of highly coherent radiation that interact with the material to produce Laue diffraction patterns for each micron-size area. Because of the low scattering properties of such beams, it is possible to derive much more accurate information about crystal properties, such as deviations from the "classic" properties of the unit cell that exist only in perfect crystals. From deviatoric changes to the lengths of the unit cell, one can determine the spatial distribution (residual) elastic strain tensors in a material; from the deviatoric changes to angles in the unit cell, information about changes to the plastic deformation can be derived including the local density and direction of dislocations [1–5]. Each Laue X-ray pattern can now be more accurately assessed than is possible with related electron beam techniques like Electron Backscattering Diffraction (EBSD).

The authors' initial research in this area was to study local strains in Alloy 600 (16%Cr 8%Fe balance Ni) that had been deformed plastically [6, 7]. Since then, a reproducible electrochemical protocol has been refined that has caused Alloy 600 tubing C-rings to undergo accelerated SCC in an aqueous caustic solution. Alloy 600 tubing has been widely used to transport high temperature coolant in nuclear power steam generators. Such tubing has frequently been found to undergo stress corrosion cracking in the steam generators of nuclear power stations particularly in areas where the tubing is supported mechanically [8, 9].

The present study has concentrated on the detection of the initial steps in the process of SCC: the accumulation of stresses at the surface of the alloy sample and the first formation of surface and internal oxides that could initiate a crack. Using experiences from previous experiments, corrosion conditions have been chosen that would allow a measured stepwise approach to crack initiation that would show incremental changes to the strain and dislocation distributions. Two different sample geometries have been used: a pre-stressed C-ring [6] and a mill annealed polished flat with little stress. In both samples cracking has been induced that could be considered to be at the incipient stage. Laue studies of both have explored the local elastic strain fields near the crack and, in the case of the flat, strain changes in both the surface oxide and the metal substrate have been followed during three steps in the corrosion process.

2 Experimental details

The C-ring was prepared from Alloy 600 nuclear grade tubing according to ASTM standard G-38. The stress applied was equivalent to a calculated strain of 2%. The flat was prepared from a mill annealed ingot of Alloy 600. The surface was mechanically polished to a 0.5 micron diamond grit finish. Then an area of several hundred square microns was cleaned and polished with a focused gallium ion beam (FIB). The corrosion test of the C-ring was conducted so that, based on past practice [10, 11], any stress-corrosion cracking of the sample would be slight. The C-ring was exposed in its stressed condition in an autoclave containing a solution of 10 wt.% NaOH in deionized water at a temperature of 315°C. All tests were conducted in a 0.5 l nickel beaker (liner) containing 230 ml of solution inside an alloy-625 autoclave. Prior to heating the autoclave, the caustic solution was de-aerated by nitrogen for at least 12 h. After temperature stabilization, a potential of 150 mV vs. pseudo-RE (Alloy 600) was applied to the whole sample (C-ring, bolt and washer assembly) and the total current density was monitored. After initial decay from ~150 $\mu A \cdot cm^{-2}$, the current densities finally reached values of ~30 $\mu A \cdot cm^{-2}$ after an exposure time of 120 minutes. For corrosion of the flat, a potential of 100 mV was used; three separate exposures were carried out for periods of 6, 12 and 24 h with a microscopic Laue diffraction study of the same area after each exposure. Following the corrosion test the sample microstructure was studied by SEM and was sectioned by a FIB to obtain a better view of the grain structure and crack propagation.

Laue studies were carried out at beamline 12.3.2 at the Advanced Light Source synchrotron using conditions reported in a previous paper [7]. The polychromatic beam size was 1 micron diameter. The Laue images were processed using FOXMAS software (Fast On-line X-ray Microdiffraction Analysis Software) [12]. FOXMAS is parallelized software that is partially based on the XMAS [2] source code and is capable of processing the same images nearly two orders of magnitude faster than XMAS. The software has been coded to run on IBM InfoSphere Streams computing software (version 2.0) on processors at the University of Western Ontario. Indexing was done using a centroid fit of the detected spots and was also re-indexed to show all results with 7 or more spots indexed. The spots were indexed for fcc nickel with a cell parameter of 0.35564 nm. in Alloy 600 [13]. Angular mis-orientation information about plastic deformation was also collected using Laue diffraction and these data were computed and displayed using software developed by this group called “Skew”.

3 Results

3.1 C-ring stress-corrosion cracking

The apex of the outside tube diameter proved to have a significant fraction of grain boundaries that appeared to be beveled due to enhanced oxidation. The sub-surface structure of such boundaries was explored using a FIB to cut a cross section (see Fig. 1). A lateral 3 kV SEM image of a FIB cut is shown in Fig. 1(a), revealing the oxide metal interface, the beveled grain boundary/surface interfaces and detail of the surface oxide. A crack pathway can be seen in the oxide surface, as well as through the compact oxide in the bevel and along the boundary.

It appears that the enhanced oxidation resulted in crack propagation that extended through the compact oxide in the bevels, the oxide crystals on the outer surface and into the grain boundaries below the bevels. This appears to be good circumstantial evidence for the initiation of cracking from the expansive growth of oxides within the bevels.

The strains associated with this cracking process were studied with the aid of a Laue diffraction map that was done on a narrow (21 micron) strip along the top of the apex in the vicinity of a number of the cracks. A limited width was required because of the stigmatic limitations of the X-ray optics. No exact match was

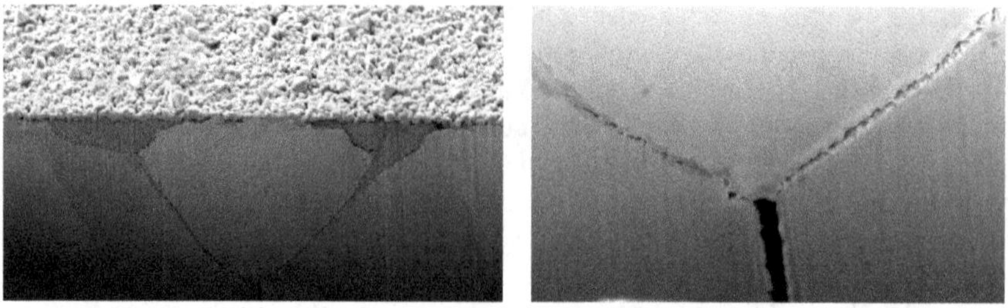

(a) (b)

Figure 1: SEM images of grain boundary structure along the apex of the outside diameter of the C-ring (a) lateral view of a FIB cut revealing the oxide metal interface; (b) detail of the triple point shown in the lower part of (a).

possible between the coordinates of an individual crack and mapped result. However, an overlap between a physical crack and the PXM map was made possible from the use of imaging X-ray absorption measurements as outlined in a previous publication [7]. Fig. 2 shows maps of the imaging X-ray image along with those of orientation, y strain tensor, composite strain and absolute mis-orientation. In Fig. 2(a), the light streaks are ascribed to near surface cracks.

By comparing the locus of this discontinuity with the orientation map in Fig. 2(b), it appears that the crack is aligned with the boundary between two grains. A line of boundaries continue up the centre of the map and along this line a number of pixels cannot be indexed, suggesting significant disruption of the crystal structure along this line, some of which must be due to plastic strain caused by the cracking. The map of y tensor strain in Fig. 2(c) shows an accumulation of compressive elastic strain in the narrow band in grains to the left of the boundary and lying close to it. Fig. 2(e) shows the distributions of mis-orientation near the crack region. Compared to the y tensor elastic strain map, there is little evidence of correlation between the crack locus and incidence of plastic strain as measured by absolute mis-orientation.

A magnification in Fig. 2(f) of the presumed crack locus and the strain pattern shows that most of the compressive elastic strain is found within several microns to one side of the crack. This pattern is also observed in the composite strain map of Fig. 2(d) but not as clearly-defined as in the y tensor map. The compressive elastic strain intensity reaches a maximum of $\sim 5 \times 10^{-3}$ within a few microns of the crack itself. The immediate region of the crack has a low crystal order as diffraction patterns there cannot be indexed and appear as black in the figure.

3.2 Mill annealed polished flat

A small region on the polished surface of the flat was pre-treated with a focussed ion beam. This was done to allow minute changes in the physical structure of the

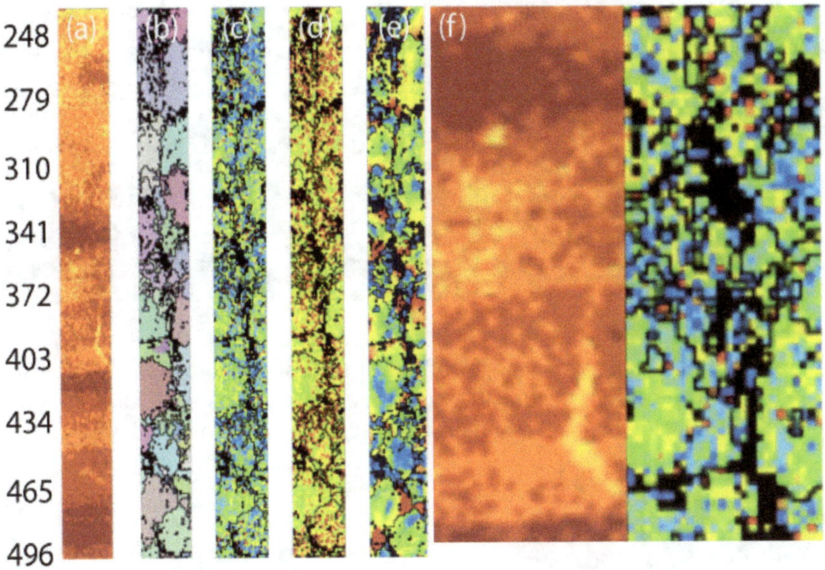

Figure 2: Laue diffraction maps of a narrow portion of the apex of the outside diameter of the C-ring; (a) absorbed X-ray image; (b) crystallographic orientation image with grain boundaries outlined by black lines; (c) *y* elastic strain tensor image; (d) composite elastic strain image; (e) absolute mis-orientation image; (f) expanded view of absorbed X-ray map and *y* strain tensor map.

flat to be followed during the three corrosion treatments. This proved to be very useful in identifying the onset of cracking after the third electrochemical treatment. SEM images of the surfaces are shown in Fig. 3 after the second and third treatments. After the third treatment bevelling of the grain boundaries became extensive as shown in Fig. 3(b) and cracks even became apparent in the oxide within the FIB pre-treated area under higher resolution. The differences between the area pre-treated by FIB and that untreated area are apparent with much greater growth of micron-scale crystallites in the untreated area. A crack opening near a grain boundary is shown in Fig. 4(a) and cracks were also found to be present among the oxide crystallites outside the FIB area. FIB sectioning was then used to create a cross-section of the oxide and underlying metal in the vicinity of the crack and this is shown in Fig. 4(b). The oxide shows internal cracking as well as an extension of the crack into the grain boundary in the underlying alloy. One possibility is that expansion of the oxide because of its Pilling-Bedworth factor could have led to an initiation of a crack in the boundary. The surface oxide was depth profiled using Auger Electron Spectroscopy and was determined to be about 80% nickel oxide/20% chromium oxide. The structure of the oxides would be expected to be NiO and Cr_2O_3 based on the solution pH and temperatures used in the autoclave [14].

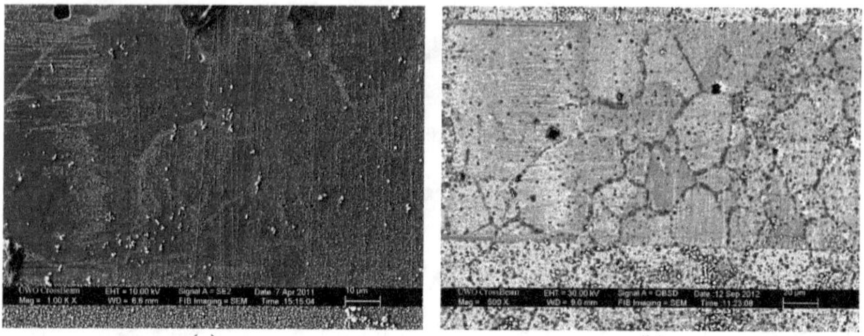

(a) (b)

Figure 3: SEM plan view images of the surface of the corrosion flat after the (a) second treatment (secondary electron display) and (b) third electrochemical treatment (backscattered electron display).

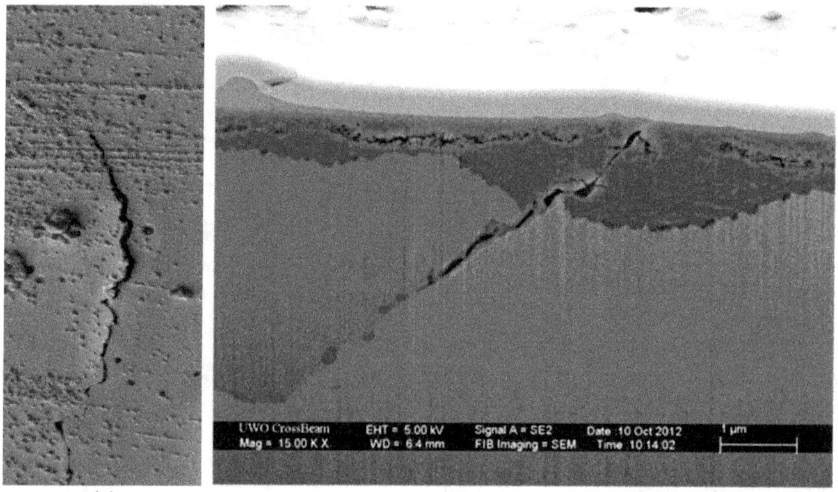

(a) (b)

Figure 4: SEM images of a crack in the surface oxide following the third electrochemical exposure (a) plan view back scattered electron image of the crack within the FIB pre-treated area (b) secondary electron image of a FIB cross section of the same crack.

The above studies took place within the small area on the flat that had been analysed on three successive occasions by PXM (after each electrochemical treatment). The Laue diffraction maps of crystallographic orientation in the same proximate areas are shown in Fig. 5. The colours shown represent crystallographic poles whose angular relationship is shown in Fig. 5(d). The black lines represent changes in angular orientation of >5 degrees. The pole figure for individual grain faces changes between exposures because the sample orientation with respect to

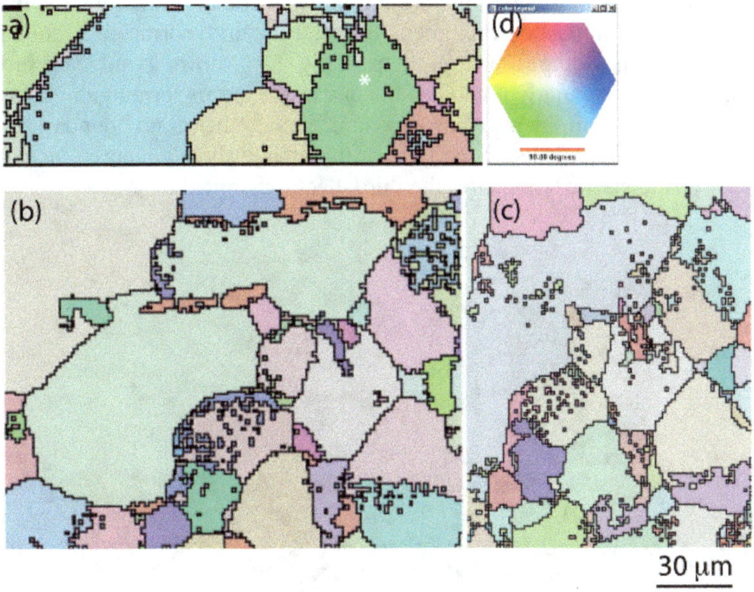

Figure 5: Crystallographic orientation maps for the same area on the polished Alloy 600 flat following: (a) 6h (b) 12h and (c) 24h of electrochemical exposure; (d) angular relationship of crystallographic poles.

the X-ray beam has changed somewhat with each analysis. The length scales are only approximately the same. A correct scale bar is shown for the map of Fig. 5(c). The same grain shapes can be recognised in these maps as well as in Fig. 3. These maps represent information from an average of several microns into the surface. The overall shape of most of the grains did not change throughout the treatment; however, there was the notable addition of small phases along many grain boundaries. Their exact origin is not understood but it is believed that they are associated with the growth of oxide phases at or near the grain boundaries.

The local elastic strain tensors are mapped for these areas in Figs. 6 and 7. Fig. 6 compares changes in the composite (von Mises) elastic strains. This strain value represents the root mean square of all 6 strain tensors that are determined for each pixel in the image; the volume element measured is actually $1 \times 1 \times {\sim}10$ microns. The regions covered by each scan were not quite the same so the figure shows only those equivalent regions. In Fig. 6(a), the central grain marked with an asterisk is also identified in the images in Figs. 3 and 5.

After the initial 6 h corrosion process, most local strains are confined to the interior of grains and probably result from forging and mechanical surface preparation. In fact, the common area shown here was unusually strained compared to other regions of the sample for this treatment. After the second 6 h corrosion process, there is a distinct tensile strain along the grain boundary near the centre of the map (see arrow in Fig. 6(b)). By contrast the areas of high strain intensity in the central grain in Fig. 6(a) diminished and the remaining regions are generally low in composite strain intensity. After the third corrosion treatment for a total of 24 h, the composite strain map in Fig. 6(c) shows an increase in local strain intensity; particularly notable is the one of the grains forming the boundary that was strained from the previous exposure. Some other grain boundaries, both within (see arrow) and outside the field of this image show a marked increase in local composite elastic strain after the third electrochemical treatment.

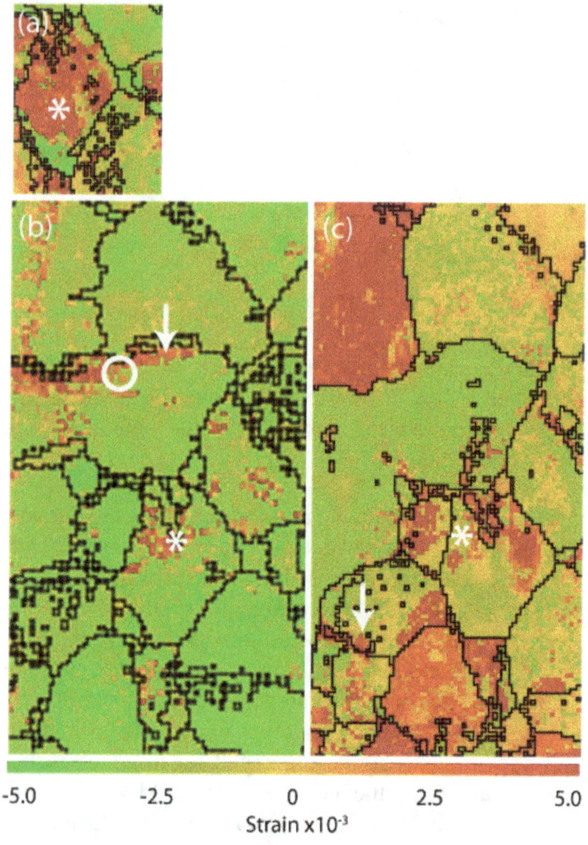

Figure 6: Composite (von Mises) elastic strain maps for the same area of the Alloy 600 flat that had been corroded for periods of (a) 6 h, (b) 12 h and (c) 24 h. The intensity scale is unidirectional. The dark lines indicate changes to orientation by > 5 degrees as in the previous figure. An asterisk identifies the same grain in each figure as well as in fig. 3. A plus sign indicates the region where the FIB section shown in fig. 4 was done. Arrows indicate regions of high grain boundary strain in (b) and (c).

The elastic tensile strain along the z axis (into the plane of the paper) is compared in Fig. 7 for approximately the same region of the flat for the second and third corrosion treatments. After the second treatment tensile strain is confined to the grain boundary region discussed above (see arrow). After the third treatment much of the entire region (and beyond) exhibits tensile strain as well as the same boundaries as before. This could possibly result from stresses caused by the growth of the oxide overlayer. Some strains in other directions within the boundary region are compressive and are probably caused by local oxide growth within the grain boundary or above it.

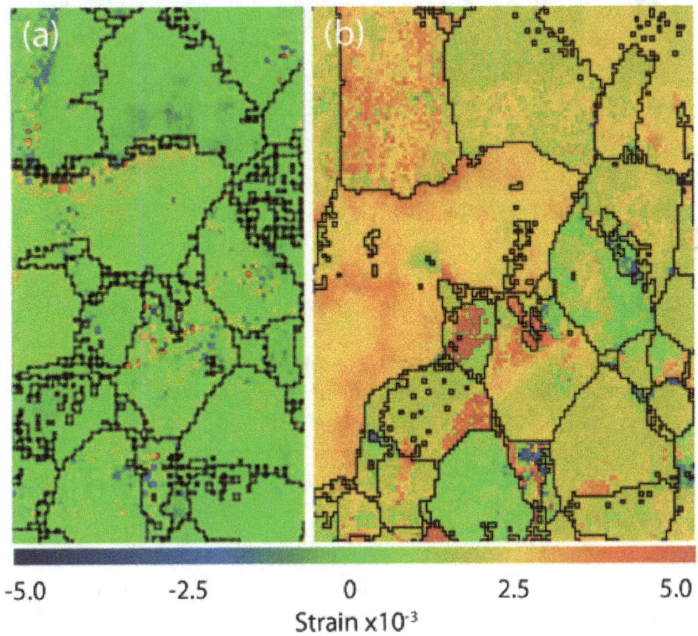

Figure 7: Comparison of strain intensity in the z direction after the second and third electrochemical treatments. Blue coloration indicates and compressive strain and red coloration a tensile strain.

The absolute mis-orientation maps after the second and third treatments are found in Fig. 8. Of particular note is the similarity in the distributions of low levels of mis-orientations thus suggesting that these features are much less altered by the processes of corrosion than the elastic strain distributions. Moreover, there is little evidence of elevated mis-orientation in the grain boundary region that exhibits high elastic strain in Figs. 6 and 7. While the distribution of the mis-orientation is maximized by proximity to grain boundaries and particularly to triple points, no unusual levels of mis-orientation were found near the grain boundary that displayed high elastic stains.

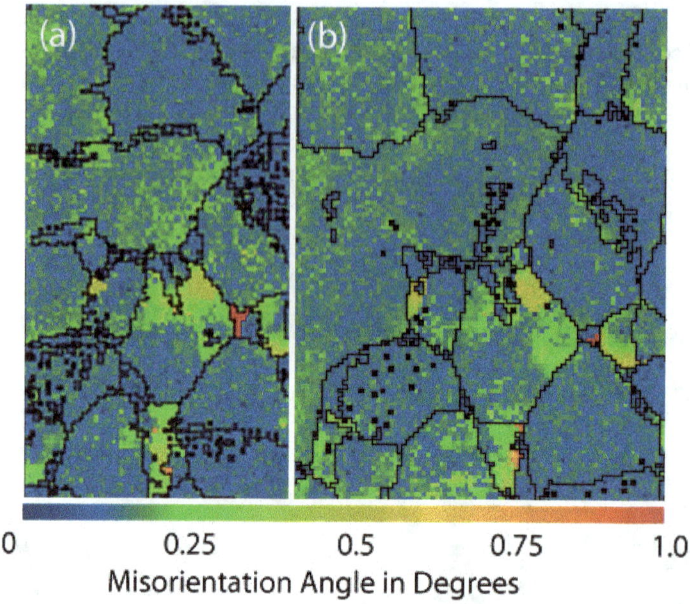

Figure 8: Comparison of absolute mis-orientation distributions for the same area of the flat Alloy 600 surface after the second and third corrosion treatments: (a) second treatment (b) third treatment.

4 Conclusions

This contribution has been primarily concerned with the relationship between the process of stress-corrosion cracking and its detection by an X-ray diffraction technique that provides a measure of elastic strain within the nexus of the crack origins. Our previous analysis of crack propagation conditions have suggested that changes in local elastic strain were easier to associate with crack opening than the plastic strain changes as measured by Laue-based mis-orientation Electron Beam Scattering Diffraction [15] with its shorter information depth probably being more sensitive to crack opening. In this study, two different forms of cracking were examined: one where the metal was pre-stressed; the other where all stresses originated from the chemical products of the corrosion process itself. In both cases it has been possible to suggest the presence of proto-cracks and actual cracks whose strain properties relate to the measured accumulation of unusual levels of stress/strain by Laue micro-diffraction.

The possible roles of surface oxide growth in the initiation of inter-granular cracking have been advanced here by the SEM observations of a compact surface oxide on the corroded alloy that undergoes some degree of stress-related cracking itself; this is believed to contribute to, if not initiate, the inter-granular cracking in the underlying alloy.

Acknowledgements

We are grateful for the financial assistance of NSERC and the UNINE program of the CANDU Owners Group. The Advanced Light Source is supported by the Director, Office of Basic Energy Sciences of the USDOE under Contract No. DE-AC02- 05CH11231. We thank Dr J. Chao for some experimental assistance and Dr S. Ramamurthy for providing the Auger depth profile.

References

[1] Chung, J.-S. & Ice, G.E., Automated indexing for texture and strain measurement with broad-bandpass X-ray microbeams. *Journal of Applied Physics*, **86**, pp. 5249–5261, 1999.

[2] Tamura N., McDowell, A.A., Spolenak, R., Valek, B.C., Bravman, J.C., Brown, W.L., Celestre, R.S., Padmore, H.A., Batterman, B.W. & Patel, J.L., Scanning X-ray microdiffraction with submicrometer white beam for strain/stress and orientation mapping in thin films. *Journal of Synchrotron Radiation,* **10**, pp. 137–143, 2003.

[3] Barabash, R.I., Ice, G.E. & Walker, F.J., Quantitative microdiffraction from deformed crystals with unpaired dislocations and dislocation walls. *Journal of Applied Phys*ics, **93**, pp. 1457–1465. 2003.

[4] Levine, L.E., Larson, B.C., Yang, W., Kassner, M.E., Tischler, J.Z., Delos-Reyes, M.A., Fields, R.J. & Liu, W., X-ray microbeam measurements of individual dislocation cell elastic strains in deformed single-crystal copper. *Nature Materials,* **5,** pp. 619–622, 2006.

[5] Ohashi, T., Barabash, R.I., Pang, J.W.L., Ice, G.E. & Barabash, O.M., X-ray microdiffraction and strain gradient crystal plasticity studies of geometrically necessary dislocations near a Ni bicrystal grain boundary. *International Journal of Plasticity*, **25(5),** pp. 920–941, 2009.

[6] Chao, J., Suominen Fuller, M.L., McIntyre, N.S., Carcea, A.G., Newman, R.C., Kunz, M. & Tamura, N., The study of stress application and corrosion cracking on Ni-16 Cr-9 Fe (Alloy 600) C-ring samples by polychromatic X-ray microdiffraction. *Acta Materialia*, **60(3)**, pp. 781–792, 2012.

[7] Chao, J., Suominen Fuller, M.L., Sherry, N., Qin, J., McInyre, N.S., Ulaganathan, J., Carcea, A.G., Newman, R.C., Kunz, M. & Tamura, N., Plastic and elastic strains in short and long cracks in Alloy 600 studied by polychromatic X-ray microdiffraction and electron backscatter diffraction. *Acta Materialia*, **60(15)**, pp. 5508–5515, 2012.

[8] Stahle, R.W. & Gorman, J.W., Quantitative assessment of submodes of stress corrosion cracking on the secondary side of steam generator tubing in pressurized water reactors: Part 1. *Corrosion*, **59(11)**, pp. 931–994, 2003.

[9] Rebak, R.B. & Szklarska-Smialowska, S., The mechanism of stress corrosion cracking of alloy 600 in high temperature water. *Corrosion Science*, **38(6)**, pp. 971–988, 1996.

[10] Kamaya, M., Measurement of local plastic strain distribution of stainless steel by electron backscatter diffraction. *Materials Characterization*, **60(2)**, pp. 125–132, 2009.
[11] Hur, D.H, Kim, J.S., Baek, J.S. & Kim, J.G., Inhibition of stress corrosion cracking of Alloy 600 in 10% sodium hydroxide solution at 315°C. *Corrosion,* **58(12)**, pp. 1031–1038, 2002.
[12] Sherry, N., Qin, J., Suominen Fuller, ML, Xie, Y., Mola, O., Bauer, M, McIntyre, N.S , Maxwell, D., Liu, D., Matias, E. & Armstrong, C., Remote internet access to advanced analytical facilities: A new approach with web-based services. *Analytical Chemistry,* **84(17)**, pp. 7283–7291, 2012.
[13] Chao, J., Mark, A., Suominen Fuller, M.L., McIntyre, N. S., Holt, R.A., Klassen, R.S. & Liu, W., Study of residual elastic- and plastic-deformation in uniaxial tensile strained nickel-based Alloy 600 samples by polychromatic X-ray microdiffraction (PXM) and neutron diffraction methods. *Materials Science and Engineering,* **A524(1-2)**, pp. 20–27, 2009.
[14] Payne, B.P., Keech, P.G. & McIntyre, N.S., X-ray photoelectron study of the oxides formed on nickel metal and nickel-chromium 20% alloy surfaces under reducing and oxidizing potentials in basic, neutral and acidic solutions. *15th Int. Conf. on Environmental Degradation*, eds. J.T. Busby, G. Ilevbare & P.L. Andresen, The Minerals, Metals & Materials Society, pp. 1065–1081, 2011.
[15] Shidmore, T., Buchheit R.G. & Juhas, M.C., Grain boundary energy vs. misorientation in Inconel® 600 alloy as measured by thermal groove and OIM analysis correlation. *Scripta Materialia*, **50(6)**, pp. 873–887, 2004.

Long-lasting outdoor atmospheric corrosion tests

M. Leiro & E. S. Ayllón
CITEFA, Argentina

Abstract

Since its introduction in 1933, the concept of patinas as a corrosion-controlling mechanism has been playing an important role in many structural applications. The analysis of the initial stages of the corrosive attack showed that this phenomenon, which was considered uniform at first, is actually a moderate one, non-metallic inclusions in structural steels being the preferred points of attack. Long-lasting tests were carried out in Argentina, at the urban site of Villa Martelli's outdoor test station. Specimens were obtained from a hot-rolled structural steel sample of 3.5 mm thickness, 45 mm wide and 300 mm long. The outdoor tests were carried out over 35 years. The results obtained by optical microscopy, electron scanning, and electron probe microanalysis showed that the addition of micro-alloying elements modifies steel behaviour. This modification means an increase in steel resistance to atmospheric corrosion.
Keywords: atmospheric corrosion, weathering steel, micro-alloying elements, patina, outdoor tests.

1 Introduction

The term "weathering steels" describes a class of low-alloy structural steels that develop an adherent protective rust layer, called patina, during exposure to the atmosphere. As a result of this, weathering steels, unlike ordinary plain carbon steels, do not require painting, which brings important advantages, namely avoiding associated costs of initial painting, periodic repainting, containment and disposal of blasting debris during repainting.

Since the beginning of the study of atmospheric corrosion, the latter was basically considered as an electrochemical phenomenon. The analysis of the initial stages of the corrosive attack showed that this phenomenon which was considered uniform at first is actually of a moderate extent, the non-metallic inclusions in steels being the preferred points of attack. Evans electrochemical model for the atmospheric corrosion shows the reactions that explain the behaviour of corrosion product films as well as the influence of micro-alloying elements on the size of colloidal aggregate particles that form patinas. There have been many propositions to explain the protection mechanism. However, there seems to be no generally accepted theory up to now [1]. Initially there were two tendencies in research: the first one was to attribute the protective nature to the chemical characteristics of the species present, and the second one was to attribute the said protective nature to the morphology of the protective films. Many techniques were used for the characterization of patina corrosion products, which were formed because of exposure to a natural environment. Non-metallic inclusions of MnS with or without Fe have a principal role at the beginning of the corrosion process. Klund [2] found that around the MnS inclusions there is an area, attributed to the steel cooling rate from its tapping temperature, along the inclusion contour. This is called MnS micro inclusions infected area and is very reactive. While the MnS inclusion works as a cathodic area, the said infected area works as an anodic one. The progress of this process leaves the non-metallic inclusion not providing support to the steel matrix, since the infected area is removed. Therefore, a hole or pit filled with corrosion products is originated and acts as an anodic area. Sometimes it is the inclusion itself the one that dissolves. This occurs when the inclusion has other chemical elements, like the lanthanum, which are easily attacked.

Previous results verified by optical microscopy, electron scanning, and electron probe microanalysis showed that the addition of rare earths modifies steel behaviour towards atmospheric corrosion because of the different chemical composition of the non-metallic inclusions developed in the process.

Szummer and Janik-Czachor [3] point out that the presence of non-metallic inclusions, which were morphological modified by rare earths, is detrimental to corrosion resistance, and that the CeS attack nucleating effect is bigger than the MnS one, usually found in steels.

2 Evans electrochemical model

Outdoor tests carried out during long periods of exposure have revealed that weathering steels (WS) with small additions of Cu, Cr, and P have much better atmospheric corrosion resistance than those steels without micro-alloying elements. Laboratory research by Pourbaix [4] and Pourbaix and de Miranda [5] has clearly shown that the said difference only occurs when steel samples are subject to repeated cycles of drying out and humidification. These tests like the ones carried out in a natural environment with a relative humidity always near saturation [6] have shown that there is no advantage in substituting ordinary steels for WS under the said environmental conditions. Many authors [7, 8] have pointed out the key influence of Cu as the most significant micro-alloying element to enhance atmospheric corrosion resistance, and they relate Cu to the formation of an inner denser and tightly adherent layer, which coats steel without cracks or defects. According to this, the tests carried out by Electron Probe Microanalysis (EPMA) [9] show that Cu concentrates mainly in those cavities resulting from the initial attack to non-metallic inclusions, and in the space among them, which is in contact with steel. Infrared (IR) spectroscopy tests show that the main components of protective patinas are first the δ ferric oxyhydroxides (δ-FeOOH), called lepidocrocite, and then the α and β oxyhydroxides and magnetite (Fe_3O_4) in smaller amounts.

The reduction of the particle size resulting from an enrichment of Cu occurs with a reduction of the pore size and a greater tortuosity in the aggregate. According to Evans model, the reduction conditions exist over a large part of the film if the pores are filled with water (from rain or mist). However, the oxidation conditions prevail during the drying out periods. The balance between these two conditions moves to the steel during the drying out periods and to the atmosphere during the humidification periods.

The excellent atmospheric corrosion properties of WS cannot entirely be explained by the presence of alloying elements. Although it seems that the phosphorus in steel improves the atmospheric corrosion resistance, it also seems that sulphur or sulphur oxides in air have an effect on protective patina formation on the steel.

Electrochemical curves show that at the potential at which the maximum anodic current is produced, the oxidised element is now the Fe^{+2} intermediary on the surface of FeOOH globules instead of the magnetite [10]. Moreover, the patina inner layer would be protected due to an increase of its compactness, according to test results by Moroishi and Satake [9], Inouye *et al.* [11] and Rosales [12]. Where there are neither micro-alloying elements or long humidification periods, the δ-FeOOH would have been transformed into Fe_3O_4, because there is no controlling effect on the Fe_3O_4 production, which does not have a protective attribute, as reported by Kihira *et al.* [13].

3 PA steel formulation

According to what has been mentioned in previous work [14], the conclusion is that a formulation containing the necessary micro-alloying elements (Cu, Cr, and Si) to produce the film of corrosion products on steel and non-metallic inclusions – modified by rare earths additions (particularly Ce) – is considered to meet both criteria: weatherability and better mechanical properties.

Following these criteria, a weathering steel experimental tapping was planned with the addition of rare earths as modifiers of non-metallic inclusions from their morphological as well as from their chemical reactivity point of view.

A minimum C content was planned to provide the mechanical strength and ferritic attribute of structural steels. The percentage of the main micro-alloying elements, Cu and Cr, were fixed on a practical basis. The available ferroalloys for steel production were considered as well as their use under the fusion conditions to be adopted. The recommendations given by specialised literature were adopted for the rare earths percentage and addition process to the ladle [15].

The percentage of Ni added – almost similar to the Cu's percentage – is the suggested one to neutralise the detrimental effect of the latter in a hot-rolled process, widely known as hot shortness. The vanadium – together with the aluminium – is a ferritic grain-refining agent. Due to its availability and ease of handling, the rare earths silicide was adopted (Fe: 35%, Si: 31% and rare earths: 34%; the latter with a composition of Ce: 50%, La: 25%, Nd: 10% and Pr: 5%) in a percentage of 1.0 g/kg of melting steel as a means of containing these rare earths.

The chemical composition of steel – analysed in the hot-rolled product – in comparison with the one of the currently produced Cor-ten A and the QST-52 non-alloyed steel is shown in Table 1.

Table 1: Chemical composition of hot-rolled PA steel, as well as Cor Ten A steel and QST-52 non-alloyed steels used as a reference samples.

Steel	C	Si	Mn	P	S	Cu	Cr	Ni	Al	Sn	As	V
PA	0.15	0.41	1.01	0.086	0.105	0.46	1.0	0.38	0.021	0.01	0.013	0.006
Cor-ten A	0.10	0.36	0.34	0.093	0.018	0.31	0.60	0.12	------	-----	0.005	0.008
QST-52	0.16	0.36	1.36	0.013	0.035	0.05	0.07	0.06	------	-----	-------	-------

An induction furnace was used and 4 ingots of 60 kg net were tapped (without dead heads). The final temperature of the process after a careful bath deoxidisation by Mn, Al and Si addition was 1673°C in the furnace and 1620°C in the ladle. The temperature of this process was between 1590 and 1600°C.

After the tapping, the ingots were heated again at a rolling initial temperature of 1150°C. The final rolling temperature was between 950–1000°C. The final thickness was 3.5 mm and the hardness was H_B = 283. Upon completion of the rolling, the steel underwent a thermal treatment process to set the mechanical properties at the usually recommended ones under the Cor-ten A steel specifications (ASTM A-588 and A-242):

- Austenitised treatment: T = 870/880°C for 10 minutes.
- Water cooling at room temperature.
- Tempering at 750°C for 1 hour.
- Cooling inside furnace down to 400°C, then in the air.
- Final hardness of H_B = 160, corresponding to a resistance to rupture of 53 kg·mm^{-2} (52 MPa).

4 Open air exposure test

Long-lasting tests were carried out at the urban site of Villa Martelli's outdoor test station [16]. Specimens were obtained from 45 mm wide and 300 mm long hot-rolled material of 3.5 mm thickness and from the same cold-rolled of 2 mm thickness. Holes of 22 mm diameter were made to set the samples in the same frame insulators, which were used in the previous tests (Fig. 1) [17]. The superficial preparation was the sanding up to Sa3 grade of SIS 05 59 00 standard (1967) and weighing the samples. Three samples exposed over 3, 6, 9, 12, 18, 24,

30, 36, 42, 48, and 60 months were removed each time, and after measuring their electrochemical potential in situ [18], they were weighed (for weight increase).

Besides there were samples exposed over 35 years and the results were obtained by optical microscopy, electron scanning and electron probe microanalysis, following the same techniques which were applied to the those samples exposed over 60 months.

Then, acid pickling was carried out at 80°C in a HCl solution (50%) with urotropine (0.1%) as inhibitor of steel attack. The samples were weighed at intervals of 10 minutes each until achieving almost linear weight variations between 30 and 50 minutes, according to DP/8407 ISO standard. The ΔW (g) value was determined by the intersection of the line obtained by linear regression by minimum squares with the vertical axis (Fig. 2) [19].

Figure 1: PA steel outdoor test at Villa Martelli's urban site.

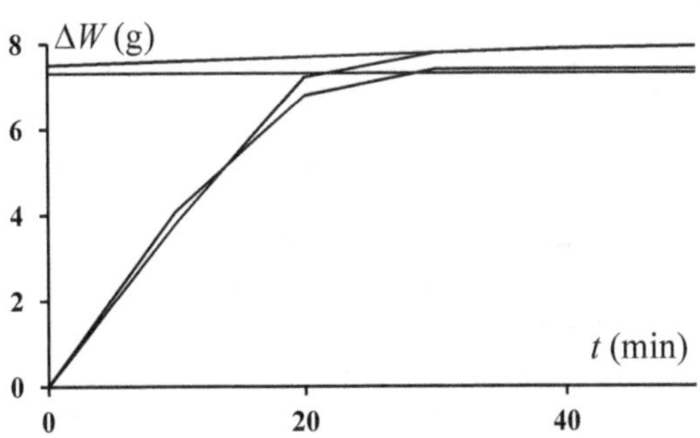

Figure 2: Weight loss based on pickling time and pertinent linear regression.

Three hot-rolled PA steel specimens were exposed for 5 years in Villa Martelli's station. The results were graphically represented by the equation

$$\Delta W = kt^n \tag{1}$$

The constants k and n were determined from experimental data by regression analysis; the results are shown in Table 2.

Table 2: Constants of the linear regression.

Steel	Thickness		Values
PA hot-rolled	3 mm	k	1.2177
		n	0.1652
PA cold-rolled	2 mm	k	1.9232
		n	0.3641

As shown in Fig. 3, the values of exponent n for hot-rolled PA steel are lower than those for Cor-ten A steel and the QST 52 non-weathering steel, which were determined in previous tests [17].

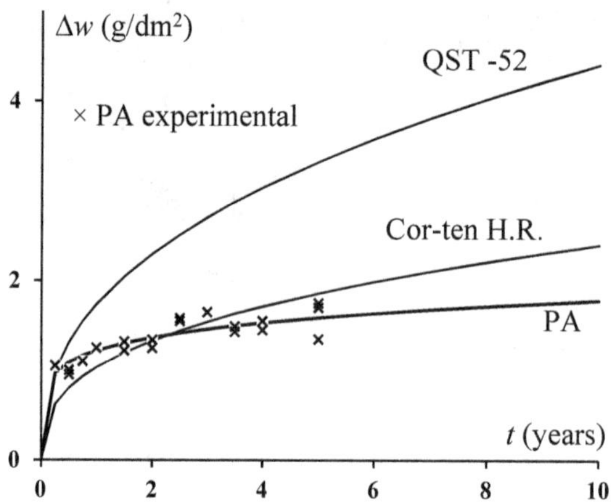

Figure 3: Time-based weight loss of hot-rolled PA steel, extrapolated at 10 years.

Figs. 4–7 show the weight loss ΔW (g) in relation to the weight of the corrosion products that form the patina as well as the line of experimental data obtained by linear regression at Villa Martelli's urban site. In those same figures, two additional lines can be seen passing through the coordinate origin; these correspond to two extreme cases of patina chemical composition. If all the Fe dissolved in the corrosive process had combined and formed FeOOH, the representative points would have been located on the line with the lowest slope; if it had been fully combined forming magnetite, they would have been located on the line with the highest slope relative to the horizontal axis.

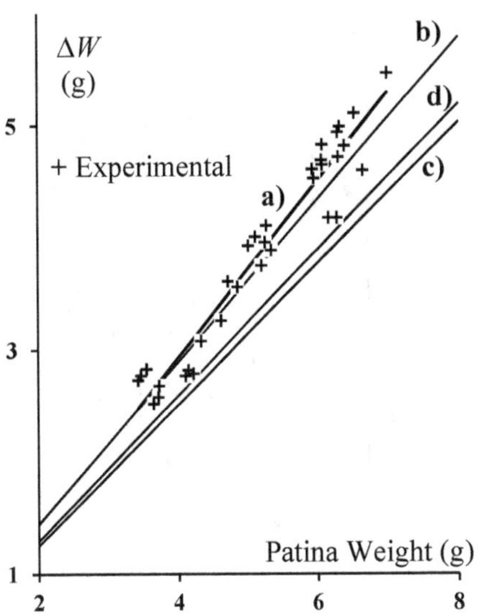

Figure 4: Hot-rolled PA steel in urban site: a) Regression; b) Fe3O4; c) FeOOH; d) 0.925 FeOOH + 0.075 Fe3O4.

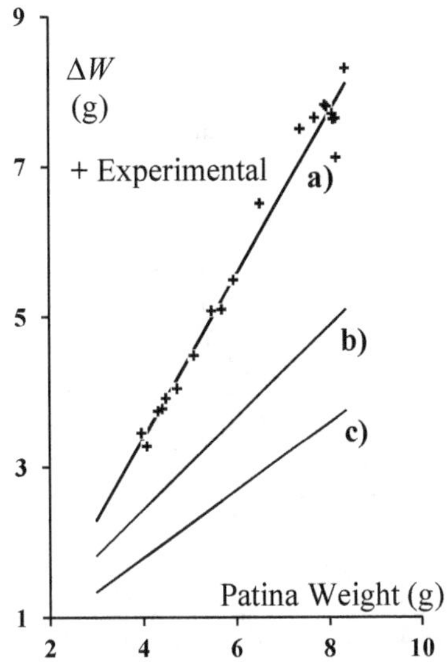

Figure 5: Cold-rolled PA steel in urban site: a) Regression; b) Fe3O4; c) FeOOH.

Since a mixture of the two compounds forms the patinas, the representative points should be located in the area between these two boundaries.

The abscissa difference between the line resulting from linear regression of experimental data and the lines pertaining to Fe_3O_4 and FeOOH provides the extreme values of corrosion products weight loss by rain wash. If the experimental result for hot-rolled PA steel is compared with those for Cor-ten A steel and QST-52 steel (Figs. 6 and 7) [17], it will be noticed that the corrosion products that are let loose from PA steel are decreased relative to those from Cor-ten A steel and non-weathering steel, when they are all exposed to open air.

The metal weight swept away by dew or rain-wash may be estimated as follows:

The difference between the abscissa W_P of the line obtained by regression from experimental data for ΔW (g) and the one corresponding to the FeOOH (W_c) – as the most unfavourable case – represents the weight loss by wash ΔW_L:

$$\Delta W_L = W_c - W_P$$

Considering the abscissas as functions of the common ordinate ΔW (g), the weight loss of the specimens is given by

$$\Delta W_L = (b_c - b_P)\,\Delta W \tag{2}$$

where b_c and b_p are, respectively, the gradients of the FeOOH line and the line obtained by regression from experimental data relative to the ΔW vertical axis.

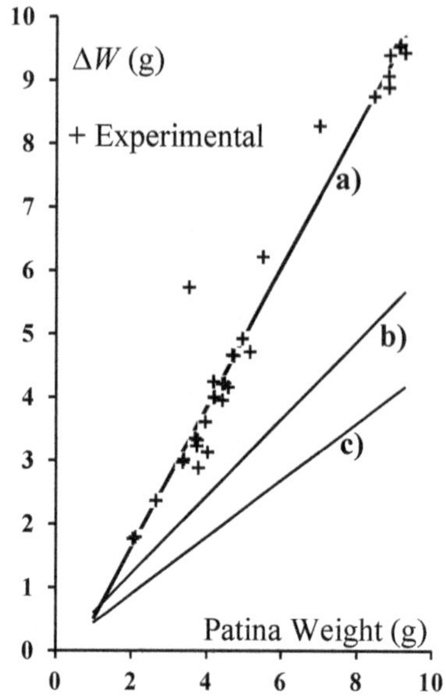

Figure 6: Hot-rolled Cor Ten A steel in urban site: a) Regression; b) Fe3O4; c) FeOOH.

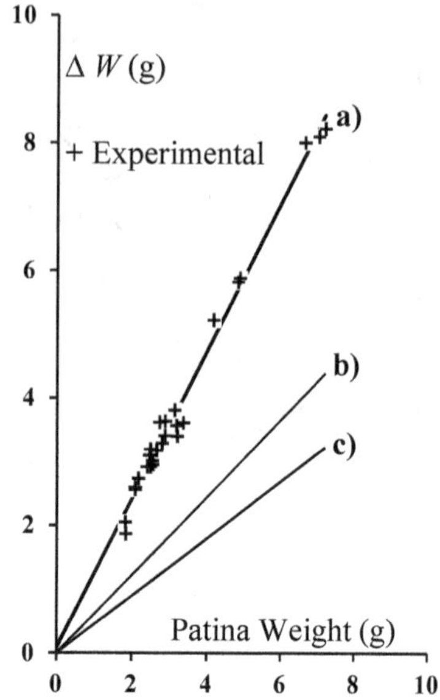

Figure 7: Hot-rolled QST – 52 steel in urban site: a) Regression; b) Fe3O4; c) FeOOH.

According to eqn (1)

$$\Delta w(\mathrm{g}\cdot\mathrm{dm}^{-2}) = \frac{\Delta W(\mathrm{g})}{A(\mathrm{dm}^2)} = k't^n$$

where A (dm^2) represents the specimen area exposed to the open air; k and n are the regression coefficients depending on the outdoor exposure site and the material. Substituting from the above equation into eqn (2)

$$\Delta W_L = (b_c - b_p)Ak't^n = Kt^n \qquad (3)$$

The values of the parameters in eqn (3) for the calculation of ΔW_L are shown in Table 3 and results are shown in Fig. 8.

Table 3: Values for parameters in the relations of eqns (3) and (4).

Steel	k'	n	A (dm^2)	b_c	b_p	K	nK
	Urban site						
PA H.R.	1.2173	0.1652	2.87	1.5910	1.2834	1.0746	0.1775
PA C.R.	1.9232	0.3641	3.01	1.5910	0.9522	3.6979	1.3664
Cor-ten A H.R.	1.2734	0.1901	3.00	1.5910	0.8598	0.8598	0.5310

H.R.: Hot-rolled; C.R.: Cold-rolled.

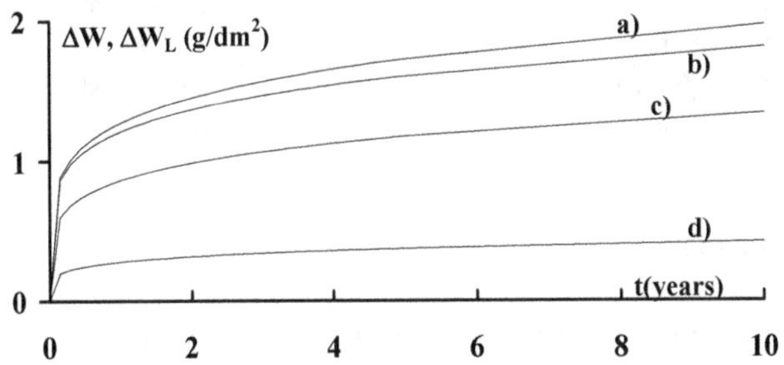

Figure 8: Comparison between the total weight loss and the corresponding one due to dew and/or rain-wash at Villa Martelli's urban site. a) ΔW Cor-ten A; b) ΔW PA H.R; c) ΔW_L Cor-ten A; d) ΔW_L PA H.R..

The rate of change of the weight loss by wash ΔW_L with respect to time is expressed by the function derivative:

$$\frac{\mathrm{d}}{\mathrm{d}t}\Delta W_L = nKt^{n-1} \tag{4}$$

with the value of nK given in Table 3 while the rate of change of ΔW_L can be seen in Fig. 8.

5 Corrosion potential in-situ measurements

The purpose of this technique is to use a non-destructive methodology to evaluate the characteristics of the corrosion products, which are formed on steels in the open air.

An agar KCl bridge-electrode holder is used to set the calomel-saturated reference electrode, and also the potential recording equipment, i.e. a millivoltmeter or high impedance recorder. As the heading suggests, the technique can be used to record the potential evolution of the specimens exposed to the open air, without removing them from the site.

In order to establish electric continuity on its corners, the sample is polished as little as possible on both sides, and contact is made through a small clamp, wire and plug to the recording equipment. The other equipment input is connected to the calomel-saturated reference electrode, which has been already set in the agar KCl bridge-electrode holder. The electrode holder open bottom is the specimen area, whose corrosion potential or open circuit (E_c) is to be measured. This is done in three different spots of the specimen starting in 10 minutes upon the closure of the circuit. Readings every 30, 60 and 90 s are recorded keeping the electrode holder in contact with every area of the specimen to measure. The potential is the result of the readings average at a same interval in the three chosen areas of the sample.

In the case of the Cor-ten A weathering steel exposed at Villa Martelli's outdoor test station for 15 years, the value obtained was +217 mV_{sce}, the same value as the one given by Pourbaix [18] for well weathered steels in industrial sites. According to the same technique, the average value of 3 samples at 35 years was +150 mV_{sce}.

6 Conclusions

Based on the long time experimental results, lasting 35 years, the conclusions are that a formulation containing the necessary micro-alloying elements (Cu, Cr, and Si) produces a protective corrosion film on the steel.

Experimental samples of PA steel were tested outdoors in an urban exposure site which produced a very effective passive patina. This aim was achieved through an increase of the density, great chemical reactivity and very small size of non-metallic inclusions with spherical morphology.

Steel with non-metallic inclusions modified by rare earth additions, particularly Ce, was considered to combine weatherability and better mechanical properties.

The outdoor tests were carried out over 35 years. The results were obtained by optical microscopy, electron scanning and electron probe microanalysis showing that the addition of micro-alloying elements modifies steel behaviour. This modification results in increased steel resistance to atmospheric corrosion.

References

[1] Merino, P., Sanchez, A. & Porto E., *Mapas de España de Corrosividad Atmosférica*, eds. M. Morcillo Linares & S. Feliú Matas, Programa CYTED, p. 319, 1993.

[2] Klund, E., *Scandinavian Journal of Metallurgy*, **1**, p. 331, 1972.

[3] Szummer, A. & Janik-Czachor, M., Effect of rare earth metals and the stability of sulphide inclusions on localized corrosion of low alloy steels. *Werkstoffe und Korrosion*, **33(3)**, pp. 150–154, 1982.

[4] Pourbaix, M., Theoretical and experimental considerations in corrosion testing. *Corrosion Science*, **12**, pp. 161–190, 1972.

[5] Pourbaix, M. & de Miranda, L., *Passivity and its Breakdown on Iron and Iron Base Alloys*, eds. R.W. Staehle & H. Okada, USA-Japan Seminar, NACE, p. 47, 1976.

[6] Bohnenkamp, K., Tribune du CEBEDAU, Tech. Rep. 324, 1970.

[7] Mizawa, T., Kyono, T., Suëtaka, W. & Shimodaira, S., The mechanism of atmospheric rusting and the effect of Cu and P on the rust formation of low alloy steels. *Corrosion Science*, **11(1)**, pp. 35–48, 1971.

[8] Suzuki, I., Masuko, N. & Hisamatsu, Y., Electrochemical properties of iron rust. *Corrosion Science*, **19(8)**, pp. 521–535, 1979.

[9] Moroishi, T. & Satake, J., Water permeability of rust layer formed by atmospheric corrosion. *Transactions of the Iron and Steel Institute of Japan*, **11**, p. 397, 1971.

[10] Stratmann, M., Bohnenkamp, K. & Engell, J., An electrochemical study of phase-transitions in rust layers. *Corrosion Science*, **23(9)**, pp. 969–985, 1983.
[11] Inouye, K., Ichimura, K., Kaneko, K. & Ishikawa, T., The effect of copper (II) on the formation of y-FeOOH. *Corrosion Science*, **16(8)**, pp. 507–517, 1976.
[12] Rosales, B.M., Superficial films behaviour during atmospheric corrosion. *Scientific Meeting on Fe Physicochemical and its Oxides*, August 1986.
[13] Kihira, H., Ito, S. & Murata, T., Quantitative classification of patina conditions for weathering steel using a recently developed instrument. *Corrosion*, **45(4)**, pp. 347–352, 1989.
[14] Ayllón, E.S., Special weathering steel, a contribution to environmental protection. *WIT Transactions on Engineering Sciences*, **54**, pp. 329–338, 2007.
[15] Zenarruja, R., Moral, H, Spira, H., Waellkens, G., Ribera, E., Borras, A., Ferro, L. & Reggiardo, H., Comité de Acería (Steel Mill Committee), IAS, Chapter 2, p. 55, 1983.
[16] Rosales, B.M. & Leiro, M.C., Jornadas SAM′95, Córdoba, Argentina, p. 295, May 16–20, 1995.
[17] Ayllón, E.S., Granese, S.L. & Rosales, B.M., Electrochemical response of weathering and plain C steels in different environments. *Corrosion Reviews*, **9(3-4)**, p. 245–268, 1990.
[18] Pourbaix, M., Une method electrochimique rapide de predeterminatión de la corrosion atmospherique, CEBELCOR, R.T. 160, **109**, pp. 1–14, 1969.
[19] Fernández, A., Navarro, J.A. & Rosales, B.M., *Proc. 4° Congreso Iberoamericano de Corrosión y Protección*, (4° Latin American Congress on Corrosion and Protection), Mar del Plata, Argentina, **1**, p. 143, October 25–30, 1992.

Accelerated corrosion tests on waste-gated turbocharger's adjustable and fixed end-links

K. Wilton-Smith, Z. Khan, A. Saeed & M. Hadfield
Engineering and Computing, Bournemouth University, UK

Abstract

A turbocharger which has been mounted to a heavy duty automotive diesel engine can be subjected to various aggressive environments and a range of temperatures varying from ambient to 670°C. Extreme operating environments and temperatures can initiate corrosion leading to the degradation of the components. The present research evaluates the corrosion resistance of the zinc-coated and stainless steel waste-gate end-links based on accelerated corrosion and dust particles tests, using an environmental test rig. Samples were exposed to the tests for 504 hours. Images of the specimens were captured to record corrosion build-up and its progression on the surfaces during the exposure time. These results were then used to evaluate the corrosion resistance ability of both, zinc coated fixed and stainless steel adjustable end-links. Accumulation of corrosion and the build-up of salt deposits were observed on both end-links during and after the experiments.
Keywords: accelerated corrosion, turbocharger, end-links, extreme environment.

1 Introduction

Mechanical component failures through surface degradation are often the result of corrosion (chemical/electrochemical) and wear mechanisms. Surface degradation due to corrosion arises through an attack of reactive compounds on surfaces and the subsequent product removal through a mechanical action. Mechanical degradation through wear is often the result of friction, abrasion, impact and fatigue. Failures because of corrosion and wear are time dependent and nonlinear. Surface degradation results in partial or complete component malfunction, higher maintenance costs and risks to the structural integrity [1–3].

This contribution presents the effects of corrosion and dust particles on the external turbocharger components stainless steel adjustable and zinc coated steel fixed end-links. This research is motivated by the turbocharger failure which occurs in service because of corrosion in the end-links. An end-link is a threaded rod integrated into the turbocharger waste-gate assembly. The end-link is threaded to the end of the waste-gate's actuator rod with the opposite end being attached to lever linkage of the waste-gate system, which is placed in a circlip shown in Fig. 1.

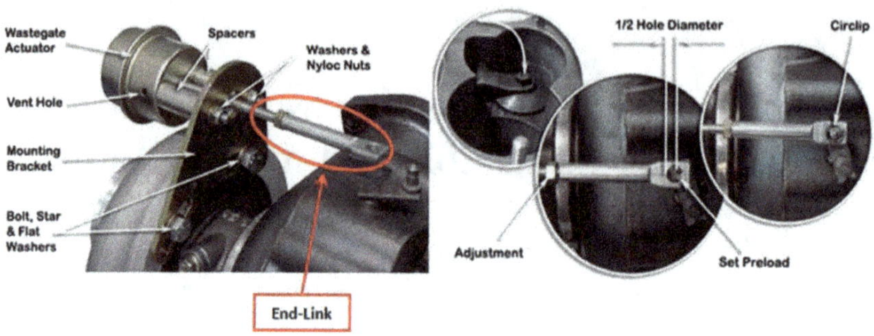

Figure 1: Turbocharger waste-gate assembly illustrating end-link locations (image courtesy of Safari Turbo).

The waste-gate assembly is a boost-controlling bypass value situated on the exhaust-end of the turbocharger. Waste-gate functions by limiting the exhaust gas passing through the turbine wheel of a turbocharged engine system. Waste-gate regulates the maximum boost pressure produced by the turbocharger to prevent and protect the turbocharged engine system from over-boosting. The energy produced is transferred back into the exhaust system of the turbine which in turn reduces the power driving the turbine wheel. This process helps the turbine wheel to match the required power for a given boost level. Turbocharger's turbine rotates at around 200k rpm, which is approximately 30 times faster than normal car engines. Since the turbine end is connected to the exhaust manifold of the engine, the temperature in the turbine reaches a value as high as 670°C. Higher temperatures are a common cause of accelerated corrosion. During turbocharger operation, if the temperature is high, the materials will oxidise faster resulting in scaling. A thick scale formed because of overheating is critical to component degradation. As end-links are assembled within close proximity of the turbine housing, they are also subjected to high temperatures, as well as aggressive corrosive environments; these factors lead to accelerated component degradation [4, 5]. For the prolong performance and lifetime of the turbocharger components, it is important to understand the materials utilised and the environment to which they are exposed. Environment plays a vital role in corrosion; the main factors responsible for corrosion are relative humidity (RH), temperature, salts, pollutants concentration and time of wetness [6, 7].

To assess corrosion, field tests and laboratory based tests are performed. A field test involves exposure of the test specimen to the environment for a long period of time and that is the reason for being undesirable. Laboratory based corrosion tests involve increased corrosion activity to obtain results faster than field corrosion tests. Examples of laboratory based corrosion tests are simulated environmental and salt-spray cabinet corrosion test. These laboratory tests must be able to keep the corrosion mechanism identical to the field corrosion tests [8–10]. Results of accelerated corrosion tests on the waste-gated adjustable and fixed end-links are reported in this contribution.

2 Experimental methodology

2.1 Samples exposure

The experiment was conducted for 504 hours in total. 45 samples of each zinc coated and stainless steel end-links were exposed to the accelerated corrosion test. All the samples were inserted in the corrosion chamber simultaneously. Images of the samples were recorded at 18th, 26th, 92nd and 115th hour of exposure. The corrosion test was suspended at the 252nd hour, when 15 samples of each zinc coated and stainless steel were removed from the chamber in order to evaluate corrosion accumulation and their operating performance. The test was then resumed for another 252 hours as normal and images were captured at 278th, 305th, 352nd, 447th, 471st and 504th hour of exposure. The corrosion test was completed after 504 hours and all the samples were removed to evaluate corrosion build-up and their operational performance.

2.2 Corrosion chamber and test conditions

Accelerated corrosion tests were performed utilising a -330 environmental chamber shown in Fig. 2.

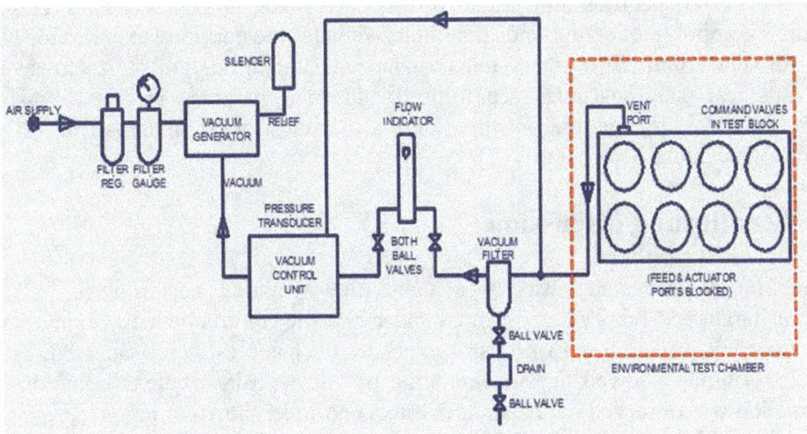

Figure 2: Schematic of the -330 environmental chamber.

The environmental chamber test was run under salt-spray accelerated corrosion (ASTM B 117) and Arizona test dust (ISO 12103 – A1) conditions simultaneously. Non-metallic separators were used to avoid contact among the samples during the test, as illustrated in Fig. 3.

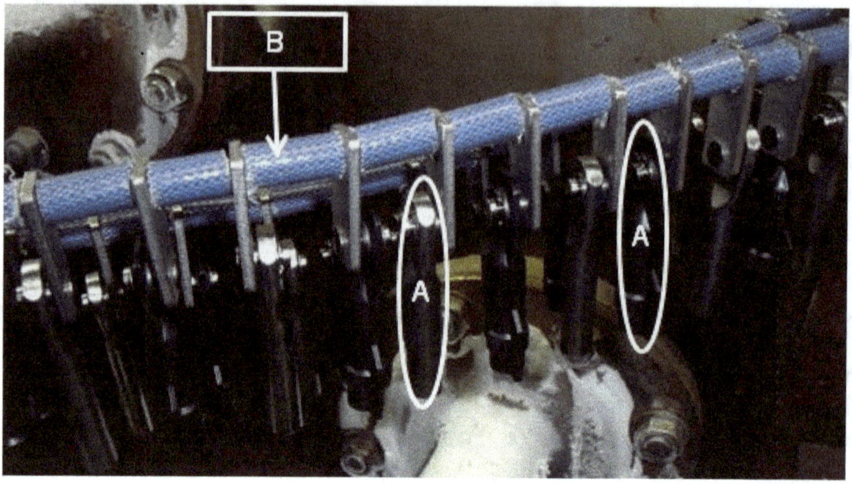

Figure 3: End-links exposure during corrosion test in the chamber (A) shows end-links; (B) shows none-metallic separator among the samples.

ASTM B117 is mainly used to evaluate corrosion resistance of the painted or metal coated samples. The salt solution contains dissolved 5% sodium chloride in distilled water in a pH range of 6.5–7.2 atomised as a fog into the chamber kept at 35°C. Sodium chloride should be free from copper and nickel and may not contain more than 0.1% of sodium iodide [11].

ISO 12103 specifies four grades of test dusts made form desert sand. Desert sand is composed of compounds that motor vehicles are normally exposed to. The performance of the filtration systems can be evaluated through this test. However, in this test, sand particles are utilised and sand particles possess abrasive characteristics; for that reason, this test is also used in wear studies of bearings, seals, windshield wipers, etc. [12].

3 Results and discussion

The aim of this study was to evaluate the operating performance of the turbocharger end-links when operated under extreme environment and exposed to dust particles. Just after 18 hours of exposure the samples were inspected. Zinc coated samples showed the accumulation of salt deposits whereas initiation of corrosion was observed on the stainless steel end-links shown in Fig. 4.

After 138 of exposure, thicker salt deposits on zinc coated end-links were observed. Stainless steel samples exhibited good resistance to salt accumulation however a slight increase in corrosion build-up was noticed. 15 samples of each kind of end-links were removed from the chamber after 252 hours of exposure in order to evaluate their operating performance and subsequent salt and corrosion accumulation. After 252 hours of exposure, the zinc coated end-links showed

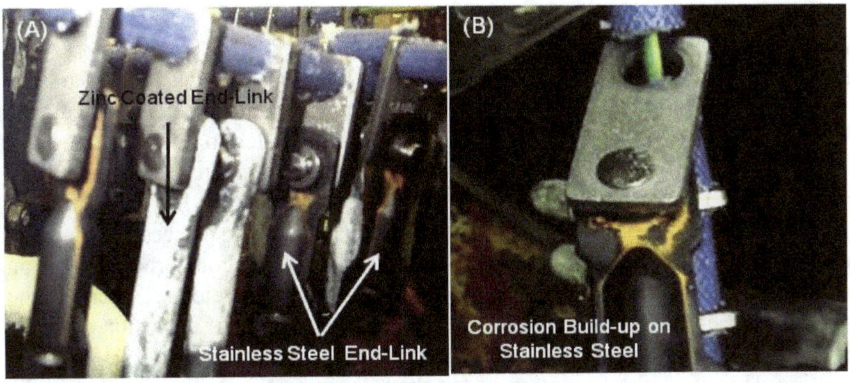

Figure 4: (A) Sample condition after 18 hours of exposure illustrate salt deposits on zinc coated and corrosion build-up on stainless steel end-links; (B) close up of the stainless steel end-link.

accumulation of thicker salt deposits and the initiation of corrosion. However, stainless steel end-links had no salt deposit formation on their surfaces. The performance of both kinds of end-links was evaluated through a functionality test; the movement of the zinc coated ones was observed to be restricted because of the salt and corrosion deposits. The stainless steel ones, however, showed free movement and were fully functional. Formation of red rust was observed after 447 hours on almost every sample exposed to the test shown in Fig. 5.

Figure 5: Sample exposure for 447 hours of the test; (A) and (B) indicate zinc coated and stainless steel end-links, respectively.

The test was terminated at 504 hours of operation. All the samples showed formation of red rust on their surface, however no salt deposits were observed on the stainless steel ones. The operation of both kinds of end-links was evaluated after test completion. The stainless steel samples exhibited good performance and were fully functional. However zinc coated end-links failed to operate.

Zinc coating provides good protection for iron and steel products from corrosion in various environmental conditions. Zinc coating has the ability to form a dense, adherent corrosion product film on metal surfaces keeping the corrosion rate considerably low. Zinc is more reactive compared to iron on ambient conditions but a thin film of corrosion products develops resulting in the reduction of corrosion rate of iron [13, 14]. The thick layer of corrosion products on the zinc coated end-links could be attributed to such characteristics of the zinc coating. However, the formation of the thick layer on the surface has led to the component failure in-terms of operating performance.

Stainless steel is widely in use due its ability to resist corrosion. For that reason the use of stainless steel is growing in both domestic and industrial applications. Stainless steel forms a thin invisible passive layer in oxygen-containing environments. This passive film is normally chromium oxide, which protects the surface against corrosion. For the creation of chromium oxide layer on the surfaces, chromium in the range of 16–28 wt% is added to stainless steel. Complete or partial breakdown of the chromium oxide layer leads to the corrosion of stainless steel. Localised breakdown of the passive film causes pitting or crevice corrosion while failures in the passive film on a larger scale often cause uniform corrosion. Solutions with low or high pH result in the breakdown of passive layer on a larger scale and the subsequent corrosion propagation is determined by the corrosive environment [15, 16]. During the salt fog test, the pH was kept between 6.5 to 7.2 at 35°C and 5% salt solution. Corrosion accumulation on stainless steel end-links is attributed to the conditions when the pH was less or greater than pH 7. Environments containing chlorides make the metals more susceptible to corrosion [17]. 5% of salt is more than found in sea water and therefore has increased the conductivity of the electrolyte on surfaces of the samples resulting in accelerated corrosion. Salts provide points of corrosion nucleation, keep the surfaces wet for longer time at lower relative humidity, and salt makes the electrolyte conductive thus increasing corrosion [18–20]. Higher salt containing environments have resulted in accelerated corrosion on both end-links.

4 Conclusion

The purpose of these tests was to evaluate the corrosion resistance of end-links when exposed to extreme environmental conditions and their ability to function when operated in an environment containing dust particles. Both zinc coated and stainless steel end-links accumulated corrosion during the test. Nonetheless, in terms of operation, stainless steel showed enhanced performance compared to zinc coated end-links. The performance of zinc coated end-links was compromised by exposing them to extreme environmental and working conditions.

References

[1] Wood, R.J.K., Tribocorrosion. *Shreir's Corrosion*, ed. J.A.R. Tony, Elsevier: Oxford, pp. 1005–1050, 2010.

[2] Saeed, A., Khan, Z., Garland, N. & Smith, R., Material characterisation to understand various modes of corrosion failures in large military vehicles of historical importance. *WIT Transactions on Engineering Sciences*, **72**, eds. A.A. Mammoli, C.A. Brebbia & A. Klem, pp. 95–106, 2011.

[3] Scott, D. Wear analysis. *Physics in Technology*, **14**, pp. 133-139, 1983.

[4] Cummins Turbo Technologies, http://www.holset.co.uk/mainsite/files/3_4-Technical%20Papers.php

[5] Safari 4×4 Engineering, Safari Power. http://www.safarisnorkel.com/turbo/ turbo_product.htm

[6] Schweitzer, P.A., *Fundamentals of Corrosion: Mechanisms, Causes and Preventative Methods*, CRC: Boca Raton, Fla., 2010.

[7] Ahmad, Z., *Principles of Corrosion Engineering and Corrosion Control*, Basic concepts in corrosion, Butterworth-Heinemann: Oxford, pp. 9–56, 2006.

[8] Schmidt, D.P., Shaw, B.A., Sikora, E., Shaw, W.W. & Laliberte, L.H., Comparison of testing techniques used to analyze the corrosion resistance of sacrificial coating systems. *Corrosion*, **63**, pp. 958–974, 2007.

[9] ASTM International, ASTM G85-11, *Standard Practice for Modified Salt Spray (Fog) Testing*.

[10] Boelen, B., Schmitz, B., Defourny, J. & Blekkenhorst, F., A literature survey on the development of an accelerated laboratory test method for atmospheric corrosion of precoated steel products. *Corrosion Science*, **34**, pp. 1923–1931, 1993.

[11] ASTM International, ASTM B 117-16, *Standard Practice for Operating Salt Spray (Fog) Apparatus*, pp. 1–12, 2011.

[12] BS ISO 12103-1:1997, *Road Vehicles – Test Contaminants for Filter Evaluation -Part 1: Arizona Test Dust*, BSI – British Standards Institution, pp. 1–6, 1997.

[13] Corrosion Doctors, Rust chemisty. Online: http://corrosion-doctors.org/Experiments/rust-chemistry.htm

[14] Marder, A.R., The metallurgy of zinc-coated steel. *Progress in Materials Science*, **45**, pp. 191–271, 2000.

[15] Iversen, A. & Leffler, B., Aqueous corrosion of stainless steels. *Shreir's Corrosion*, vol. 3, eds. B. Cottis, M. Graham, R. Lindsay, S. Lyon, T. Richardson, D. Scantlebury & H. Stott, Elsevier: Amsterdam, pp. 1802–1878, 2010.

[16] Truman, J.E., Coleman, M.J. & Pirt, K.R., Note on influence of nitrogen content on resistance to pitting corrosion of stainless steels. *British Corrosion Journal*, **12**, pp. 236–238, 1977.

[17] Nishikata, A., Yamashita, Y., Katayama, H., Tsuru, T., Usami, a., Tanabe, K. & Mabuchi, H., An electrochemical impedance study on atmospheric

corrosion of steels in a cyclic wet-dry condition. *Corrosion Science*, **37(12)**, pp. 2059–2069, 1995.

[18] Corvo, F., Mendoza, A.R., Autie, M. & Betancourt, N., Role of water adsorption and salt content in atmospheric corrosion products of steel. *Corrosion Science*, **39**, pp. 815–820, 1997.

[19] Saeed, A., Khan, Z. & Montgomery, E., Corrosion damage analysis and material characterization of Sherman and Centaur – The historic military tanks. *Materials Performance and Characterization*, **2**, pp. 30–44, 2013.

[20] Neufeld, A.K., Cole, I.S., Bond, A.M. & Furman, S. A., The initiation mechanism of corrosion of zinc by sodium chloride particle deposition. *Corrosion Science*, **44**, pp. 555–572, 2002.

Investigation on contact surface damage of an automotive connector by fretting corrosion

M. A. Beloufa
PRISME Institute EA 4229, ENSI de Bourges, France

Abstract

The main cause of electrical contact resistance degradation by corrosion is the vibration of contact interfaces. The purpose of this article is to analyse the change of contact resistance under vibration testing for uncoated sphere/plane contact made of new high-copper alloys. The influence of electrical and mechanical properties of materials and mainly the hardness on contact resistance was studied in this work. During the fretting test, the contact point was submitted to 16,000 vibration cycles under a fretting amplitude of 50 µm and 1 Hz frequency. The spherical part was fixed, while the plane part was submitted to a relative motion. At the end of the test, the fretted surfaces and the wear debris were analysed by scanning electron microscope and energy dispersive X-ray spectroscopy to evaluate damage, oxidation and elemental composition present in the wear surfaces. In addition, the measurement of the wear track profile using a 3D surface scanning system was introduced. Increases in contact resistance and contact temperature have been examined during the fretting test. The results showed that the contact resistance for the harder alloy was higher than that obtained for the other materials. In addition, topographic measurement showed that the small wear track corresponds to the harder material. Finally, depending on the hardness and the resistivity, these uncoated materials influence fretting appearance and its level.
Keywords: contact resistance, EDX and SEM analysis, fretting corrosion, high-copper alloys, wear surface, 3D scanning profiler, crack, delamination.

1 Introduction

Electrical connectors used in telecommunication, automotive and many other applications are exposed to degradation due to the vibration of mated contact surfaces. External mechanical vibration is the main cause of the fretting motion between contacting surfaces which is manifested by the mechanical and electrical damage of contact interfaces [1]. The failure process begins by wear of contact surfaces then oxidation of these surfaces or of wear debris which can cause rapid and dramatic increases in electrical contact resistance, proceeding to virtual open circuits [2–4]. Low-level relative movement between mated contacts can also be induced by differential thermal expansion of materials [5].

Determination of contact resistance changes during fretting of coated contact materials has been the subject of many investigations [6–10]. Unfortunately, contact materials with different coating (silver, palladium or tin, among others) have many disadvantages. Silver is prone to tarnish in atmospheres containing certain sulphur or chlorine compounds, which limits its application in electronic connectors while palladium will catalyse the formation of insulating frictional polymers which originate in absorbed organic air pollutants on the fretting surface. Wear was responsible for the insulating materials at the interface causing contact resistance to rise. When the contact finish is gold, a reduction in its wear is beneficial but gold is more expensive. An improvement due to the harder nickel under-plate has also been detected. It has been shown that the thicker layer leads to decreasing fretting contact resistance [6].

Commonly, the main degradation of electrical connectors due to oxidation is prevented by coating. Unfortunately, after a certain number of vibration cycles, the fretting wear removes this protective metal layer leading to exposure of the base metal substrate to fretting corrosion. Regarding the disappearance of protective coating, uncoated high-copper alloys were studied in this work. The aim of this article is to analyse the effect of these new materials on the appearance and increases in fretting contact resistance. Fretting corrosion behaviour has been discussed and investigated using scanning electron microscope (SEM) and energy dispersive X-ray spectroscopy (EDX) in many papers [11–13]. In this study, oxidation of the contact surfaces and the wear debris was extended and confirmed with the EDX technique. Moreover, fretting damage on the contact zone by several wear mechanisms such as abrasion, delamination wear and accumulation of wear debris were observed with the SEM technique. Analysis of the contact surface profile with the help of a profiler has been the subject of numerous recent research studies [14, 15]. In this work, the surface profile of the fretted zone was examined by a profiler and then the wear volume loss and the fretted contact area were calculated.

2 Contact geometry

The U-shaped samples shown in Fig. 1 were made from a sheet of 20 mm width, using the processing technique of stamping and bending. The two parts of the sample (plane part or lower part and a spherical part or upper part) have the same thickness i.e. 0.8 mm. The radius R of the spherical segment was 3 mm (Fig. 1). It has been demonstrated by Beloufa [16] and Queffelec *et al.* [17] that the contact resistance for sphere/plane contact was lower than the cylindrical/plane contact. Therefore, the tested U-shaped samples have a contact segment with spherical shape.

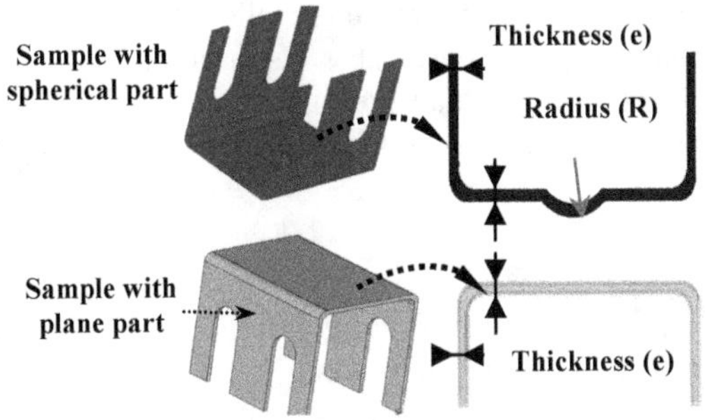

Figure 1: U-shaped samples.

3 Material studies

Previously tested U-shaped samples were made by recent high-copper alloys. Four types of contact materials have been selected for this study. These materials will be used in automotive power connectors and they combine excellent contact performance with good mechanical and electrical properties. Table 1 shows, in descending order of electrical conductivity, some of the electrical, thermal and mechanical properties of the selected alloys [18].

Table 1: Characteristics for different used copper alloys.

Copper alloy (UNS designations)	C10100	C19210	C12200	C70250
Composition	Cu 99.99	CuFeP	Cu 99.9% + 0.008% P	CuNi3SiMg
Yield stress (MPa)	200	322	200	514
Vickers hardness (HV)	65–95	100–130	65–95	180–220
Young's modulus (MPa)	$1.2\ 10^5$	$0.9\ 10^5$	$1.2\ 10^5$	$1.22\ 10^5$
Tensile strength (MPa)	240–300	360–440	240–300	620–760
Thermal conductivity (W/mK)	401	350	340	190
Electrical resistivity (Ω·mm)	$1.68\ 10^{-5}$	$1.88\ 10^{-5}$	$2.12\ 10^{-5}$	$4\ 10^{-5}$
Electrical conductivity (%IACS)	100	91	80	43

The primary selection criterion for these alloys was the variation on their electrical and mechanical properties. Table 1 shows that C70250 alloy is harder and less electrically conductive than the other copper alloys. So, the aim is to determine the relative susceptibility of these materials to fretting corrosion.

4 Test apparatus

Fig. 2 illustrates schematically a fretting test apparatus for measurement of electrical contact resistance during vibration.

The two parts of the contact specimen were fixed in Teflon insulator supports. A DC power supply was used to apply a constant current which was equal to 2 A with an open circuit voltage of 2 V. This electrical current was transmitted from the flat sample to the spherical sample through the contact zone. When high currents are employed, there can be a profound effect on interfacial topography, film formation, contact resistance and resistance heating which accelerates oxidation and micro-arcing.

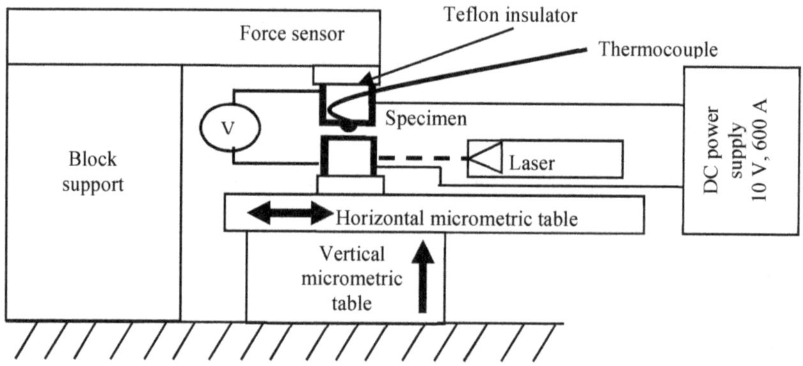

Figure 2: Schematic of the fretting apparatus and the circuit used to measure the contact.

A micro-voltmeter with 1 μV resolution was used to measure the voltage drop when passing a DC current through the contacts and then deduct the dynamic contact resistance. Contact resistance determinations were made using a DC four-wire dry circuit technique [19]. Constant contact force F_c was applied with the help of the vertical displacement of the micrometric table. This force was equal to 10 N and was controlled and maintained constant throughout the testing. The contact force was measured by a force sensor which was connected to a voltmeter through a conditioner. A horizontal micrometric table was used for producing fretting motion. These micrometric tables were driven by a DC stepping motor through a micrometre screw and were controlled by a motion controller. The upper sample or the spherical sample was fixed while the flat sample was submitted to a relative movement with a frequency of 1 Hz and fretting amplitude of 50 μm; this amplitude was measured by a laser with 0.1 μm resolution. The minimal distance between the tested flat sample and laser head was 3 cm. Bibliographical sources quote different values of displacement amplitudes as a border between fretting and reciprocating sliding motion. This value is variously interpreted and contained within a wide range of amplitudes between 50 and 300 μm [20]. Three specimens were used for each test. For each series of tests the fretting motion was carried for 16,000 cycles where the contact voltage attained the open circuit value. The laboratory air temperature ranged from 20 to 25°C and its relative humidity from 40 to 50%. Contact resistance and contact temperature data were collected by a personal computer through a general purpose interface bus (GPIB) IEEE-488 and then stored for subsequent statistical analysis.

Before each experiment, the samples surfaces were wiped with an antioxidant paste and then the samples were cleaned in an ultrasonic alcohol bath, dried by a wiper paper and carefully mounted in the fretting test assembly. Since the geometries of the flat and upper specimen were relatively small, thermo-couples with high response time (J type) were used to measure the temperature variation near to the contact zone during the test. The collected temperature data were collected with the help of a voltmeter through a conditioner. For security reasons, the test was stopped when the contact temperature exceeds 50°C. In order to avoid any accidental arcing between the contacts, the connecting members were brought into contact with the current power supply OFF. The power supply was switched to ON only once the contact force was established and then the contact resistance and contact temperature were measured at each end of a cycle and plotted according to the number of fretting cycles.

5 Experimental results and discussion

5.1 Electrical contact resistance

Fig. 3 shows the results in logarithmic scale of the contact resistance as a function of fretting cycles only for the high-copper alloy C10100. The dotted curve represents the obtained results of contact resistance for three tested specimens. For all materials, the maximum contact resistance which corresponds to the open circuit voltage of 2 V occurs at 16,000 cycles when insulating materials tend to accumulate.

This is manifested by a rapid increase of the contact resistance stability as the fretting progresses. For all materials, the contact resistance behaviour seems to be characterized by three distinct stages. In the first stage, which represents 500 cycles of fretting, the contact resistance decreases with very low fluctuation. Relative movement leads to a sliding/wiping action which facilitates the escape of the wear debris and cause burnishing with rupturing the asperities, ensures good metallic contact and increases the conductive contact surfaces. Hence, the contact resistance decreases. It is well known that the current passes through micro asperities or spots present on the contact area. For metallic contact of rough surfaces [21, 22], it has been shown that contact resistance R_c can be approximated by the formula:

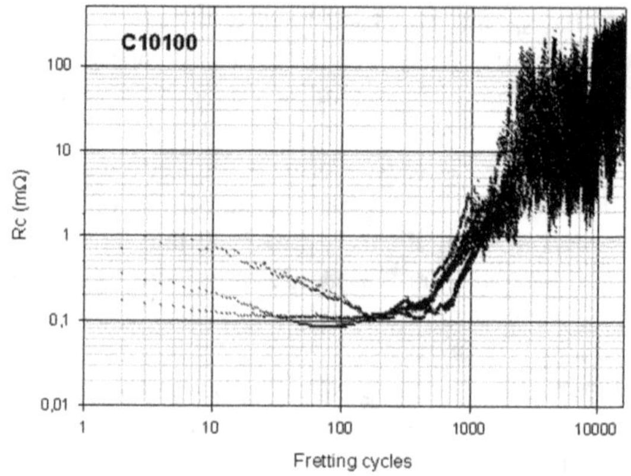

Figure 3: Change in contact resistance of copper alloy C10100 according to the fretting cycles.

$$R_c = \frac{\rho}{2}\left(\frac{1}{na} + \frac{1}{\alpha}\right) \tag{1}$$

where ρ is the resistivity of contacting materials and α is the radius of a circle containing a cluster of n uniformly distributed contact spots, the radius of each spot is a.

In the second stage, apparently between 500 cycles to 2500 cycles of fretting, the fretting was initiated and the contact resistance of all materials increases progressively up to 2 orders of magnitude. However, as the fretting progresses, wear debris were generated and accumulated between the contacting surfaces and then oxidized. Removal of oxide exposes the original surface which then oxidizes in a repeating process. Hard oxidized wear debris accelerates the progress of wear. As a result, contact resistance enhances continuously, this phenomenon is known as fretting corrosion.

In the third stage or the saturation stage, between 2500 and 16,000 cycles, the contact resistance increases sharply and is very unstable as characterized by large fluctuations between 1 and 3 orders of magnitude and sometimes by an open circuit. The peak and bottom values of contact resistance were speculated to be formed by accumulation of wear debris and oxidized wear debris in the fretting area, and subsequently, by extrusion of the accumulated wear debris on the contact area, respectively. Then, a hypothesis was set up that wear debris and copper-alloy have a different hardness from each other and that hard materials crush soft materials and crack the oxides which leads to a lower contact resistance. In this third stage, a thick insulating layer containing oxides and wear debris will form and any remaining metallic contact will be lost, causing a sharp increase in the contact resistance. In addition, prolonged exposure to fretting action will create the conditions for fatigue to occur and cause softening and progressive separation of contacting metals. When the current is confined to flow through the conducting spots (a-spots), the temperature of the point of contact (T_c) may be higher than that of the bulk (T_b). Hence, the increase in constriction resistance over the resistance that would exist (R^i_c) if the metal-metal contact were continuous across the entire contact area can be expressed as:

$$R_c = R^i_c\,[1 + 2/3(T_c - T_b)] \tag{2}$$

The term ($T_c - T_b$) is called the super-temperature and is related to the voltage drop across the contact interface (U) by:

$$T_c^2 - T_b^2 = \frac{U^2}{4L} \tag{3}$$

where L is the Wiedemann-Franz Lorenz number (2.45×10^{-8} V^2/K^2). Hence, a small increase in the contact voltage (U) can raise the super-temperature high enough to produce metallurgical changes such as softening or even melting of the conducting areas [23]. Accumulation of fretting debris and oxides in the contact zone reduces the size of contact spots surrounded by oxidized debris thus increasing the current density and the super-temperature of these conducting paths [11]. The contact temperature measured after 16,000 cycles for all materials was about 30°C.

5.2 Comparison of contact resistance results

Fig. 4 shows a comparison between the smoothing curves of contact resistance for the all tested materials. A running average smoother was applied with sampling proportion of 0.01 and an interval of 100 points. Contact resistance was slightly higher for the harder material C70250 than the other softer materials. It has been shown in an earlier paper [24] that static contact resistance measured under indentation test for C70250 alloy was higher than that of the other alloys. C70250 copper alloy which contains the hard Nickel element has been hardened by precipitation and offers interesting mechanical properties with high strength, high Young's modulus, high hardness and a low electrical conductivity (Table 1) [18]. This causes the highest contact resistance to be obtained with lower contact area.

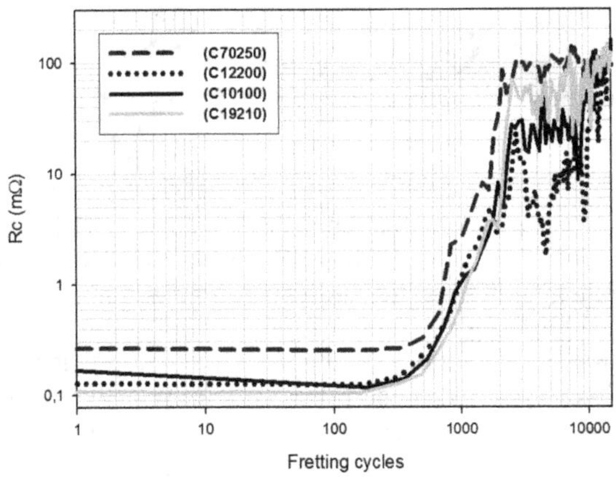

Figure 4: Comparison of smoothing contact resistance results for tested materials.

6 Surface studies of fretted contact zone

6.1 SEM observation and EDX analysis

Fretting damage of contact surfaces was examined by Scanning Electron Microscopy (SEM). Some representative examples of the SEM examination of the fretting damage in contact zones are shown in Figs. 5 and 6. SEM micrographs were taken at 16,000 cycles before and after cleaning of the fretted surface for the harder material C70250 and softer material C12200. Fig. 5(a) shows the accumulated wear debris at the edges of the fretted zone for the softer alloy C12200 and the zoom of wear particles with ×50,000 of magnification. Wear debris seems to be more compact than the observed wear debris for the harder alloy C70250 (Fig. 6(a)).

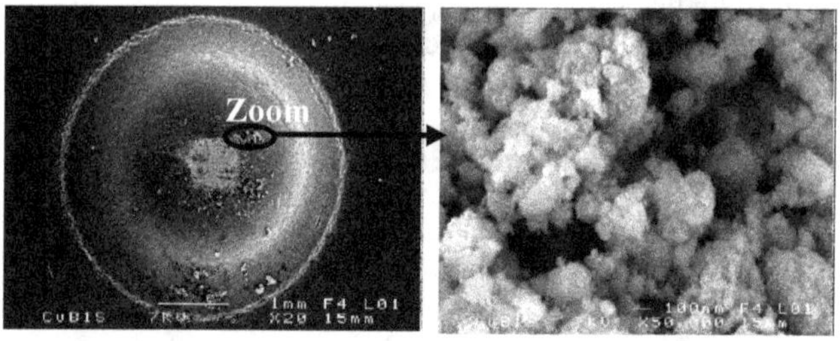

(a)

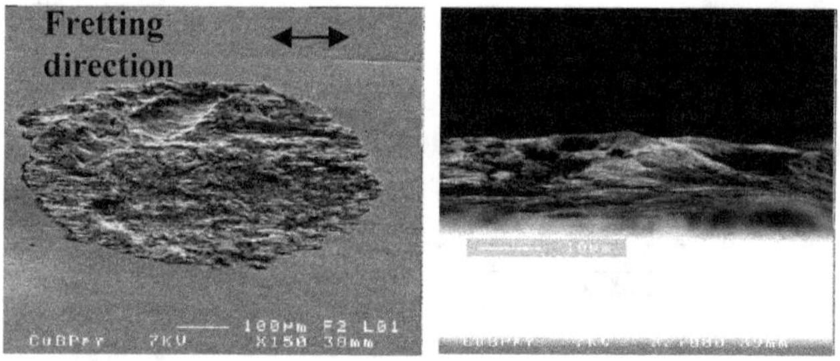

(b)

Figure 5: SEM micrograph of the fretted contact zone for C12200 copper alloy. (a) Spherical sample before cleaning of the fretted surface; (b) Flat sample after cleaning of the fretted surface.

Fig. 5(b) shows delamination wear and wear marks following the direction of the fretting motion caused by the abrasive particles remaining between the two contacting areas. A pushed material at the edge of the wear track was observed on the cross section in Fig. 5(b). The characteristic features of plastic deformation, abrasion and delamination wear were observed in the contact zones (Fig. 6(b)). The micrograph of the cross section presented in Fig. 6(b) shows a surface degradation which is characterized by extensive delamination and cracking as a result of contact fatigue [25]. It is clear that fretting causes serious damage in the contact areas. Typical damage being severe as abrasion, delamination and accumulation of the wear debris at the edges of the wear track leads to an increase of contact resistance. Elemental compositions of the fretted contact zones were determined by the Energy Dispersive X-rays analysis (EDX).

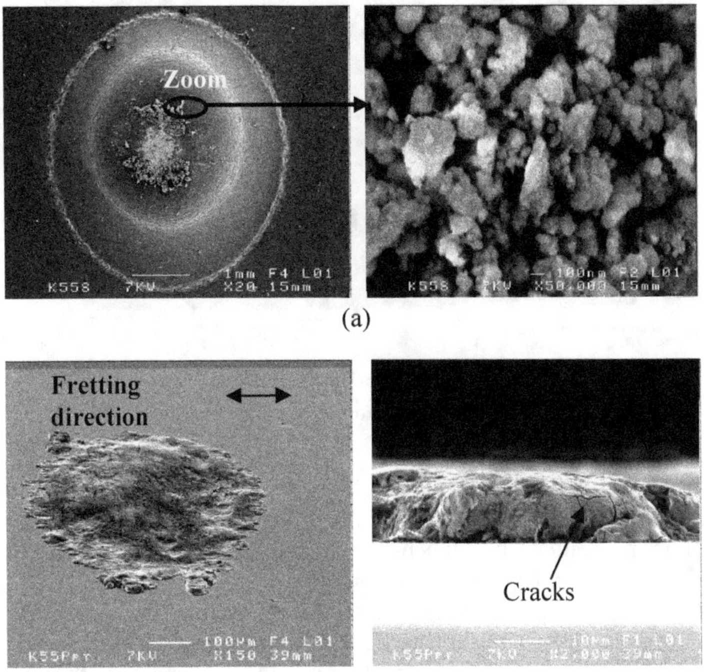

(a)

(b)

Figure 6: SEM micrograph of the fretted contact zone for C70250 copper alloy. (a) Spherical sample before cleaning of the fretted surface; (b) Flat sample after cleaning of the fretted surface.

The main advantage of this type of X-ray fluorescence analysis is the ability to identify a wide range of elements simultaneously and to obtain a semi-quantitative measure of their concentrations in the contact regions under examination. Fretting contact areas have been examined by EDX analysis after the end of 16,000 fretting cycles inside the wear track and on the wear debris for the C12200 and C70250 alloys. The elemental composition spectre measured from a spot across the wear track for the two materials is shown in Fig. 7. The spectre presented in Fig. 7 (a) shows the presence of strong peak of oxygen (26.5%) for the harder alloys C70250 and 20.7% of oxygen for the softer material C12200 (Fig. 7 (b)). The presence of strong peaks of oxygen and copper indicates that wear tracks were composed mainly of these elements.

X-ray elemental mapping clearly reveals the nature of chemical species present in the contact zone and the extent of oxidation. The X-ray elemental mapping confirms the oxidation of the removal copper.

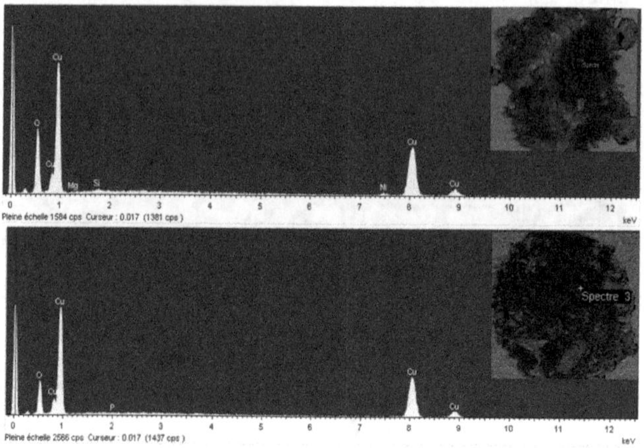

(a)

(b)

Figure 7: EDX spectra taken inside the fretted surfaces. (a) C70250 copper alloy; (b) C12200 copper alloy.

6.2 Profile measurement

The cleaned wear track of the tested alloys was examined by a 3D scanning profiler [26] to evaluate the dimensions of the wear tracks. Fig. 8 shows the smoothing initial profiles and the smoothing profiles of the wear tracks for the C70250 and C12200 alloys obtained after 16,000 fretting cycles. The *X*, *Y* grid resolution for the collected data was 1.2 μm over a measurement area of 1.2 mm × 1.2 mm. The 3D sub-micrometric profiler method can determine the parameters of wear in fretted surfaces.

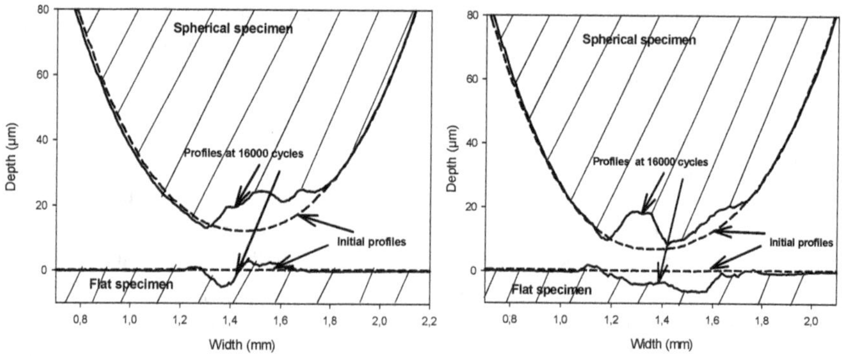

(a) (b)

Figure 8: Scanning fretted surfaces and profiles of the wear track after 16000 fretting cycles (a) C70250 copper alloy; (b) C12200 copper alloy.

The real contact area, volumetric wear removal, width and length of the wear track are presented in Table 2 for the flat specimen and for all tested materials. The measured wear sizes for the harder material C70250 were lower than the softer materials C10100, C12200 and C19210. The softer materials suffer more surface damage, material loss and generation of wear debris.

Table 2: Wear sizes for all tested high-copper alloys.

	C10100	C12200	C19210	C70250
Depth (μm)	7.7	6.54	7.8	4.94
Width (μm)	593	600	659	450
Fretted area (mm^2)	0.37	0.34	0.35	0.28
Fretted volume loss (mm^3)	$5.9\ 10^{-4}$	$4.7\ 10^{-4}$	$6.1\ 10^{-4}$	$1.4\ 10^{-4}$

The harder materials suffer very little surface damage and not much wear debris generation. It is well known that the wear volume is proportional to load F_c and fretting amplitude x and is inversely proportional to hardness H; the wear volume V can be calculated by the following equation of Archard [27]:

$$V = \frac{2\,K\,F_c\,N\,x}{3H} \tag{4}$$

where N is the total number of cycles and K is a wear coefficient.

The radius of contact area (a) can be calculated by the following equations:

$$a = \left(\frac{3F_c R}{4E^*}\right)^{1/3} \tag{5}$$

$$\frac{1}{E^*} = \frac{2(1-\nu^2)}{E} \tag{6}$$

where R is the radius of spherical specimen (R = 3 mm), E the Young modulus and ν the Poisson's ratio. The diameter calculated by the eqns. (5) and (6) for the C12200 alloys was found equal to 140 μm. Fig. 9 gives a schematic of the fretted

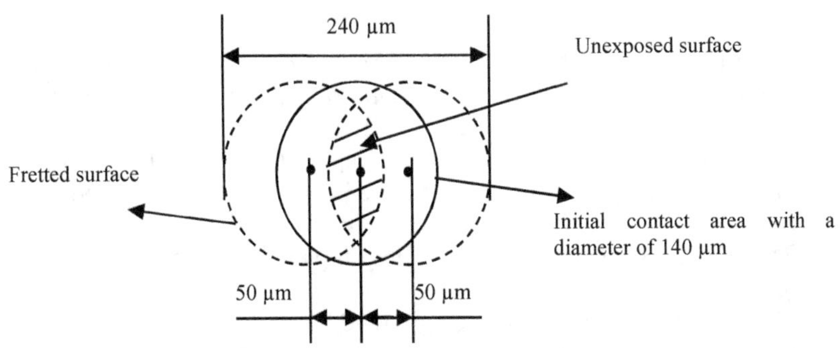

Figure 9: Fretted surface for one fretting cycle (fretting amplitude of 50 μm, C12200 alloys).

surface for one fretting cycle. The fretted area after 16,000 fretting cycles was higher than the fretted area for one cycle. The measured width of the track for the C12200 after 16,000 fretting cycles was equal to 600 µm (Table 2) and was higher than the initial width of 240 µm. The width and depth increase during the fretting test. This is due to the progressive wear of the spherical specimen; the initial contact area was spherical and then increased and became flat.

7 Conclusion

Resistance increases noted during fretting action are felt to be due to trapped wear debris composed mainly of oxides. Based on the variation in contact resistance as a function of fretting cycles, wear debris, elemental composition of the wear track, surface profile and wear track sizes, were assessed by surface analytical techniques (SEM, EDX and profile-meter). SEM results showed that fretting causes serious damage in the contact areas as abrasion, delamination wear, fatigue cracks and accumulation of the oxidized wear debris which can lead to an increase of contact resistance. EDX analysis confirmed the presence of a strong percentage of oxygen in the wear track for the tested copper alloys. 3D scanning profiler showed that softer materials suffer more surface damage, material loss and generation of wear debris than the harder material. The hard material was more wear resistant than soft materials and leads to high electrical contact resistance during fretting test. These studies underline the importance and necessity of using different surface characterization techniques in order to obtain an unambiguous explanation and information of the phenomena occurring at the contacting interfaces. The obtained results could help the connector manufacturers in the choice of contact materials and were a useful tool to evolve compromise between the electrical and mechanical aspects for connectors.

Some proposed solution could lead to a decreasing of the contact resistance and increasing the contact durability as the use of a lubricant between the two contacting surface and increasing of the contact load which will reduce the micro-displacement of the contact interface by increasing the friction force.

References

[1] Fagerstrom, W.B. & Nicotera, E.T., Fretting corrosion in connectors. *Proc. Connectors and Interconnections Symposium. Electronic Connector Study Group*, Philadelphia, pp. 303–312, 1961.

[2] Sauger, E., Fouvry, S., Ponsonnet, L., Kapsa, Ph., Martin, J.M. & Vincent, L., Tribologically transformed structure in fretting. *Wear*, **245**, pp. 39–52, 2000.

[3] Braunovic, M., Fretting damage in tin-plated aluminium and copper connectors. *IEEE Transactions on Components, Hybrids, and Manufacturing Technology*, **12**, pp. 215–223, 1989.

[4] Malucci, R.D., Fretting corrosion degradation, threshold behavior and contact instability. *Proc. of the 49th Int. Conf. on Electrical Contacts IEEE Holm*, Washington, pp. 1–15, 2003.

[5] Bock, E.M. & Whitley, J.H., Fretting corrosion in electric contacts. *Proc. of the 20th Int. Conf. on Electrical Contacts IEEE Holm*, Chicago, pp. 128–138, 1974.

[6] Antler, M., Survey of contact fretting in electrical connectors. *Proc. of the 30th Int. Conf. on Electrical Contacts IEEE Holm*, Chicago, pp. 1–21, 1984.

[7] Noël, S., Lécaudé, N., Correia, S., Gendre, P. & Grosjean, A., Electrical and tribological properties of tin plated copper alloy for electrical contacts in relation to intermetallic growth. *Proc. of the 52nd Int. Conf. on Electrical Contacts IEEE Holm*, Montreal, pp. 1–10, 2006.

[8] Ito, T., Takata, K., Hattori, Y., Iida, K. & Saitoh, Y., Microscopy study of fretting corrosion of tin plated contacts. *Proc. of the 53rd Int. Conf. on Electrical Contacts IEEE Holm*, Pittsburgh, pp. 216–221, 2007.

[9] Park, Y.W., Sankara Narayanan, T.S.N., Lee, K.Y., Fretting corrosion of tin-plated contacts. *Tribology International*, **41**, pp. 616–628, 2008.

[10] Van Dijk, P., Rudolphi, A.K. & Klaffke, D., Investigations on electrical contacts subjected to fretting motion. *Proc. of the 21st Int. Conf. on Electrical Contacts (ICEC)*, Zurich, pp. 1–7, 2002.

[11] Gagnon, D. & Braunovic, M., Effect of fretting in lead-free systems. *Proc. of the 50th Int. Conf. on Electrical Contacts IEEE Holm*, Seattle, pp. 248–254, 2004.

[12] Braunovic, M., McIntyre, N.S., Chauvin, W.J. & Aitchison, I., Surface analysis of fretting damage in electrical contacts of aluminum with different contact materials. *Proc. of the 29th Int. Conf. on Electrical Contacts IEEE Holm*, Chicago, pp. 231–242, 1983.

[13] Kassman, A. & Jacobson, S., Surface damage, adhesion and contact resistance of silver plated copper contacts subjected to fretting motion. *Wear*, **165**, pp. 227–230, 1993.

[14] McBride, J.W., Developments in fretting studies applied to electrical contacts. *Proc. of the 52nd Int. Conf. on Electrical Contacts IEEE Holm*, Montreal, pp. 170–180, 2006.

[15] McBride, J.W., The loaded surface profile: A new technique for the investigation of contact surfaces. *Proc. of the 23rd Int. Conf. on Electrical Contacts (ICEC)*, Sandai, pp. 150–156, 2006.

[16] Beloufa, A., Influence of shapes, contact forces and high copper alloys on the contact resistance and temperature. *Proc. of the 2nd Int. Conf. on Engineering Mechanics, Structures, Engineering Geology (EMESEG '09-WSEAS)*, Rodos Island, pp. 139–144, 2009.

[17] Queffelec, J.L., Ben Jamaa, N., Travers, D. & Pethieu, G., Materials and contact shape studies for automobile connector development. *Proc. of the 36th IEEE Conf. on Electrical Contacts*, pp. 225–231, 1990.

[18] Zauter, R. & Kudashov, D.V., Precipitation hardened high copper alloys for connector pins made of wire. *Proc. of the 23rd Int. Conf. on Electrical Contacts (ICEC)*, Sandai, pp. 257–261, 2006.

[19] Standard Methods for Measuring Contact Resistance of Electrical Connections (Static Contacts), B53g-80, American Society for Testing and Materials, Philadelphia.

[20] Ohmae, N. & Tzukizoe, T., The effect of slip amplitude on fretting. *Wear*, **27**, pp. 281–293, 1974.

[21] Holm, R., *Electric Contacts: Theory and applications*, Springer-Verlag: Berlin, 1999.

[22] Greenwood, J.A., Constriction resistance and the real area of contact. *Journal of Applied Physics*, **17**, pp. 1621–1632, 1966.

[23] Timsit, R.S., On the evaluation of contact temperature from potential-drop measurements. *IEEE Transactions on Components, Hybrids, and Manufacturing Technology*, **6**, pp. 115–121, 1983.

[24] El Abdi, R., Beloufa, M.A. & Benjemaa, N., Contact resistance study of high copper alloys used in power automotive connectors. *Journal of Automobile Engineering*, **222**, pp. 1375–1383, 2008.

[25] Antler, M., Sliding wear of metallic contacts. *Proc. of the 26th Int. Conf. on Electrical Contacts IEEE Holm*, Chicago, pp. 1–24, 1980.

[26] TaiCaan Technologies Ltd, http://www.taicaan.com.

[27] Archard, J.F., Contact and rubbing of flat surfaces. *Journal of Applied Physics*, **24**, pp. 970–981, 1953.

Evaluation of the fretting corrosion mechanisms on the head–cone interface of hip prostheses

I. Caminha[1], C. R. M. Roesler[2], H. Keide[1], C. Barbosa[1], I. Abud[1] & J. L. Nascimento[1]
[1]*National Institute of Technology, Rio de Janeiro, Brazil*
[2]*University Hospital, Federal University of Santa Catarina, Brazil*

Abstract

Fretting corrosion is one of the main concerns in the application of hip prosthesis. The residual corrosion products of stainless steels are associated with the occurrence of several adverse reactions in the human body. An investigation into these corrosion products is extremely important in the phase of designing hip prostheses, aiming at the minimization of the amount of exposure of the organic tissues to corrosion products released during the permanence of the prosthesis inside the patient. In this contribution, the mechanical stability and fretting corrosion resistance of a modular hip prosthesis, fabricated from ASTM F 138 austenitic stainless steel, were evaluated according to the criteria of ASTM F 1875 standard, method I, which prescribes a long-term test, with the purpose of determining the amount of damage through the quantification of the corrosion products and debris which result from the fretting corrosion conditions. The mechanical tests were performed in a servohydraulic mechanical testing machine and the modular interfaces were exposed to an electrolytic 0.9% NaCl in distilled water solution and subjected to a minimum load of 230 N and a maximum load of 2.3 kN, frequency of 5 Hz for 10 million cycles, according to ASTM F 1875 standard, thus simulating real conditions of use. The load ratio was specified according to the ASTM F1440 standard. A significant amount of fretting-corrosion products was observed: 0.22 g resulted from five specimens after 10^7 cycles and the diffraction X-ray tests showed the presence of the crystalline phases Fe_2O_3, Fe_3O_4 and Cr_2O_6. It can be concluded that the mass loss in the head–cone connection allowed the entrance of the physiological saline solution in the inner region of the head, increasing the predicted micro movement in the head–cone

interface, resulting in an accelerated process of fretting corrosion and, consequently, the liberation of debris and corrosion products that could lead to adverse biological reactions.
Keywords: fretting corrosion, hip prosthesis, modular components, mechanical properties.

1 Introduction

The increasing life expectancy in the last decades and the trauma risk due to the more frequent exposure of individuals to different types of accidents and the currently insecure life conditions has led to an increased search for orthopedic implants with high performance, capable of resisting even more severe loads for longer times. For this reason, metallic materials, which were developed for applications in orthopedic implants, must present some specific properties, such as biocompatibility, strength and resistance to degradation by wear or corrosion.

In spite of the increasing use of titanium and its alloys for this application, in the last decades, in Brazil, ASTM F138 austenitic stainless steel [1, 2] is still very frequently used in the fabrication of modular components for total hip arthroplasty due to its lower cost, together with good mechanical properties and satisfactory corrosion resistance, since the main customer is the Health Ministry, which supplies the demands of public hospitals in Brazil.

The use of modular components in the total hip arthroplasty presents several advantages, such as a great variety of stem geometry, which allows a better choice of modular component for the surgeon, according to the patient disease, mainly in the case of revision surgeries, thus minimizing the risk of inadequate procedures.

However, the modular interfaces are subjected to micro movements, which can result in fretting and corrosion, leading to the release of debris, which can cause adverse reactions and accelerated wear in the articulation interface. "Fretting" can be defined as a wear phenomenon which occurs when two solids in contact are subjected to tangential oscillatory movement with small displacement amplitude. It can be described more accurately as a movement whose amplitude is smaller than the contact extent. The presence of a corrosive environment contributes to the acceleration of the wear process and this condition is defined as "fretting corrosion". The occurrence of corrosion results in degradation processes which reduce the structural integrity of the implant and release products whose reaction can be harmful in contact with organic tissue.

When the prostheses are fabricated from ASTM F138 stainless steel, the generation of a passive barrier on the surface (passive layer) is the main obstacle to corrosion process. This film, which is composed of metallic oxides formed spontaneously on the metal surface and whose thickness can be increased with surface treatments, avoids the migration of metallic ions from the metal to the solution and the migration of anions from the solution to the metal through the metal-solution interface (physiological environment). For being effective barriers, these films must be compact and cover completely the metallic surface, they must have an atomic structure which limits the migration of metallic ions and/or electrons through the metallic oxide/solution interface, and they must be capable of remaining on the surface of these alloys even when subjected to mechanical and abrasive stresses in the clinical use of prostheses.

The most frequently mentioned mechanisms of failure in these components are aseptic loosening, periprosthetic osteolysis and metallosis and are directly related to the debris released by the prosthesis components, mainly the non cemented prostheses [3].

According to the literature [4], the residual products of corrosion in stainless steel based implants are associated to tissue necrosis, inflammation, allergenic reactions and even cancer.

The aim of the present work is an evaluation of fretting corrosion in modular prostheses fabricated with ASTM F 138 stainless steel, in order to minimize the adverse reactions during the permanence of the prosthesis in the patient.

2 Material and methodology

In this work five modular hip prostheses, composed of heads and stems, that is, five stems and five heads, fabricated from the same batch of ASTM F 138, were analyzed. A SZX16 model Olympus stereomicroscope was used for comparing the surfaces of the modular components before and after a fretting corrosion test. In Fig. 1, an as-received representative sample of the batch can be observed.

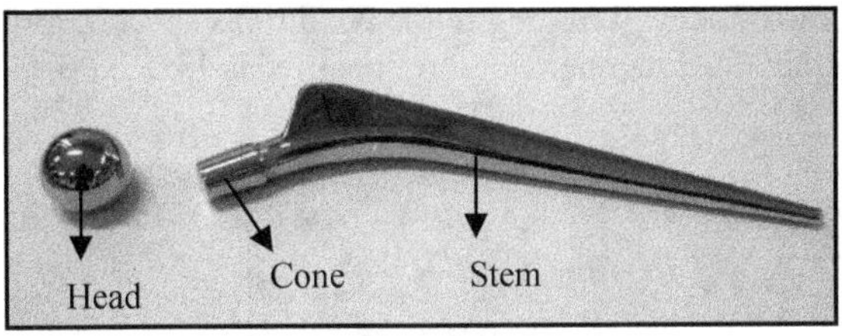

Figure 1: Head and stem of the hip prosthesis fabricated from ASTM F 138 stainless steel.

The fretting corrosion tests on the cone-head interface were performed according to method I of ASTM F 1875 standard, which prescribes a 0.9% NaCl physiological saline solution, thus simulating in vivo conditions, since these components are in contact with body fluids when implanted in patients. With the purpose of analyzing both fatigue resistance of hip stems and mechanical stability of modular components, the experimental procedures described in ASTM F 1440 [5] and ISO 7206-4 [6] were also adopted.

These tests were performed in two universal machines for mechanical testing (Instron model 8872) with a 25 kN load cell with the following characteristics, described in the ASTM F 1875 standard [7]:

- Maximum load: 3.3 kN;
- Frequency: 5 Hz;
- Test finish after 10 million cycles;
- Error of applied load lower than 1% of the maximum load;

- Control function of cyclic load (sinusoidal);
- Monitoring of vertical displacement of the prosthesis head and registering of cycles counting;
- Computerized operation;
- Equipment for recirculation and heating of solution test with temperature control within 37°C ± 1°C.

Before starting the test, each modular component composed by head and stem was prepared according to the following procedure:

- Assembling the modular component and checking the fittings;
- Placing the modular component in a device which allows alignment of angles mentioned in the standard, with respect to the load direction;
- Drying the mounting;
- Introducing the environment chamber together with the metallic cup with the mounted prosthesis;
- Adding the saline solution (0.9% NaCl in distilled water) to the environment chamber;
- Starting the pump for recirculation and heating the test solution;
- Fitting the complete assembly in the test machine as shown in Fig. 2.

The machine was programmed for the application of the load in the center of the prosthesis head, with minimum 0.3 kN and maximum 3.3 kN loads, with sinusoidal 5 Hz frequency according to the conditions specified in ASTM F1875 standard. Maximum load requirements are specified in ISO 7206-4-2010 standard.

Figure 2: Complete assembly before starting test in INSTRON 8872 machine.

A limiting condition was imposed for automatically stopping the machine: a vertical displacement of 5 mm. If this value is not reached, the test is finished after completing 10 million cycles of loading and unloading. The parameters of sample placement follow the requirements of the ASTM F1440 standard.

X-ray diffraction analysis was carried out under the following conditions: anode material: Cu Kα, angle between 10 and 100°, step size 0.05°, scan step time: 1 s, receiving slit size: 0.03 mm, current: 40 mA, voltage: 40 kV.

3 Results and discussion

The images of the modular surfaces (head and cone), which were obtained by optical microscopy, did not present surface defects such as cracks other types of flaws, introduced by the fabrication process, which could influence the results of this test. In Figs. 3 and 4, the views of the surfaces of the modular components (head and cone) of a hip prosthesis, before and after testing, are presented. The other prostheses have similar appearances.

A significant corrosive attack inside the head and in the cone can be observed after 1 million cycles with increasing rates along the test, releasing corrosion products which resulted in mass loss in the head–cone connection. The mass loss, which was observed in the test solution, allowed the introduction of saline solution inside the head, increasing the predicted micro movements, leading to an accelerated wear (friction), with tearing of debris, as shown in Fig. 5. No cracks of any type and mechanical failure could be observed in the five hip prostheses, which were analyzed; this means that the mechanical stability of the modular component was preserved during the test.

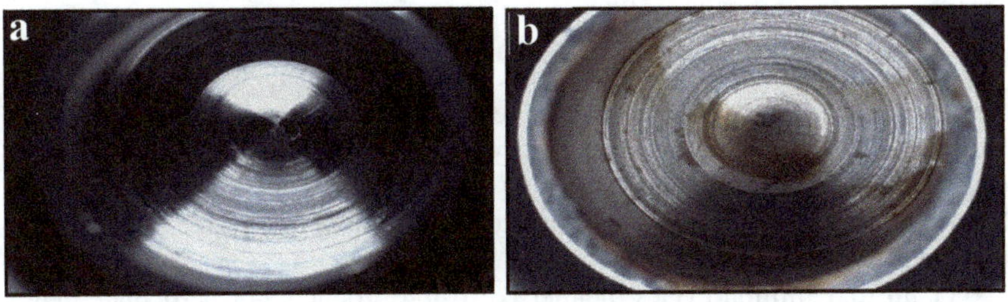

Figure 3: Detailed view inside the heads: (a) before and (b) after the test. Magnification: 10×.

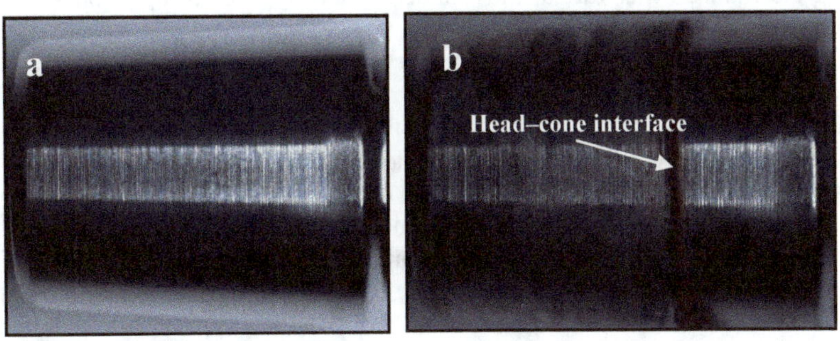

Figure 4: Detailed view of the cone region: (a) before (b) after the test. Magnification: 10×.

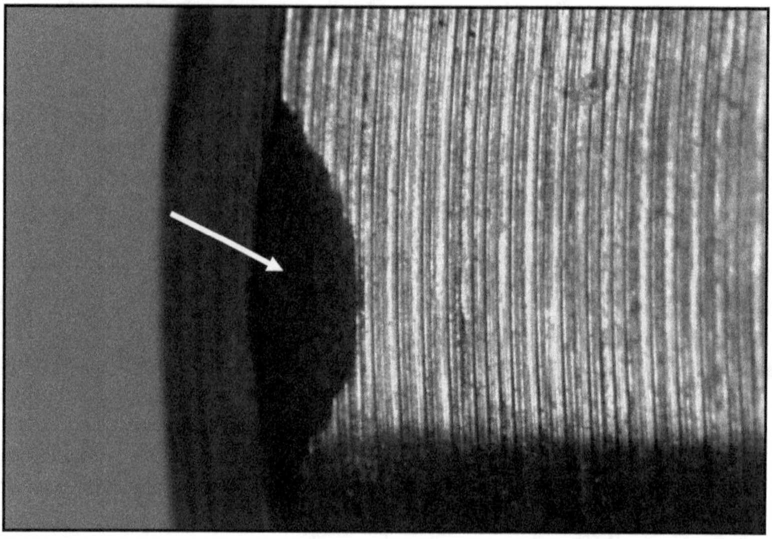

Figure 5: Detailed view of tearing which was observed on the cone. Magnification: 60×.

In Figs. 6 and 7, the corrosion products and particulate residues (debris), resulting from the fretting corrosion test can be observed at the bottom of the environmental chamber (indicated by an arrow in Fig. 7). These ones are a consequence of the damages caused by fretting and corrosion fretting at the head–cone interface of the primary hip prosthesis. These images represent the five hip prostheses subjected to this analysis.

For up to 1 million cycles, neither corrosion products nor particulate debris could be observed but, after this period, the degradation of the modular component was increasing and localized on the head–cone, as can be seen in detail in Fig. 8.

After 10 million cycles, the corrosion products and particulate debris were quantified, resulting in an average value of 0.22 g. In the literature, there is only one citation of fretting-corrosion products quantification, but in their experiment the cycle number was varied, thus hindering any type of comparison with the results presented in this work.

For better comprehension of fretting corrosion mechanisms in hip modular components, the variation of load and number of cycles is necessary. The characterization of the corrosion products and particulate residues by X-ray analysis is also very important, in order to identify the phases and type of metallic particles which were released, with the aim of minimizing the incidence of aseptic loosening, periprosthetic osteolysis and metallosis in the surgeries of total hip arthroplasty. In the present work, a complete characterization of the fretting corrosion products was not concluded, but it can be observed that they presented a predominantly ellipsoidal shape with size distribution approximately uniform, whose larger diameter was around 500 μm.

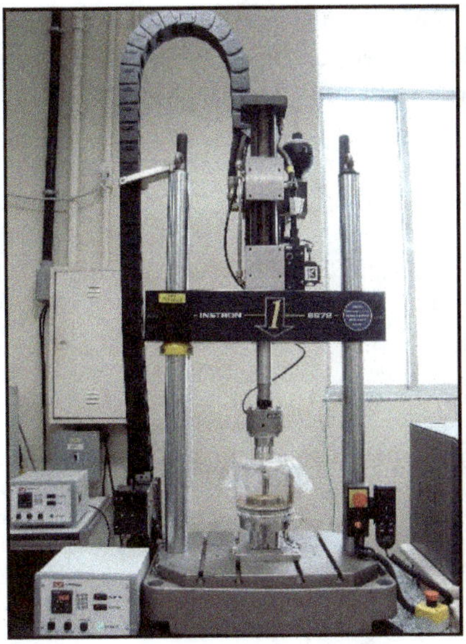

Figure 6: Test assembly after 10^6 cycles.

Figure 7: Detail.

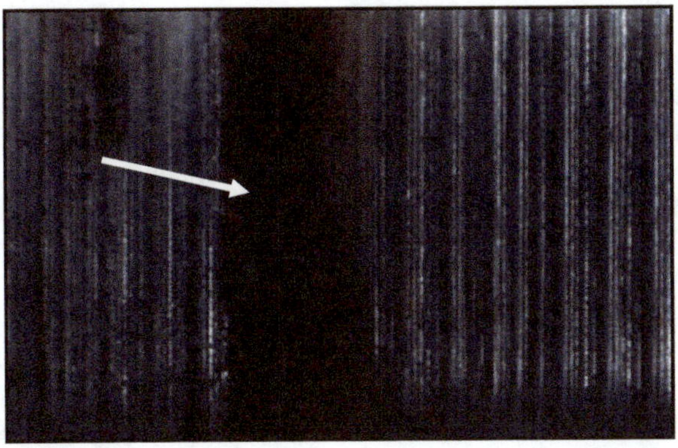

Figure 8: Detailed view of the head–cone interface connection shown in Fig. 4(b). Magnification: 60×.

Furthermore, the X-ray diffraction analysis performed on the corrosion products showed the presence of the following crystalline phases: Fe_2O_3, Fe_3O_4 and incipient Cr_3O_8, compatible with the ASTM F138 stainless steel.

4 Conclusions

The evaluation of fretting corrosion on the head–cone connection interfaces of five modular hip prostheses fabricated of stainless steel led to the following results:

- A significant degree of localized corrosion in the inner region of the head and in the head–cone interface was observed, originating from the fretting corrosion and fretting during 10 million cycles, resulting in particulate debris and corrosion products.
- In order to achieve a better comprehension of the fretting corrosion mechanisms, more tests are necessary, including changes in the applied load and number of cycles.
- At first glance, fretting corrosion does not affect the mechanical stability of the modular components, at least for up to 10^7 cycles.

References

[1] Cavalcanti, E.H. de S., Souza, S.M.C. de, Ferreira, C. de A., Campos, M.M., Abud, I. de C. & Palmeira, L., Avaliação da resistência à corrosão de prótese total de quadril de aço inoxidável austenítico removida de paciente. *22° CONBRASCORR – Congresso Brasileiro de Corrosão*, Salvador, Bahia, 2002.

[2] Giordani, E.J., Ferreira, I. & Balancin, O., Propriedades mecânicas e de corrosão de dois aços inoxidáveis austeníticos utilizados na fabricação de implantes ortopédicos. *REM: R. Esc. Minas, Ouro Preto*, **60(1)**, pp. 55–62, 2007.

[3] Crestani, M.V., Boschim, L.C. & Schwartsmann, C.R., Metalose simulando tumor abdominal. *Revista Brasileira de Ortopedia*, Nov/Dec., 2004.

[4] Merritt, K. & Brown, S.A., Biological effects of corrosion products from metals. *Corrosion and Degradation of Implant Materials: Second Symposium, ASTM STP 859*, eds. A.C. Fraker & C.D. Griffin, American Society for Testing and Materials, Philadelphia, pp. 195–207, 1985.

[5] ASTM F 1440 – 92, Standard Practice for Cyclic Fatigue Testing of Metallic Stemmed Hip Arthroplasty Femoral Components Without Torsion, reapproved 2008.

[6] ISO 7206-4-2010 – Implants for surgery – Partial and total hip joint prostheses -- Part 4: Determination of endurance properties and performance of stemmed femoral components.

[7] ASTM F 1875 – 98, Standard Practice for Fretting Corrosion Testing of Modular Implant Interfaces: Hip Femoral Head-Bore and Cone Taper Interface, reapproved 2004.

Electrochemical analysis of corrosion products from long-lasting atmospheric tests

M. Leiro & E. S. Ayllón
CITEFA, Buenos Aires, Argentina

Abstract

A traditional electrochemical cell was used to test the corrosion products from PA steel, a weathering steel containing non-metallic inclusions, exposed for 1–9 years to atmospheric conditions. Polarization curves were obtained and compared to those from a weathering steel without non-metallic inclusions. The weight loss and polarization resistance were extracted from these curves and compared with values previously estimated. Electrochemical analysis also provided the characteristics of cathodic reduction of corrosion product films from both non-weathering and PA steel after a 6-year open-air exposure.
Keywords: atmospheric corrosion, electrochemical analysis, non-metallic inclusions.

1 Introduction

When submerging a pure metal or an alloy in an aqueous solution, a migration of metallic ions between the solid phase and the aqueous phase starts immediately. In fact, it is a constant exchange of metallic ions in which the prevailing speed will be the one leading to a decrease of the free energy of the system. The passing of the metallic ions to the solution will create a potential difference in the metal-solution interface, which will hinder the passage of new ions. A potential difference will be reached, at which the free energy variation will be null and a thermodynamic equilibrium will exist. If a metal is submerged into a solution that has its own ions, the electrochemical potential will be different for each metal. Noble metals are those like gold (Au), and platinum (Pt), whose ions are deposited on the metallic solid phase thus increasing their potential with respect to the solution. The metal will be traditionally more positive than the solution. Those metals like iron (Fe) and zinc (Zn) will have a tendency to pass ions to a solution, no matter what its ionic concentration is, getting the metallic solid phase lower potentials than the solution, or negative potentials in contrast to the noble metals.

The direct measurement of the potential difference between the metal and the solution is experimentally impossible. However, a cell with two electrodes can be formed to measure the potential difference between the two. One of the electrodes will be the pure metal or alloy, whose potential needs to be found and the other will be an electrode with an easy-to-reproduce potential to take it as a pattern. The electrode taken as a reference is the one called hydrogen (H) electrode, which is a platinised platinum electrode submerged in an acid solution (pH = 0) with a hydrogen activity equal to 1.0 M, over which hydrogen bubbles at a specified atmospheric pressure. Traditionally, zero potential is assigned to every temperature.

Two different metallic electrodes together form a cell, for example, metallic Cu and Zn in solutions containing their own ions. The potential difference between these metals may be calculated from previous data. Cu will be +0.34 V more positive than the solution, and Zn will be –0.76 V. The union through a salt bridge of both solutions leads to a potential measurement of 1.1 V between electrodes. The measurement may be carried out through a very high impedance instrument.

If two electrodes are joined with an electron conductor – a copper wire, for instance – a current flow *I* will result, which will depend on the electrolyte resistance and the potential difference between the solutions adjacent to the electrodes. Cu^{++} metallic ions will be deposited on the Cu electrode and Zn will dissolve with Zn^{++} ions forming part of the solution. This latter is an anodic reaction:

$$Zn \rightarrow Zn^{++} + 2e^- \quad (1)$$

The Cu reaction is a cathodic one or a reaction in which electrons produced in the anodic reaction are consumed:

$$Cu^{++} + 2e^- \rightarrow Cu \quad (2)$$

This is the electrochemical corrosion pattern or model. The Zn electrode is corroded and the Cu electrode gives the reduction reaction. The electric current generated will be a measure of the reaction rate that will be the same in both electrodes unless other reactions appear.

The potential measured before the connection is a thermodynamic value informing about the possibility of a reaction and the current *I* is a measure of the reaction kinetics.

If the potential of the electrodes with respect to the solutions in which they are submerged is measured when the current flows, both of them will be observed to vary from their initial value. The Cu electrode will be less positive and the Zn one will be less negative. In order for the corrosion current to flow, the electrodes have to depart from their thermodynamic equilibrium; they have to polarise. The difference between the new potential *E*, which depends on the current *I*, and the initial one given by Nersnt equation, is called overpotential or overvoltage.

The oxidation or anodic reaction produces electrons:

$$Fe \rightarrow Fe^{2+} + 2e^- \quad (3)$$

The main cathodic or reducing reactions, which consume electrons, are:

$$2H^+ + 2e^- \rightarrow H_2 \quad (4)$$

and

$$O_2 + 4H^+ + 4e^- \rightarrow 2H_2O \quad (5)$$

These equations represent the Fe electrochemical corrosion process. In general, the anodic reaction is the transfer of a metallic atom from the crystalline lattice to an aqueous solution as a Fe^{+2} cation.

Likewise, the cathodic reaction is basically the consumption of electrons generated by different ways of anodic reactions with only one reducing equation or more. Here, the electron conduction is through the Fe mass generated by the Fe weight since it is a sample in which the anodic and cathodic areas are spontaneously produced. Both reactions may be studied individually through electrochemical techniques. The latter apply a particular potential to the sample; the current flow is measured or controlled and the resulting potential is measured with respect to a reference electrode – in volts or millivolts, generally. These are called polarisation techniques. Diagrams of the applied potential E are obtained in terms of log$i$ where i is the current density, that is, the measured current divided by the tested sample area ($A \cdot cm^{-2}$).

The corrosion potential may be measured with a reference electrode by means of a high impedance instrument. In general, a saturated calomel reference electrode is commonly used in all electrochemical laboratories.

When it is necessary to express the values obtained in the hydrogen scale, these are transformed as follows:

$$\text{Hydrogen scale} = \text{calomel saturated scale} + 240 \text{ mV} \quad (6)$$

2 Equipment

A traditional Pyrex glass electrochemical cell – as shown in Fig. 1 – with a magnetic stirrer was used to keep the hydrodynamic conditions of the system constant. The cell – of around 120 ml – has six inputs. The main one, which coincides with the symmetry axis of revolution of the cell, is for the work electrode, made up of a sample of the material to be tested, and whose exposed surface is around 1 cm^2. The remaining four top inputs are used for the reference electrode, the platinum (Pt) wire counter-electrode, as well as the output and input gases tubes. Laterally, and at 45° from the cell axis, there is another input for a glass electrode to measure the pH of the solution, which can be done continuously during the test.

The reference electrode used was a saturated calomel one. It had an agar/ Na_2SO_4 0.1 M salt bridge – through a Luggin capillary with ϕ diameter in its end, and located at distance ϕ from the work electrode surface. The current density in the counter-electrode surface must be low in order not to exceed the potentiostat capacity. Therefore, its area must exceed the one of the sample by at least ten times. The work electrode was prepared by setting the sample in a glass tube end with epoxy resin, which was also used to protect the edges and back. A Cu wire with a 1 mm diameter was used inside the glass tube to connect it with the cell outer circuit.

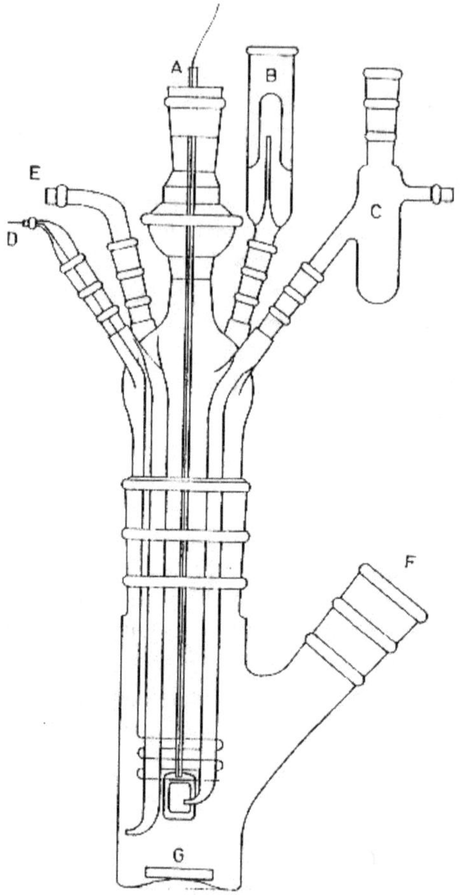

Figure 1: Traditional glass cell for electrochemical tests. A: Work electrode; B: Gases output with water trap; C: Reference electrode holder; D: Pt counter-electrode; E: Gases input; F: Input for complementary measurements of T, pH, $[M^{+n}]$, etc.; G: Submerged device to magnetically stir the solution.

The curves were drawn with a Tacussel PRT-20-2X potentiostat with a Servovit 13 scanning generator. This equipment keeps constant the specimen potential E with respect to the reference electrode. The resultant cathodic (I_c) or anodic (I_a) current – which can be measured in the cell outer circuit – closes with ions by means of the Pt counter-electrode. The E potential may be constant or may vary, according to a preset law. Here, a linear variation law with a 10 $mV \cdot min^{-1}$ slope was adopted. The curves started from the equilibrium potential – where the over-potential and the current are null – scanning in anodic or cathodic direction up to +1200 mV and –1300 mV, respectively. The curves were drawn in triplicate at environmental temperature, showing their anodic and cathodic averages.

The corrosion current density i_{corr} (A·cm^{-2}) was determined as the intersection of the cathodic curve slope in the Tafel zone and the rest horizontal potential – also called open circuit since no current flows – which coincides with the corrosion potential E_{corr}; this resulted in an i_{corr} value of 5×10^{-6} A·cm^{-2}. The cathodic curve does not change substantially if drawn for the same stirred solution and if the latter is oxygen-saturated by air bubbling; this shows that oxygen reduction does not interfere and that there are other cathodic reactions involved. The passivation current value of the patina reveals its protective attribute. Polished steel up to 0.25 μm shows a constant increase of the current with the applied over-potential increase. The specimens exposed at Villa Martelli's outdoor test station for 1–9 years reveal that the passive nature of Cor-ten A steel increases continuously [1–3].

The protective attribute of corrosion products on steels depends on the climatic conditions of the place where they are formed. This was shown by the electrochemical test described involving low carbon steel without microalloying elements in different test stations in Ibero-American countries, as shown in Fig. 2 [4].

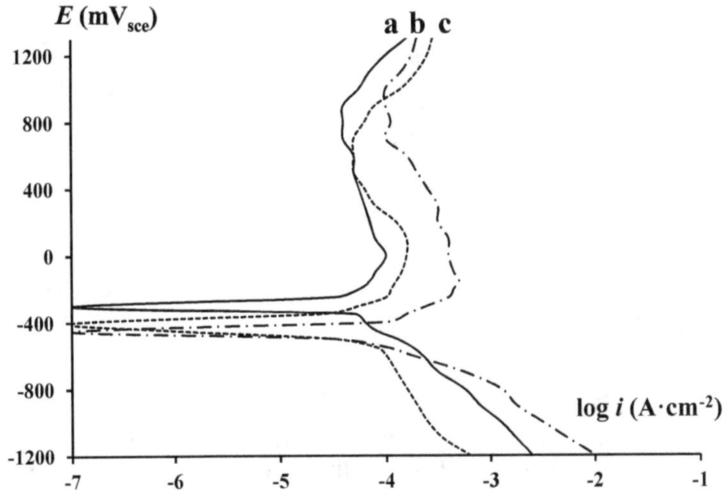

Figure 2: Polarisation curves of low carbon steel in Na_2SO_4 0.1M at 25°C, exposed at 3 outdoor test stations for 1 year [4].

3 Non-metallic inclusions reactivity test

PA weathering steel was designed to contain the necessary microalloying elements as well as non-metallic inclusions; the latter were expected to enhance both weatherability and mechanical properties [5]. Cor-ten A and PA steel specimens polished with diamond paste up to 0.25 μm were polarised anodically to determine the pitting potential (E_p) of these steels in a buffered solution at pH = 9 with NaCI 0.05 M.

The test was carried out in triplicate using specimens of 1 cm^2 area exposed to the electrolyte, which was polarised at a scanning rate of 10 mV·min^{-1} from the rest potential to the pitting potential, which was E_p = –180 mV for both steels. The said potential remained constant and the corrosion attack was nucleated, recording time-based currents. In Fig. 3, the curve log $i = f(t)$ can be observed, showing the nucleation kinetics of the PA steel attack in contrast with Cor-ten A.

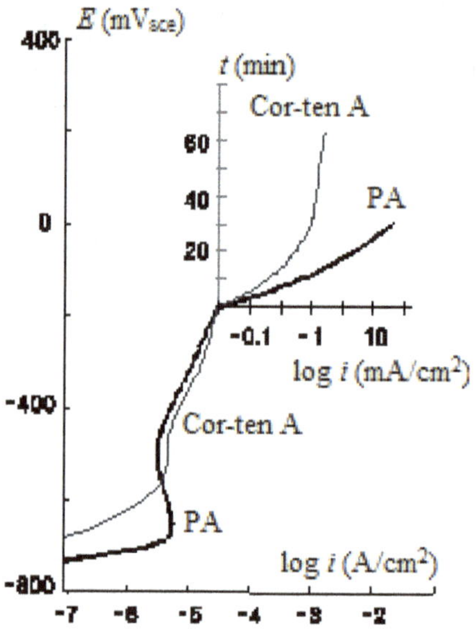

Figure 3: Reactivity inclusions test on PA and Cor-ten A steels, polished up to 0.25 μm.

3.1 Polarisation curves

Anodic and cathodic polarisations were performed on PA steel samples exposed for 6 years to the open air in Villa Martelli; the results are shown in Fig. 4. The same experimental technique described in [6], was applied to a Cor-ten A steel exposed for 15 years in the same station. The latter curve has been reproduced for the sake of comparison. The corrosion current i_{corr} and the anodic β_a and cathodic β_c Tafel slopes were calculated by the GAMMACRUNCH program [6], which was used by its authors for different corrosion problems [7, 8]. The results were as follows: $i_{corr} = 2\times10^{-7}$ A/cm^2; β_c = –201.8 mV/dec.; β_a = 218.8 mV/dec.

The result of weight loss by outdoor exposure for PA steel was compared with that for Cor-ten A, previously obtained:

$$\Delta W = \frac{i_{corr} A_r}{2F} \times 8.64\times10^8 = 0.05 \text{ g/m}^2\text{day} \tag{7}$$

where A_r = 55.85 is the Fe atomic weight and F = 96,490 C/mol is the Faraday constant. The value obtained for PA steel is a 6-year exposure average: $\Delta W \approx 0.072$ g/m^2 per day, which is of the same order of magnitude.

The results obtained by the GAMMACRUNCH program allow the calculation of the polarisation resistance R_p [6] from

$$\frac{1}{R_p} = \frac{\Delta i}{\Delta E} = 2.303 i_{corr} \frac{\beta_a + |\beta_c|}{\beta_a |\beta_c|} = 4\times10^{-9}\ \text{A/mV·cm}^2$$

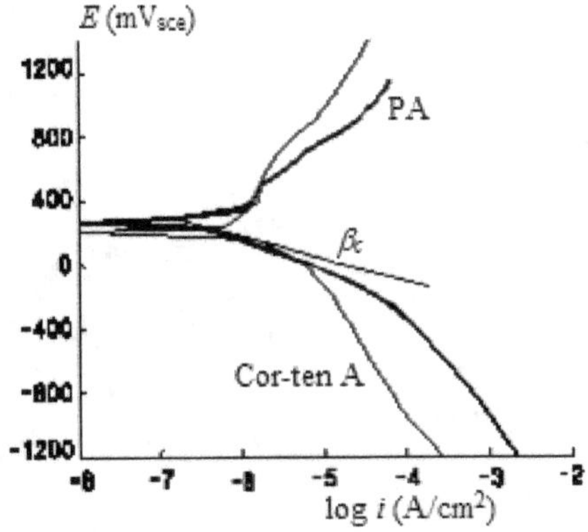

Figure 4: Polarisation curves in Na_2SO_4 0.1 M of PA steel exposed for 6 years and of Cor-ten A steel exposed for 15 years in Villa Martelli's urban site.

Thus, considering $\Delta E = 20$ mV, $\Delta i = 0.08$ µA/cm^2. The experimentally obtained value for $\Delta E = 20$ mV$_{sce}$ was $\Delta i = 0.024 \pm 0.014$ µA/cm^2, a result of the same order of magnitude as that previously calculated.

3.2 Cyclic polarisation curves

The curves were produced from the rest potential E_r or the open circuit at a scanning rate of 100 mV·min^{-1} in the anodic direction up to +1300 mV in order to start from there an anodic-cathodic cycle down to –1200 mV, and back again to +1300 mV. Results for PA steel, Cor-ten A steel and SAE 1020 steel (C: 0.2; Mn: 0.5; P: 0.01; S: 0.024; Si: 0.09), exposed simultaneously with the PA steel, are shown in Figs. 5(a), (b), and (c), respectively.

3.3 Test on corrosion products film reduction

These tests were carried out at a potential scanning rate of 2.77 mV·min^{-1}, from the rest potential and in a cathodic direction. The electrolyte used was a Na_2SO_4 0.1 M solution at 25°C with an initial pH = 6.3, and a rotation rate of the magnetic stirrer of 60 rpm, which remained constant in all the tests performed. The pH of the solution was measured all the time by an electrode and a Tacussel Minises 8000 electron equipment. The scanning was performed down to –950 mV, and samples of the solution were taken, the Fe^{+2} content of which was analysed in a Carl Zeiss M4QIII atomic absorption equipment.

Following the cathodic treatment, the specimen potential was stabilised and an anodic scanning at 100 mV·min^{-1} was performed from the said potential up to +700 mV. The results of a non-weathering SAE 1020 steel, and a PA weathering steel after six years' exposure in the urban site of Villa Martelli's outdoor test station are shown in Fig. 5 and Table 1. Both steels were exposed simultaneously at the same site.

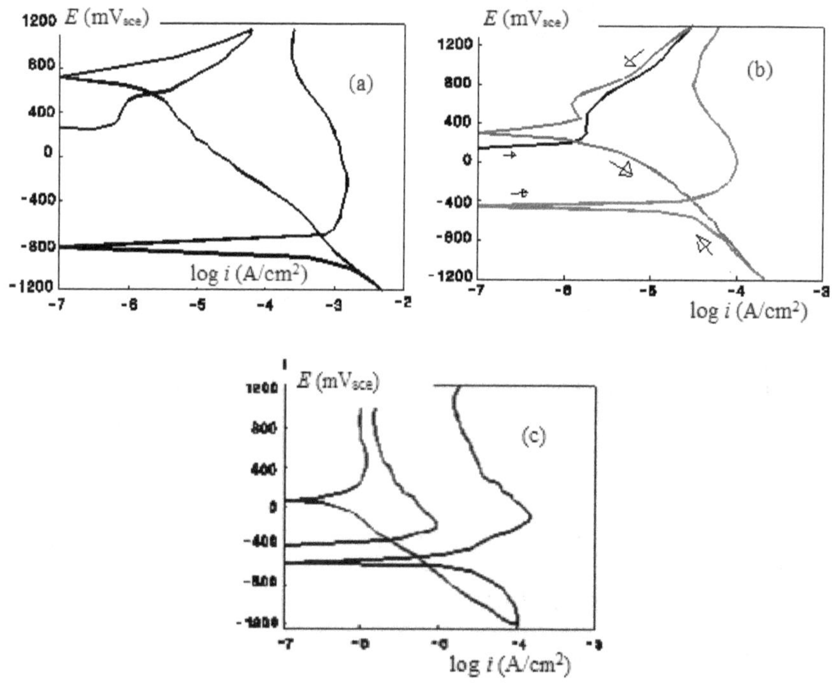

Figure 5: Steel cyclic polarisation curves; (a) PA steel exposed for 6 years; (b) Cor-ten A steel exposed for 15 years; (c) SAE 1020 steel exposed simultaneously with the PA steel in the urban site of Villa Martelli's outdoor test station.

Table 1: pH and $(Fe^{+2})_{aq.}$ based on the PA steel potential in the cathodic reduction test.

Potential (mV_{sce})	+200	–200	–700	–900	–1100
pH	6.3	7.3	5.3	5.2	6.3
$(Fe^{+2})_{aq.}$ (ppm)	----	-----	= 1	= 1	11.7

The measured areas under the anodic and cathodic curves shown in Fig. 6 are proportional to the Q charge, which has flowed during the anodic and cathodic processes. The E potential axis (mV_{sce}) may be considered as the time t axis (s), since the scanning rate is constant during the test. Thus [9]

$$Q = a \int i dt = a' \int i dE \tag{8}$$

The PA steel values evaluated using Simpson's rule show that the cathodic area is bigger than the anodic one according to the ratio $z = 1.63$.

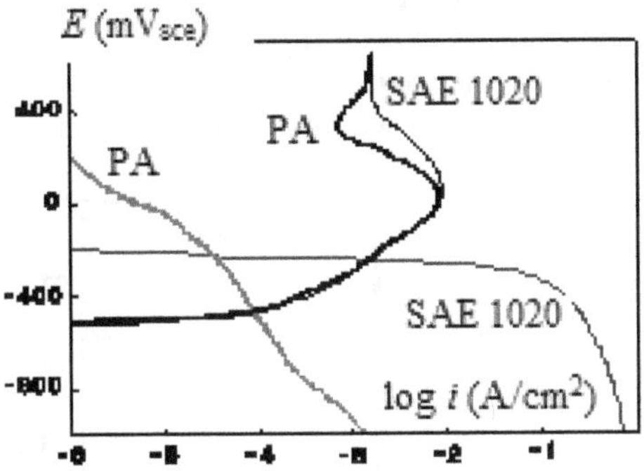

Figure 6: Cathodic reduction curves of corrosion products formed in the open air and anodic curves for the same sample after the cathodic test.

The curves in Fig. 7 show the results of a cathodic reduction at two different scanning rates: curves (a) cathodic reduction at 2.77 mV/min down to a potential of –650 $mV_{sce.}$; (b) at 100 mV/min down to –1200 mV_{sce}. The corresponding anodic curves at 100 mV/min show a maximum current at the potential E = –235 mV_{sce} and E = –300 mV_{sce}, respectively.

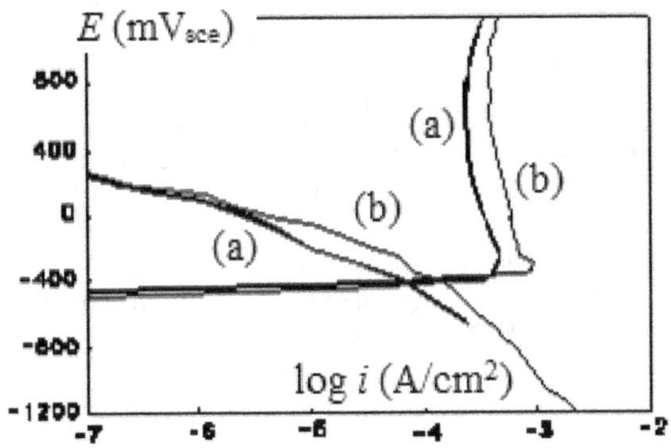

Figure 7: Curves (a) cathodic reduction at 2.77 mV/min, and following anodic reduction at 100 mV/min; curves (b) cathodic reduction at 100 mV/min., and following anodic curve at the same rate.

Fig. 8 summarises the characteristics of the three cathodic reductions of the PA steel, and the respective anodic and cathodic loads.

These results were supplemented with the measurement of the corrosion potential E_{corr} evolution of the PA steel, considering the immersion time in a Na_2SO_4 0.1 M solution (Fig. 5). Curve (a) was drawn at a pH = 6.0, which remained constant during the 900 hours of the test. On the other hand, curve (b) corresponds to a test similar to the previous one, which showed a pH = 4.0 at first, and that varied until achieving a 5.6 value after 100 hours. In curve (c), a pH = 4.0 buffer was used; the potential decreased more quickly at first, but then it stabilised at the same potential values. During this last test, no corrosion product precipitation was observed in the solution; and the test was stopped when the full loosening of the patina outer layer – without being fragmentised – occurred.

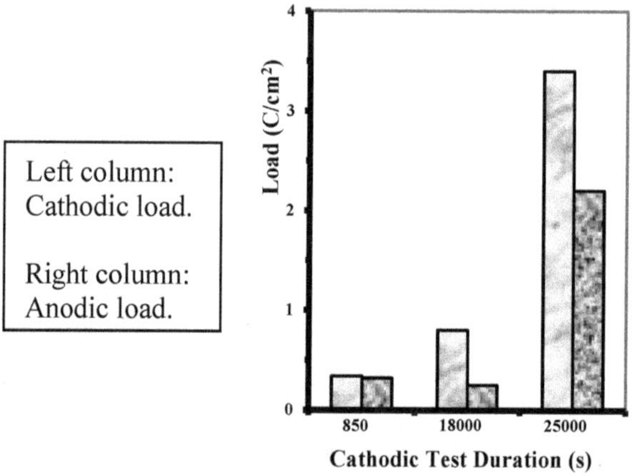

Cathodic test duration (s)	850	18,000	25,000
Curve	Fig. 7(b)	Fig. 7(a)	Fig. 5(a)
Cathodic scanning rate (mV/min).	100	2.77	2.77
Cathodic potential achieved (mV_{sce})	–1200	–650	–950
E at maximum anodic current (mV_{sce})	–300	–235	–40
Load balance (cathodic/anodic)	1.1	4.1	1.63

Figure 8: Summary of the characteristics of the cathodic reduction curves, and cathodic and anodic loads, exchanged among the PA steel corrosion products and the electrolyte.

4 Conclusion

The results from electrochemical tests, particularly those involving cathodic reactions of patina formed in the open air, show how different factors influence the magnetite formation as a final product of the said reactions. These experimental results allow an interpretation of why the PA steel cathodic curve shows a current variation by three orders of magnitude, which are less than the non-weathering reference ones, until achieving reduction potentials from those of the inner layer compounds to those of magnetite; thus, the weathering steel corrosion product film has a protective attribute similar to the one presented by non-weathering steels.

Therefore, it is inferred that the controlling mechanism of the patina protective attribute results from the compactness provided by the micro-alloying elements, which leads to a decrease of the pore and particle sizes.

References

[1] Rosales, B.M. & Ayllon, E.S., Urban industrial corrosion of a weathering steel. *Atmospheric Corrosion*, ed. W.H. Ailor, John Wiley & Sons. Inc.: New York, Ch. 28, p. 415–422, 1982.

[2] Rosales, B.M. & Ayllon, E.S., Atmospheric marine corrosion of structural steels. *Atmospheric Corrosion*, ed. W.H. Ailor, John Wiley & Sons. Inc.: New York, Ch. 29, p. 423–430, 1982.

[3] Rosales, B.M. & Ayllon, E.S., Controlling factors in the atmospheric corrosion of steel. *Atmospheric Corrosion*, ed. W.H. Ailor, John Wiley & Sons. Inc.: New York, Ch. 35, pp. 501–508, 1982.

[4] Granese, S.L., Rosales, B.M., Ayllon, E.S., Varela, F.E, Gervasi, C.A. & Vilche, J.R., Characterisation of the atmospheric corrosion products. *Proc. of the First Pan-American Corrosion and Protection Congress*, Mar del Plata, Argentina, vol. 1, p. 200, 1992.

[5] Ayllon, E.S., Special weathering steel, a contribution to environmental protection. *WIT Transactions on Engineering Sciences*, **54**, pp. 329–338, 2007.

[6] Britz, D., Programa GAMMACRUNCH (GAMMACRUNCH Program), Private Communication, April 1994.

[7] Jensen, M. & Britz, D., A linearized least-squares method of calculation of corrosion parameters. *Corrosion*, **46(2)**, pp. 111–114, 1990.

[8] Jensen, M. & Britz, D., Comparison of some methods of calculation of corrosion parameters from discretely sampled polarisation curves. *Corrosion Science*, **32(3)**, pp. 285–302, 1991.

[9] Galvele, J.R., *Corrosion*, CNEA-OEA, p. 17, 1979.

Electrochemical and corrosion resistance of new Ti-Mo-Fe alloys for biomedical applications

Y. M. Abd-Elrhman[1], M. A. Gepreel[1], A. Abd El-Moneim[1] & S. Kobayashi[2]
[1]*Egypt–Japan University of Science and Technology, Egypt*
[2]*Ehime University, Japan*

Abstract

Titanium and its alloys have been used extensively in the last few decades as materials for orthopaedic and dental implants and other medical devices due to their high strength, low modulus and high corrosion resistance in biological media. Besides other important material features, the corrosion parameters and corrosion products are responsible for limiting the biocompatibility of metallic materials and can produce undesirable reactions in implant-adjacent and/or more distant tissues. Electrochemical corrosion behaviour of novel beta titanium alloys, Ti-4.7Mo-4.5Fe, Ti-3Mo-0.5Fe and Ti-2Mo-0.5Fe was investigated in naturally aerated Ringer's solution at room temperature compared with currently used biomedical titanium alloy, Ti-6Al-4V. Very low current densities were obtained from the potentiodynamic polarization curves, indicating a passive behaviour for all the studied alloys. Electrochemical impedance spectroscopic (EIS) studies showed high impedance values for all samples indicating an improvement in corrosion resistance of the spontaneous oxide layer. The fitting data obtained suggest a single passive layer form on the virgin surfaces of the alloys. The EIS results exhibited capacitive behaviour (high corrosion resistance) with phase angles close to –80°C and high impedance values at low and medium frequencies, which are indicative of the formation of a highly stable film on these alloys in the test solution. The new alloys are promising metallic biomaterials for the future, owing to their very low elastic modulus and good corrosion resistance capabilities.
Keywords: titanium alloys, corrosion resistance, Ringer's solution.

1 Introduction

More than two thirds of implants are made of metallic biomaterials i.e. stainless steel, cobalt (Co)-chromium (Cr) alloys, and titanium (Ti) and its alloys. Among these metallic biomaterials, Ti alloys exhibit the highest biocompatibility, corrosion resistance and specific strength (ratio of strength to density), compared with the other metallic biomaterials [1]. Recently, researchers paid much attention to developing low rigidity β and near β type titanium alloys composed of non-toxic and non-allergic elements with excellent biocompatibility, superior corrosion resistance, high strength and low elastic modulus (close to that of human bone). The relatively low modulus of β and near-β titanium alloys may also reduce the "stress-shielding" effect [2–6]. Although titanium alloys possess a set of suitable properties for biomedical applications, their corrosion resistance still plays a vital role in such applications, whereas the corrosion products are largely responsible for limiting their biocompatibility and can cause inflammations with the implant-adjunct tissues. Therefore, a corrosion resistance investigation is considered one of the in-vitro evaluation procedures for the acceptance of these alloys as biomaterials. Fortunately, titanium and its alloys show high corrosion resistance due to the spontaneous formation of a protective oxide film of 1–4 nm thickness [7] on its surface at open circuit conditions. In physiological fluids that contain about 1 wt.% NaCl, the chloride present can destroy the oxide film, so it is necessary to investigate the durability and stability of these passive films in simulated body fluids [8].

Ti-4.7Mo-4.5Fe, Ti-3Mo-0.5Fe and Ti-2Mo-0.5Fe are a set of new beta titanium alloys developed to be used as biomaterials. The present work is undertaken to study the corrosion behaviour of the investigated alloys, as well as to compare their performance with that of commercial alloys such as Ti-6Al-4V thus to assess their performance and the possibility of their being used as biomaterials. Although there are many alloys used as biomaterials, they still have high cost, so the candidate materials are presented as low cost biomaterials.

2 Experimental procedure

2.1 Materials preparation

Nominal compositions of Ti-4.7Mo-4.5Fe, Ti-3Mo-0.5Fe and Ti-2Mo-0.5Fe (in atom %) alloys were arc-melted with a tungsten electrode on a water-cooled copper hearth. Titanium blocks were melted to purify the Ar gas atmosphere before the Ar-arc melting. All the ingots of the alloys were melted and inverted six times in order to promote their homogeneities. Specimens were solid-solution-treated (ST) at 1173 K in β phase field for 3.6 ks followed by rapid quenching into ice water. The designed alloys were selected to be investigated and characterized from the point view of corrosion to assess their performance as biomaterials.

2.2 Electrochemical evolution

Electrochemical experiments were conducted in two standard cells, namely, a polarization cell and a reference cell, connected by a salt-bridge probe. The polarization cell contained the working electrode and a platinum wire as a counter electrode, while the reference cell contained the standard saturated KCl Ag/AgCl reference electrode. Working electrolyte was a naturally aerated aqueous Ringer's physiological solution. The Ringer's solution has an electrolyte composition closer to that of plasma as compared to normal saline; this solution is composed of 0.32 g $CaCl_2.2H_2O$, 0.3 g KCl, 8.6 g NaCl, and quantum sufficit of water for each 1000 mL of electrolyte; its measured pH was 5.4. Prior to any immersion, the working electrodes were grinded using 2000 grade silicon carbide paper and rinsed with distilled and de-ionized water. All electrochemical measurements were managed by a Princeton Applied Research potentiostate and controlled by a personal computer with dedicated software (VersaSTAT 4). In order to verify the reproducibility of electrochemical results, all experiments were repeated three times measured at the same sample, after a new polish for the surface of the working electrode and good reproducibility was observed.

Measurements of the open-circuit potential (OCP), for each composition were performed on the polished samples, until the potential values became reasonably stable. The corrosion potential was measured as a function of time for about 116 min. with a stability of about 0.005 mV/s.

Potentio-dynamic polarization studies were carried out after 7,000 s of immersion in Ringer's solution after almost reaching the steady state. Potentio-dynamic polarization measurements were performed in the potential range from –500 mV in the cathodic direction to 1600 mV in the anodic direction from the OCP vs. the Ag/AgCl reference electrode at a scan rate of 0.167 mV/s. The corrosion potential E_{corr} and corrosion current density i_{corr} were determined from the polarization curves using the Tafel extrapolation method.

Electrochemical impedance spectroscopy (EIS) measurements were carried out at open-circuit potential and the impedance spectra were acquired in the frequency range from 100 kHz to 100 mHz. The applied alternating potential signal had an amplitude of 10 mV. EIS plots were obtained after the specimens solution was treated at 900°C. Data acquisition and analysis were performed with a personal computer. The spectra were interpreted using the ZSimpWin software. Impedance data were represented in Bode amplitude and phase angle plots. The charge transfer resistance, *Rs*, and double-layer capacitance, *CPE*, values were determined. The Bode representation shows the logarithm of the impedance modulus (log *Z*) and phase angle as a function of the frequency logarithm (log frequency).

2.3 Corroded surface

To characterize the anodic oxide films formed on the surface of the specimens, the latter were ultrasonically cleaned with methanol after the potentio-dynamic polarization tests and then examined by the optical microscope.

3 Results and discussion

3.1 Open-circuit potential analysis

The OCP is the potential at which the specimen is in equilibrium with the surrounding environment. The open-circuit potential of metals is always measured as a function of time, until it stabilizes at a stationary value after a certain period of immersion. The open-circuit potential is used as a criterion for the corrosion behaviour. This type of measurement is qualitative and still insufficient for a complete electrochemical analysis.

The results obtained from the OCP with immersion time in Ringer's solution are presented in Fig. 1 and Table 1 for the studied alloys and at the different heat treatment conditions. As can be seen from Fig. 1, the potentials increase quickly during the first few minutes and then shift slowly in the more noble direction, until they reach quasi-stationary values. This behaviour is indicative on the spontaneous formation and growth of the oxide film on the surfaces of the samples. Furthermore, the OCP in the noble direction indicates that an oxide film is passivating the metallic surface [9]; this suggests that a protective passive film acts as a barrier against metal dissolution, reducing the corrosion rate. Al-Mobarak [10], in his research, gave another interpretation to the appearance of the initial shift of potential during the first few minutes; he claimed that the first shift is due to the dissolution of the pre-immersion oxide layer that formed in air before immersing the samples in the solution.

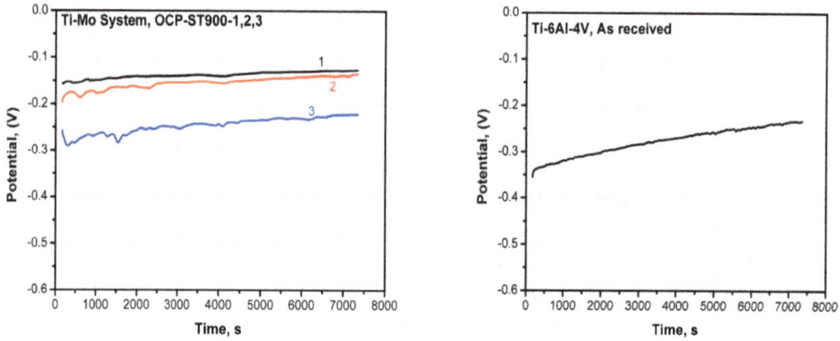

Figure 1: Open-circuit potential (OCP) versus time in Ringer's solution of Ti-4.7Mo-4.5Fe (denoted as "1"), Ti-3Mo-0.5Fe (denoted as "2"), Ti-2Mo-0.5Fe (denoted as "3"), and Ti-6Al-4V.

From the OCP profiles, a comparison among the obtained quasi-stationary OCP values for the specimens at different heat treatment conditions can be made. There is clear evidence regarding the effect of the heat treatment process from the variation of OCP values. As a result of the solution treatment process (ST900) after the hot rolling process (HR900), all the OCP values for all studied compositions were towards the noble direction; this result indicates that the spontaneous oxide film formed on the alloys due to the solution treatment process displays better corrosion protection characteristics than the oxide film formed on hot rolled samples. By comparing the obtained results of the OCP of the studied compositions with the OCP of most widely used titanium alloy in the biomedical applications, Ti-6Al-4V, one can notice that the oxide film on the studied samples formed more rapidly than on Ti-6Al-4V. Nevertheless, they are mostly in agreement regarding the direction towards the noble direction.

Table 1: Open-circuit potentials at the steady state conditions for Ti-4.7Mo-4.5Fe (denoted as "1"), Ti-3Mo-0.5Fe (denoted as "2"), Ti-2Mo-0.5Fe (denoted as "3"), and Ti-6Al-4V in ringer's solution.

Heat treatment process ST900	
Samples	[OCP/mv vs. Ag/AgCl]
(1) Ti-4.7Mo-4.5Fe	−129±7
(2) Ti-3Mo-0.5Fe	−139±13
(3) Ti-2Mo-0.5Fe	−223±17
*Ti-6Al-4V	−233±15
**Ti-6Al-4V used only for comparison and not subjected to any heat treatment process (used as received).*	

It is clear from the above OCP results that the heat treatment process is a strong factor in the electrochemical performance of the metals. These values of the OCP can be interpreted in the light of the present microstructures due to the heat treatment process. The type of the present crystal structure, i.e. hcp (α-phase), bcc (β-phase), orthorhombic α″…etc. influences the number of dislocations leading to differences in the mechanical properties of the alloys. These differences, i.e. in ductility, strength and so on, due to the heat treatment, will of course have an effect on the electrochemical properties such as OCP. The formed β-phase as a result of the solution treatment process affected the OCP values of the alloys and directed them towards the noble direction. Therefore, the denser the phase present is, the greater the reduction in the corrosion rate.

3.2 Potentio-dynamic polarization studies

Fig. 2 shows the potentio-dynamic polarization curves for the studied alloys compared with that for Ti-6Al-4V as tested in Ringer's solution at 25°C. Potentio-dynamic polarization studies were conducted in the potential range of –500 mV to +1600 mV with respect to OCP vs. Ag/AgCl for all alloys. From these curves, the corrosion parameters, i.e. the corrosion potential (E_{corr}) and the corrosion current density (i_{corr}) were obtained from the Tafel region, then the corrosion current density was used to determine the corrosion rate for all alloys. The corrosion current density is representative for the degradation degree of the alloy. A distinctive passivation region was observed on the anodic polarization curves for all the studied alloys. This region of passive film formation was studied through the passivation current density (i_{pass}) which is obtained around the middle of the passive range. It was observed from Fig. 2 that the shape of the polarization curves is quite similar, while different corrosion parameters were determined for all alloys after heat treatment; see Table 2. This qualitative analysis of the polarization curves shows that the studied alloys have good potential to be used as biomaterials. The corrosion behaviour of the present new alloys shifts towards nobility, i.e. the corrosion potentials, (E_{corr}) shift towards positive values and the corrosion currents (i_{corr}) decrease compared to those of Ti-6Al-4V. The new alloys exhibit very high resistance to corrosion compared with one of the most commonly used biomaterial, Ti-6Al-4V. The corrosion parameters of the present new alloys are about ten times better than those of Ti-6Al-4V and this clearly indicates that these alloys have excellent pitting and crevice corrosion resistance.

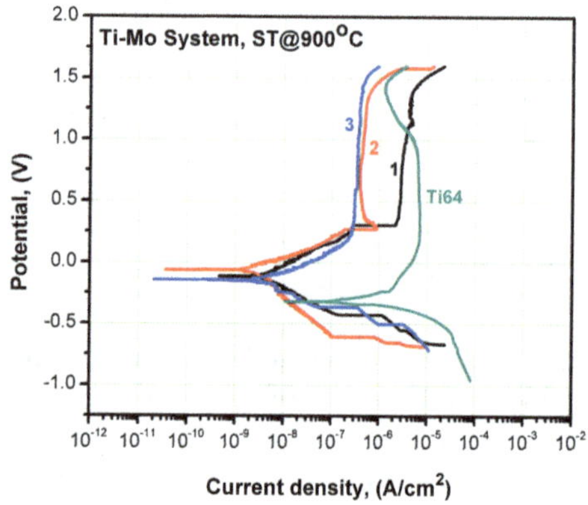

Figure 2: Potentio-dynamic polarization curves for Ti-4.7Mo-4.5Fe (denoted as "1"), Ti-3Mo-0.5Fe (denoted as "2") and Ti-2Mo-0.5Fe (denoted as "3") after the heat treatment compared with Ti-6Al-4V (as received) in Ringer's solution at room temperature.

The corrosion rates were obtained from the corrosion current densities using the following equation (ASTM Standard G102-89 [11]):

$$\text{Corrosion rate (in } \mu\text{m/year)} = 3.27\ (i_{corr}/\rho)\ EW \tag{1}$$

where, i_{corr} is the corrosion current density in $\mu A/cm^2$, ρ the density (4.51 g/cm^3) and the EW is the equivalent weight for titanium alloys:

$$EW = [\text{atomic weight (47.9)}] / [\text{the valence (4)}].$$

The calculated corrosion rates for the new present alloys at various heat treatments are tabulated in Table 2. The value of the corrosion rate is a significant parameter that gives valuable information about the real behaviour of the corrosion. The corrosion rate is directly proportional to the corrosion current density so they have the same trend. In conclusion, the present new alloys are more resistant to corrosion compared with the most widely used biomaterial, Ti-6Al-4V.

Table 2: Corrosion potential (E_{corr}), current density (i_{corr}) and passivation current density (i_{pass}) determined from the polarization curves for Ti-4.7Mo-4.5Fe (denoted as "1"), Ti-3Mo-0.5Fe (denoted as "2") and Ti-2Mo-0.5Fe (denoted as "3"), after the heat treatment process, compared with those of Ti-6Al-4V (as-received) in Ringer's solution at room temperature.

Potentio-dynamic characterization					
Compositions	Heat treatment	E_{corr} [mv vs. Ag/AgCl]	i_{corr} [nA/cm²]	i_{pass} [μA/cm²]	Corrosion rate [μm/year]
1	ST900[a]	–110	6.876	2.804	0.060
2	ST900	–64	3.338	0.3915	0.029
3	ST900	–138	5.838	0.3449	0.051
Ti-6Al-4V	As-received	–319	661.8	6.561	5.746

Impedance (EIS) results of studied alloys were represented through the Bode plots as shown in Fig. 3. All specimens exhibit no significant change in their behaviour. This result could be due to a complete coverage of the surface of newly formed spontaneous oxide layer. In this paper, a satisfactory fit of all data could be obtained using simple equivalent circuit R_s $(Q\ R_p)$, see Fig. 4, where R_s and R_p are the solution and the oxide film resistances, respectively, while, Q is the constant phase element (CPE), indicating the capacitive behaviour of the film. It can be seen in Table 3 that all compositions have very small capacitance values and the values of n are near to 1, so Q presents an almost pure capacitive behaviour. For the oxide film resistance, R_p (Table 3) is of the same order of magnitude, 10^5 $\Omega.cm^2$, and these high values demonstrated that the spontaneous formed oxide layer on the alloys is a highly stable and resistive layer.

Table 3: Values of the equivalent circuit parameters derived from the fitting of the impedance data measured for the passive film grown spontaneously on the studied alloys, using the equivalent electric circuit R_s $(Q\ R_p)$, from Fig. 4.

Compositions	R_s ($\Omega.cm^2$)	Q (S cm^{-2} s^n)	n	R_p, ($\Omega.cm^2$)
1	24.45	7.13×10^{-5}	0.844	1.54×10^5
2	22.22	4.10×10^{-5}	0.825	7.53×10^5
3	29.63	3.92×10^{-5}	0.835	4.06×10^5

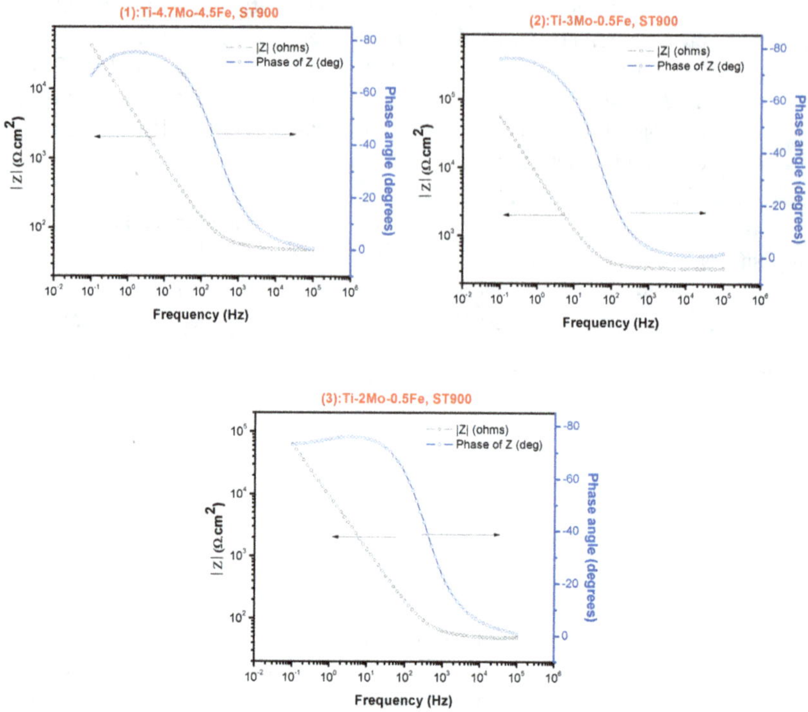

Figure 3: Bode diagrams of the studied alloys after solution treatment process at 900°C: (1) Ti-4.7Mo-4.5Fe, (2) Ti-3Mo-0.5Fe and (3) Ti-2Mo-0.5Fe.

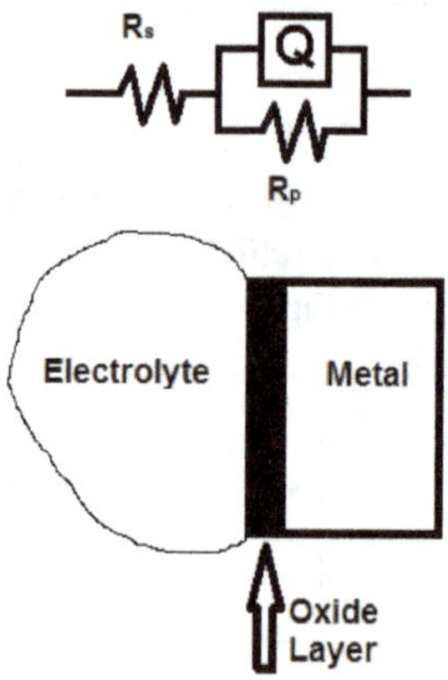

Figure 4: Equivalent circuit, used for fitting the experimental data.

3.3 Corroded surfaces evaluation

Figure 5 shows the optical microscope (OM) micrographs of the passive oxide films formed on the surfaces of the designed alloys. The OM micrographs are in agreement with the above anodic curves whereas the morphologies of anodic films critically depend on the behaviour of the oxide layer growth. The oxide layer formed on the surfaces of the specimens after solution treatment depends on the composition of the specimen. For all studied compositions, passivation occurred and the very small corrosion rates give a good indication on the promise of the candidate alloys as biomaterials.

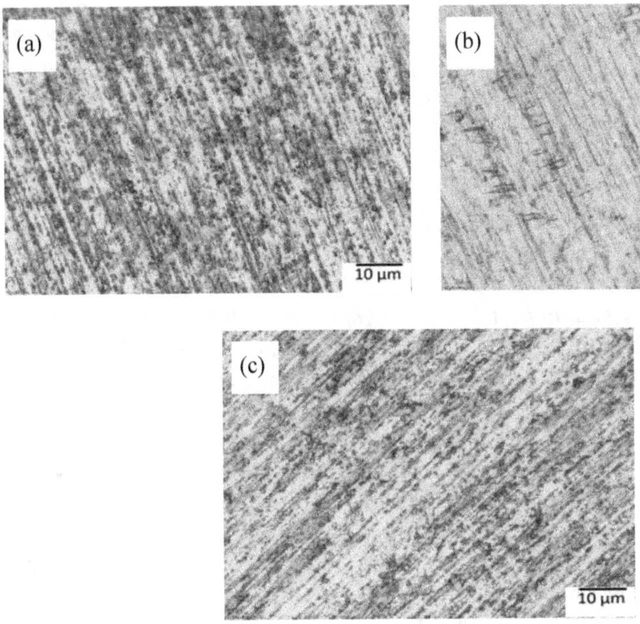

Figure 5: Optical microscope micrographs of oxide films on the surfaces of the studied alloys after anodic test: (a) Ti-4.7Mo-4.5Fe, (b) Ti-3Mo-0.5Fe and (c) Ti-2Mo-0.5Fe.

4 Conclusions

In this study, new titanium-iron based alloys were developed using a low-cost alloying element, iron, to reduce the total cost of the product. Two beta stabilizer elements were used in our alloys, Mo and Fe; the electrochemical properties of these alloys were studied. The main results are summarized as follows:

- The open circuit potentials (OCP) of the designed alloys in the Ringer's solution directed towards the positive direction and reached almost steady state condition within 30 min. The short time which has taken to reach the steady state is good evidence of the stability of the formed oxide layer.

- The potentio-dynamic tests after the solution treatment show very low corrosion rates for the present alloys.
- High impedance values were obtained for all studied alloys; the obtained fit suggests a single passive film present on the metals' surface.

All these electrochemical results are connected and suggest that Ti-Mo-Fe alloys are promising materials for biomedical applications because of their electrochemical stability.

References

[1] Niinomi, M., Nakai, M. & Hieda, J., Development of new metallic alloys for biomedical applications. *Acta Biomaterialia*, **8**, pp. 3888–3903, 2012.

[2] Lütjering, G. & Williams, J.C., *Titanium (Engineering Materials and Processes)*. Springer, pp. 5–20, 2007.

[3] Oliveira, N.T.C., Aleixo, G., Caram, R. & Guastaldi, A.C., Development of Ti–Mo alloys for biomedical applications: Microstructure and electrochemical characterization. *Materials Science and Engineering A*, **452-453**, pp. 727–731, 2007.

[4] Donachie, J.M., *Titanium: A Technical Guide*. ASM International, 2000.

[5] Dewidar, M.M., Yoon. H. & Lim J.K., Mechanical properties of metals for biomedical applications using powder metallurgy process: A review. *Metals and Materials International*, **12**, pp. 193–206, 2006.

[6] Geetha, M., Singh, A.K., Asokamani, R. & Gogia, A.K., Ti based biomaterials, the ultimate choice for orthopaedic implants – A review. *Progress in Materials Science*, **54**, pp. 397–425, 2009.

[7] Marino, C.E.B., Oliveira, E.M.D., Rocha-Filho, R.C. & Biaggio, S.R. On the stability of thin-anodic-oxide films of titanium in acid phosphoric media. *Corrosion Science*, **43**, pp. 1465–1476, 2001.

[8] Alves, V.A., Reis, R.Q., Santos, I.C.B., Souza, D.G. & Gonçalves, T., In situ impedance spectroscopy study of the electrochemical corrosion of Ti and Ti–6Al–4V in simulated body fluid at 25°C and 37°C. *Corrosion Science*, **51**, pp. 2473–2482, 2009.

[9] Oliveira, N.T.C. & Guastaldi, A.C., Electrochemical stability and corrosion resistance of Ti-Mo alloys for biomedical applications. *Acta Biomaterialia*, **5**, pp. 399–405, 2009.

[10] Al-Mobarak, N.A., In situ impedance spectroscopy study of the electrochemical corrosion of orthopedic Ti-6Al – 7Nb in simulated body fluid. *Advanced Materials Letters*, **3**, pp. 433–440, 2012.

[11] ASTM G102-89, Standard Practice for Calculation of Corrosion Rates and Related Information from Electrochemical Measurements, 2004.

Improvement of corrosion resistance and biocompatibility of medical nickel–titanium alloys through forming TiO_2 thin film

Y. Kimura
Kogakuin University, Shinjuku-ku, Tokyo, Japan

Abstract

Nickel–titanium alloys have been investigated for applications to various surgical procedures. However, little is known about the toxicity of Ni-elements and the actual conditions of surface modification of Ni-Ti alloy for preventing the toxicity of Ni. In this study, the corrosion resistance of Ni-Ti alloys was improved through producing a thin TiO_2 film by conducting in-air oxidation and reactive sputtering. These thin films were characterized using field emission type scanning electron microscopy, X-ray diffraction, X-ray photoelectron spectroscopy and atomic force microscopy. Using the polarization curves obtained by electro-chemical measurements in quasi-living body environment, the corrosion resistance of Ni-Ti super elastic alloy was evaluated. In addition, the effect of an air-oxidized Ni-Ti alloy specimen upon cell multiplication and morphological change of cell were investigated by culturing cells for 72 hrs on a Ni-Ti alloy specimen surface with TiO_2 thin film. From these considerations, some trials for establishing a method for TiO_2 thin film formation on the surface of Ni-Ti alloy were conducted from the view point of improving biocompatibility.
Keywords: nickel–titanium alloy, corrosion resistance, biocompatibility, surface modification, TiO_2 thin film, reactive sputtering, air oxidation – suppressed partial pressure, cell multiplication – morphological change.

1 Introduction

In recent years, some concern has been raised about the corrosion resistance of implant alloys within an aging society. Implant materials have been required to exhibit biocompatibility in the long term without a re-operation. The biocompatibility of implant materials means they should not adversely influence the living body nor lead to loss of function of the materials of interest. Especially, the strength degradation by corrosion and elution of a toxic metallic ion by corrosion in the severe living body environment has attracted growing interest. The shape memory effect and super elasticity effect of Ni-Ti alloys [1, 2] have received considerable attention in the medical domain [3–6]. The Ti-Ni alloy has been increasingly applied to medical and dental devices, such as coronary stents and orthodontic wires. Thus, the application of Ni-Ti alloys in medical and dental devices has been the subject of many research papers. However, in practical cases, the presence of Ni element (Ni ion) in Ni-Ti alloys for such applications has been a cause for apprehension due to carcinogenicity and allergic reaction [7]. In the present state of affairs, the view about the use of Ni-Ti alloys in the medical domain is negative. A comprehensive understanding of the surface modification of Ni-Ti alloys for use under quasi-living body environment is still lacking.

Therefore, in this paper, the corrosion resistance of Ni-Ti alloys was improved through the production of a thin TiO_2 film. The TiO_2 thin film has been studied for photocatalysis and a lot of production methods have been developed. In this study, the TiO_2 thin film was produced by air oxidation and reactive sputtering. Air oxidation is a comparatively simple thin film production method and thin films made by this method do not easily exfoliate. Reactive sputtering is the film production method that can control the forming condition precisely and thin films made by this method are highly coherent. These thin films were characterized using field emission type scanning electron microscopy (FE-SEM), X-ray diffraction (XRD), X-ray photoelectron spectroscopy (XPS) and atomic force microscopy (AFM). Using the polarization curves obtained by electro-chemical measurement in quasi-living body environment, the corrosion resistance of Ni-Ti super elastic alloy was evaluated. In addition, the effect of an air-oxidized Ni-Ti alloy specimen upon cell multiplication and morphological change of cell were investigated by culturing cells for 72 hrs on Ni-Ti alloy specimen surface with TiO_2 thin film.

From these considerations, some trials for establishing a method for TiO_2 thin film formation on the surface of Ni-Ti alloy were conducted from the view point of improving biocompatibility.

2 Experimental procedures

2.1 Test specimen

Plate specimens with sizes of $11.5 \times 11.5 \times 2.5–3.0\ mm^3$ were cut from a block of Ni-Ti alloy (44% titanium, 56% nickel by weight). The samples were finished by mechanical abrasion and then polished to a mirror finish. The test specimen was ultrasonically cleaned in ethanol. Air oxidation treatments were performed in an electric furnace in oxidizing atmosphere. Four kinds of samples (A1–A4) were prepared through changing the heat treatment temperature and heating time (A1: 673 K, 30 min, A2: 773 K, 30 min, A3: 773 K, 60 min, A4: 1073 K, 30 min) [8].

A DC reactive sputtering system (MPS-2000-HC3, ALVAC) was employed for forming the TiO_2 thin film [9]. The target is titanium with purity of 99.99%. The distance between target and substrates was fixed at 200 mm. After the chamber was evacuated to a vacuum lower than $\times 10^{-6}$ Pa, argon was introduced into the chamber. The input power was fixed at 200 W and the substrate temperature was not controlled. After the discharge colour was confirmed, oxygen was introduced into the chamber. Three kinds of samples were prepared through changing the ratio of argon and oxygen flow rate (S1: $Ar:O_2 = 5:10$ [sccm], S2: $Ar:O_2 = 20:10$ [sccm], S3: $Ar:O_2 = 50:10$ [sccm]).

2.2 Anodic polarization measurement and immersion test

The corrosion resistance was evaluated employing an anodic polarization measurement method based on JIS T 0302 (testing method for corrosion resistance of metallic biomaterials by anodic polarization measurement). The electro-chemical measurement was conducted using a three-electrode method (HZ-3000 Hokutodenkou Japan) composed of the specimen (working electrode), Pt (counter electrode) and the saturated calomel electrode (SCE) (reference electrode). The electro-chemical measurement was conducted using a plate specimen in lactated Ringer's solution at 310 K. The test solution was de-aerated by extra-high purity nitrogen for 20 minutes before testing to suppress dissolved oxygen at a level similar to that of the living body. The potentio-dynamic corrosion test was conducted with the sweep rate of 20 mV/min from –1000 mV to 2000 mV (vs. SCE).

Immersion tests were conducted as follows. The specimens were put into polyethylene vessels with 20 ml of 1.0% lactic acid (JIS T 0304). After immersion at 310 K for 30 days, quantification of released ions was carried out by means of an inductively coupled plasma emission spectrometer ICP-AES: SPS7700 (Seiko Instruments Inc.). Three specimens were tested for each condition. Amounts of released component ions per unit surface area of the specimens were then calculated.

2.3 Evaluation of bio-compatibility of air oxidized Ni-Ti alloy

As a final goal of biocompatibility evaluation, it was planned to employ the inflammatory response of macrophage. In previous research, U937 (Human Leukemic Monocyte Lymphoma Cell) was used to differentiate into macrophage [10, 11]. In this study, however, an osteoblast-like cell line MC3T3-E1 was applied because of improvement in reproducibility of test results due to its having no ability to differentiate into other cell systems.

At first, a pure Ni square plate specimen (27 mm) whose thickness is 1.0 mm was dissolved in 1.0 M phosphate buffer whose temperature is 310 K employing three-electrode anodic polarization measurement method under the constant current density of 130 mA/cm^2. Then through analysing it by ICP, test solutions with various Ni ion concentrations from 0.125 to 2.0 ppm were prepared.

The effect of Ni ion concentration upon cell multiplication and morphological change of cell was investigated. Also, the effect of air oxidized Ni-Ti alloy specimen upon cell multiplication and morphological change of cell were investigated by culturing cells for 72 hrs on Ni-Ti alloy specimen surface with TiO_2 thin film.

3 Experimental results and discussion

3.1 Characterization of modified surface

In Fig. 1, the FE-SEM morphology of Ni-Ti specimen surface, the film thickness determined by FIB processing and SIM (Scanning Ion Microscopy) observation and the structure of thin film determined by XRD are summarized.

Then, the structural analysis of the thin film layer was conducted by sputtering and XPS [12]. From the results of compositional analysis toward the depth direction by XPS shown in Fig. 2, the composition of extreme thin surface of specimen was understood. A thin TiO_2 layer exists in the vicinity of film surface.

On the surface of the thin film formed by air oxidation, a peak of Ni was observed, and it is considered that a Ni-free layer was not present on the surface of the specimen. On the other hand, a peak of Ni was not observed on the surface of the film formed by reactive sputtering; thus, it is considered that a Ni-free layer was provided on the surface of sputtered specimen.

Sugawara *et al.* [8] investigated the effect of thermal oxidation treatment on the surface characteristic and corrosion resistance of a Ni-Ti shape memory alloy.

In this study, a mechanically polished Ni-Ti alloy was subjected to isothermal oxidation (TO) in N_2-20vol.%O_2 at temperatures ranging from 300 to 800°C. TO-treated surfaces were then characterized by FE-SEM, energy dispersive X-ray spectrometer (EDX), XPS and XRD. The results showed that the Ni-Ti alloy exhibited different oxidation behaviour depending on the treatment temperatures. A Ni-free layer was observed in the oxide layer for the specimen TO-treated at temperatures 500°C and above. Element concentration and chemical binding condition from the surface to interior were investigated using XPS. As a result, oxygen diffusion was detected in the surface layer and the depth of this layer increases with increasing treatment temperature.

3.2 Evaluation of corrosion resistance by electro-chemical measurement

Figs. 3 and 4 show the polarization curves obtained in lactated Ringer's solution at 310 K using in air oxidized and reactive sputtered Ni-Ti alloy. From the polarization curves obtained from in air oxidized specimen, it can be understood that the passivation current density was extremely suppressed compared with that of non-treated materials, and improvement of corrosion resistance was confirmed. As in-air oxidation heat treatment temperature increases, film thicknesses increase and corrosion resistance improved from the view point of passivation current density. It is considered that corrosion resistance rather depends on film thickness than on crystal structure. However, the potential width of passive state remains approximately equal to that of virgin materials.

Sugawara *et al.* also carried out electrochemical corrosion tests using a three-electrochemical cell connected to a computer-driven potentiostat [8]. TO-treated specimens at temperatures ranging from 300 to 500°C showed higher corrosion resistance compared to that of the surface finished by polishing. This was because

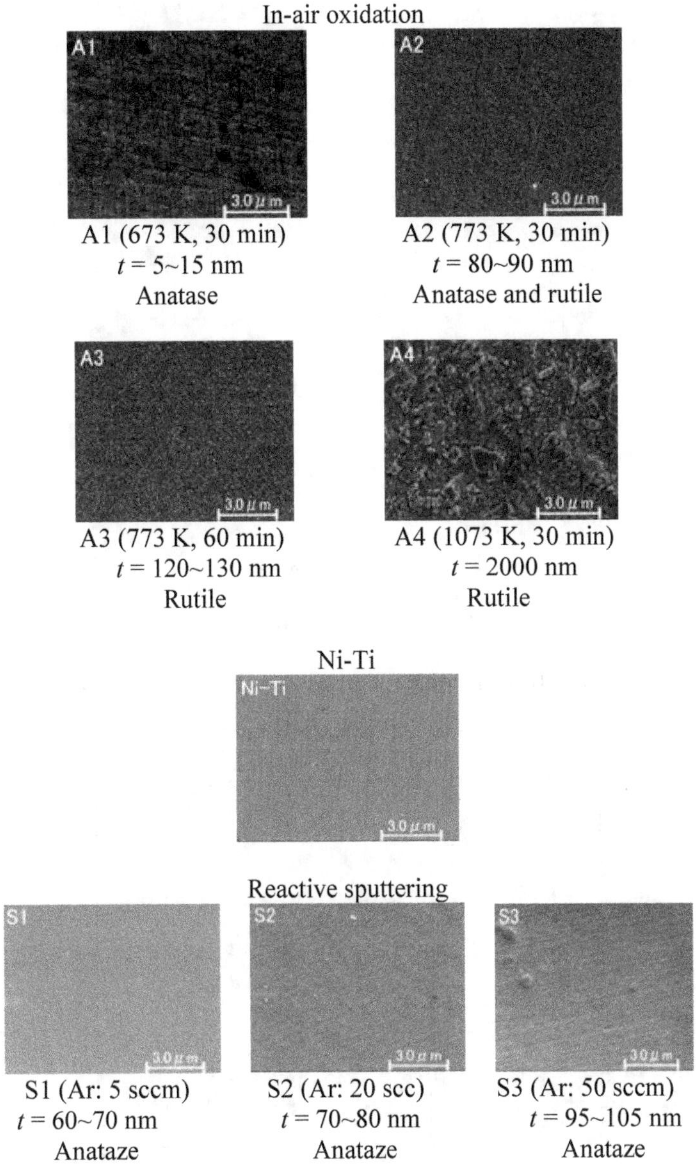

Figure 1: Morphology, film thickness and structure of surface-modified specimens.

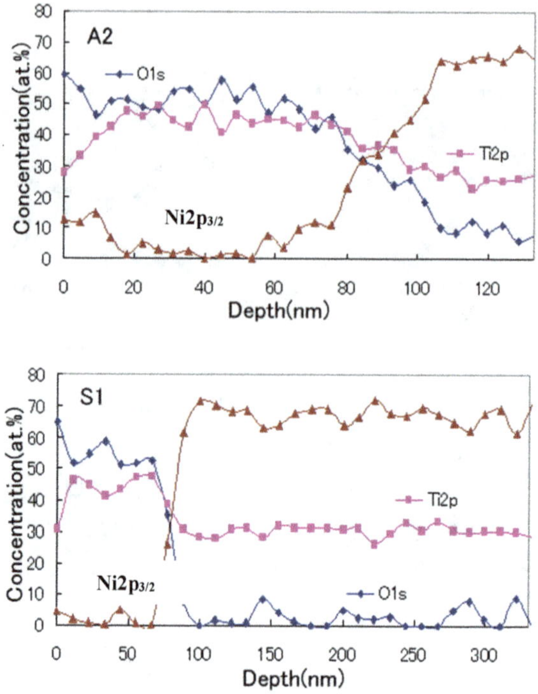

Figure 2: Analysis of composition toward depth direction by XPS.

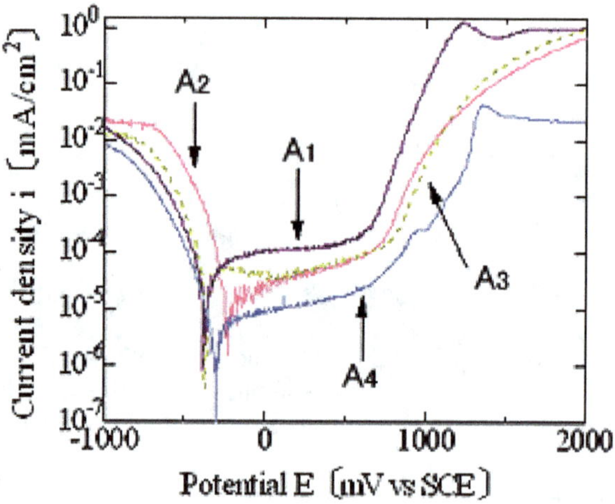

Figure 3: Polarization curves of surface-modified Ni-Ti alloy (in-air oxidation) in lactic Ringer's solution of 310 K.

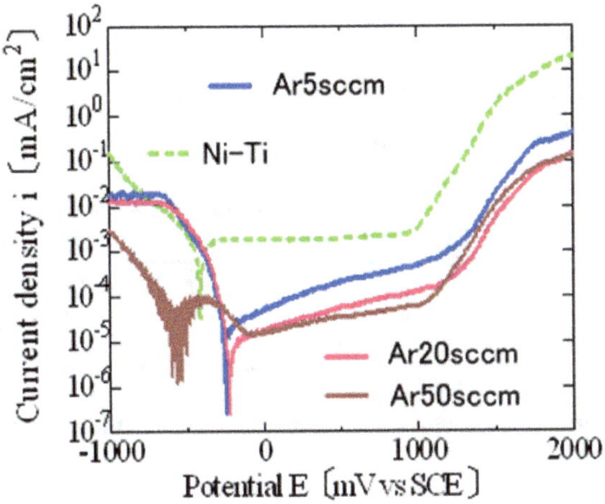

Figure 4: Polarization curves of reactive sputtered modified Ni-Ti alloy in lactic Ringer's solution of 310 K.

the TO-treatment created a thick oxide layer with an amorphous structure. TO-treated specimens at above 600°C showed very low passive current densities, but lower pitting potential. This was because these samples were covered with crystalline oxide. Consequently, the TO-treatment at 500°C produced a smooth protective nickel free oxide layer, which contributes to good biocompatibility of Ni-Ti implants. As a result, it was made clear that TO-treatment applied to Ni-Ti alloy can drastically suppress Ni ion elusion from the alloy surface.

From polarization curves of a reactive sputtered specimen, it can be stated that the passivation current density was significantly improved compared with that of non-treated ones, and the improvement of corrosion resistance was thus confirmed.

As Ar flow rate increases, film thicknesses increase and corrosion resistance improve. The potential width of the passive state is wider than that of the virgin material, thus biocompatibility without re-operation in the long term is expected.

3.3 Bio-compatibility of air oxidized Ni-Ti alloy specimen

At first, the effect of Ni ion concentration upon cell multiplication was investigated in this section. Fig. 5 shows the number of adhered cells after cultured 72 hrs in 1.0 M phosphate buffer with various Ni ion concentrations under 2 ppm or cultured 72 hrs on Ni-Ti alloy specimen surface with TiO_2 thin film. The solid line shown in this figure indicates sowed cell density, that is, 5600 cells/cm^2. From these results, cell multiplication occurred under each ion concentration condition in the same manner as control. Therefore, no significant difference in the number of adhered cells was recognized. As a result, there is no effect of Ni ion upon cell multiplication within ion concentration below 2 ppm. Also, in case of Ni-Ti alloy with TiO_2 thin film formed by air oxidization, no effect of Ni ion dissolution from extreme thin surface layer upon cell multiplication was detected.

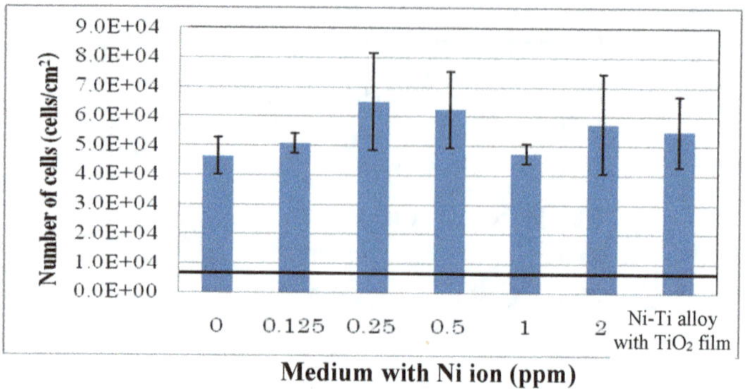

Figure 5: Number of adhered cells after cultured 72 hrs in 1.0 M phosphate buffer.

Also in Fig. 6, the number of adhered cells after cultured 24–72 hrs in 1.0 M phosphate buffer with various Ni ion concentrations under 0–100 ppm are indicated. Cell multiplication characteristics were suppressed with Ni ion concentration increasing. Then, cells died out under the Ni ion concentration condition of 100 ppm.

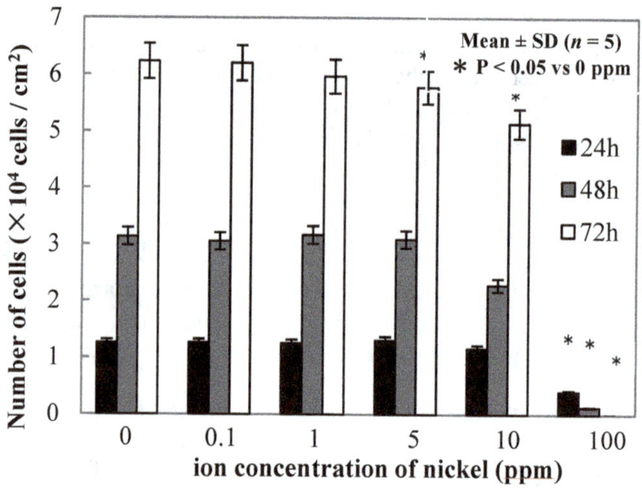

Figure 6: Changes in number of adhered cells in 1.0 M phosphate buffer with different Ni ion concentration (MC3TA-E1).

Mizutani *et al.* also investigated the effects of thermal oxidation on the biocompatibity of a Ni-Ti shape memory alloy [13]. They conducted cell culture tests with L929 fibroblast cells to examine the cell proliferation. The amount of Ni ions elusion in the culture medium used during the cell proliferation test was analysed by atomic absorption photometer. As a result, all TO-treated samples showed good biocompatibility after 3 days of immersion tests as shown in Fig. 7, and released Ni ions were significantly reduced by TO-treatment as shown in Fig. 8. These results are caused by the generated thick and stable surface-modified layer on the alloy surface by TO-treatment. In addition, the wear resistance of the Ni-Ti alloy also improved by TO-treatment. After the wear test, biocompatibility was evaluated by cell proliferation tests, and the results showed that the surface damage on the TO-treated sample did not affect the attachment cells. From these results, it was concluded that the isothermal oxidation treatment is effective as a surface modification method for a Ni-Ti alloy.

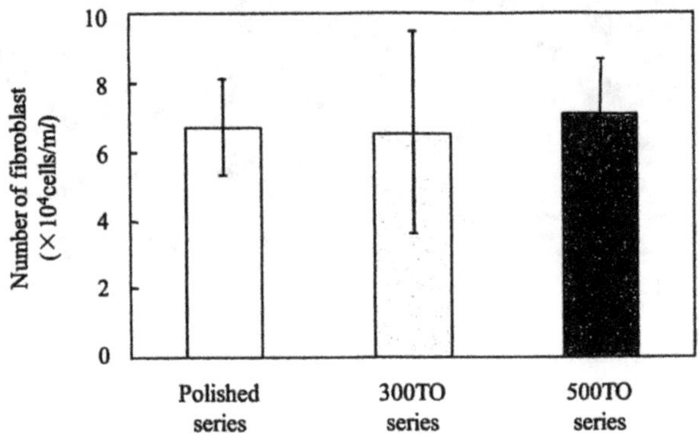

Figure 7: Results of cell proliferation tests ($n = 4$, mean ± SD) [14].

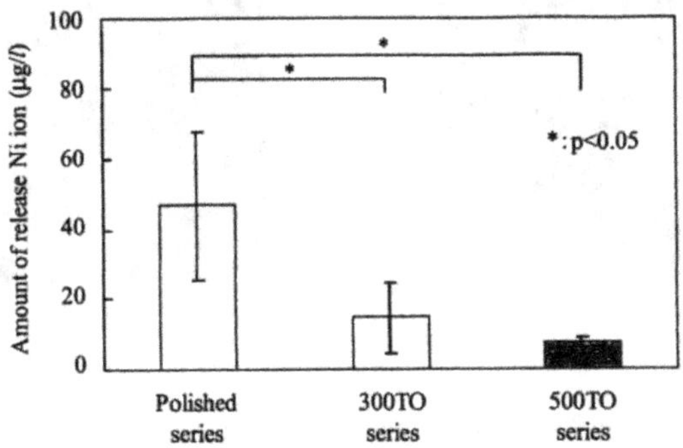

Figure 8: Comparison of released Ni ion ($n = 4$, mean ± SD) [14].

Then, the effect of Ni ion concentration upon morphology change of cell was investigated. In Fig. 9, the cell morphology after cultured 72 hrs on Ni-Ti alloy specimen surface with TiO_2 thin film is indicated. Fig. 9(a) shows the cell morphology obtained after fluorescent stain, and (b) shows the cell morphology obtained after Giemsa stain. Also, in Fig. 10(b), the fluorescent stain cell morphology after cultured 72 hrs in 1.0 M phosphate buffer with 0.5 ppm Ni ion concentration is indicated and in Fig. 10(a) that of control is indicated. In the case of cell after cultured 72 hrs in 1.0 M phosphate buffer with Ni ion concentration of 0.5 ppm, some morphological change, that is, linearization of cell was recognized. In contrast, in the case of cell after cultured 72 hrs on Ni-Ti alloy specimen surface with TiO_2 thin film, no remarkable change in cell morphology was generated just like the case of control.

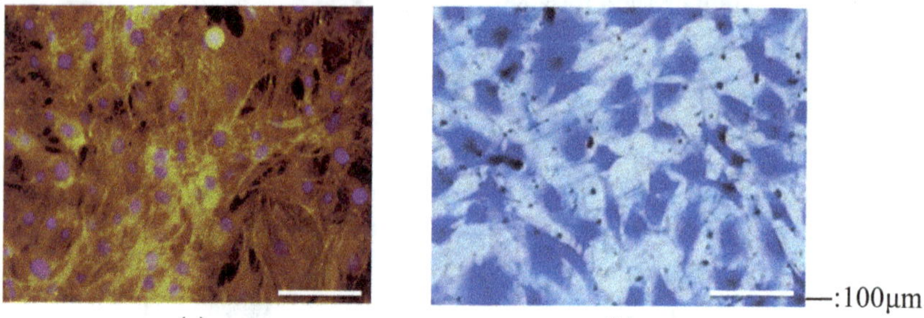

(a) (b)

Figure 9: Cell morphology after cultured 72 hrs on Ni-Ti alloy specimen surface. (a) Fluorescent stain; (b) Giemsa stain.

From these observed results, TiO_2 thin film formed by air oxidization also significantly contributes to improved biocompatibility of the Ti-Ni alloy.

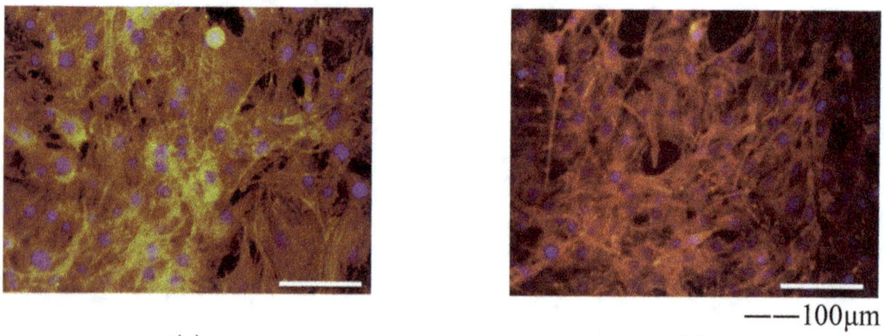

(a) (b)

Figure 10: Cell morphology after cultured 72 hrs in 1.0 M phosphate buffer with 0.5 ppm Ni ion. (a) Ni ion: 0 ppm; (b) Ni ion: 0.5 ppm.

At the same time, the effect of Ni ion concentration upon the morphology change of cell was investigated under the Ni ion concentration conditions from 0–100 ppm. As a result, some morphological change, that is, linearization of cell was recognized in phase-contrast micrographs under the Ni ion concentration conditions of 10 ppm and above as shown in Fig. 11. Then, the cells almost died out under the Ni ion concentration condition of 100 ppm. This morphological change was also detected by the change in aspect ratio of cells as show in Fig. 12.

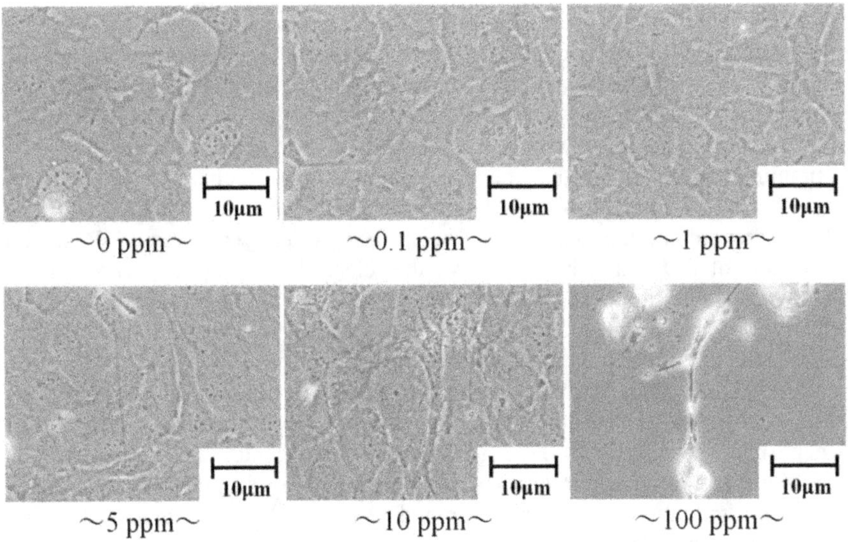

Figure 11: A phase-contrast micrograph of cells (MC3T3-E1, 72 h).

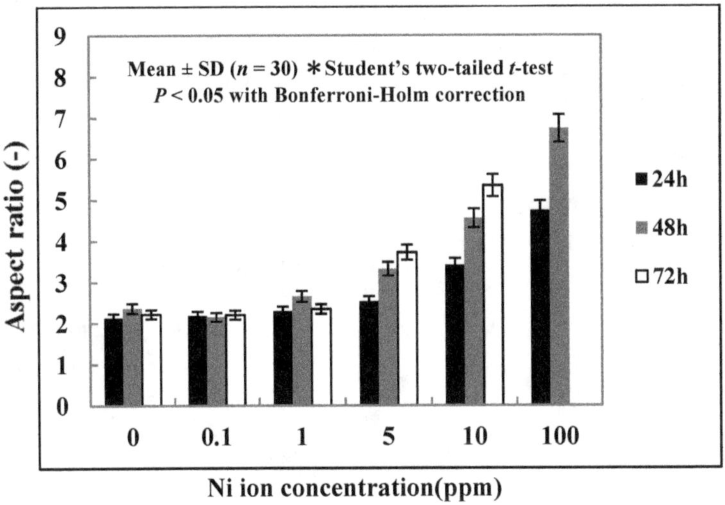

Figure 12: Aspect ratio (determined by image J) dependency of cell upon ion concentration (MC3T3-E1).

3.4 Biocompatibility improvement through optimal TiO_2 thin film

In this section, other trials of surface modifications were conducted.

Fukushima *et al.* discussed corrosion resistance and surface characterization of an electrolyzed Ti-Ni alloy [14]. This alloy contains nickel, which is known to give rise to cytotoxicity, metal allergy and carcinogenicity. Therefore, the purpose of that study was to improve the corrosion resistance of Ti-Ni alloy by electrolytic treatment; thus, an investigation was carried out using different acidic electrolyte compositions. As a result, specimens electrolyzed with lactic acid, water and glycerol were found to show higher corrosion potential and release lower amount of titanium and nickel ions than mechanically polished specimens ($p < 0.05$). With the electrolytic treatment, nickel concentration in the surface oxide layer of Ti-Ni alloy decreased and the thickness of the surface oxide layer increased. Based on the results of that study, it was shown that electrolytic treatment with suitable electrolyte could improve the corrosion resistance of Ti-Ni alloy, which is effective to produce medical and dental devices that utilize the shape memory effect or super elasticity with better biocompatibility.

On the other hand, Mizutani *et al.* examined the creation of bioactive surface by nanosecond-pulsed laser irradiation of commercially pure titanium [15]. In that paper, nanosecond-pulsed laser irradiation was conducted under various conditions, and then morphologies, chemical compositions and crystal structure of irradiated titanium surface were examined. As a result, it was made clear that nanosecond-pulsed laser irradiation can give bioactive properties on commercially pure titanium surface.

Therefore, in the present paper, the following two kinds of methods were employed. The first one is in-air oxidation under suppressed partial pressure oxygen condition using sputtering equipment. Judging from above mentioned results [15], another method is the sequential combination of electro-polishing and air oxidation. Figs. 13 and 14 show the polarization curves obtained in lactated Ringer's solution at 310 K using these two kinds of surface-modified

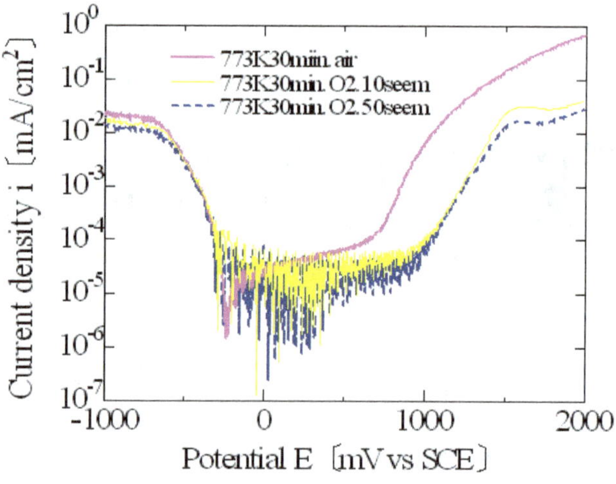

Figure 13: Polarization curves of Ni-Ti alloy oxidized in suppressed partial pressure air obtained in lactic Ringer's solution of 310 K.

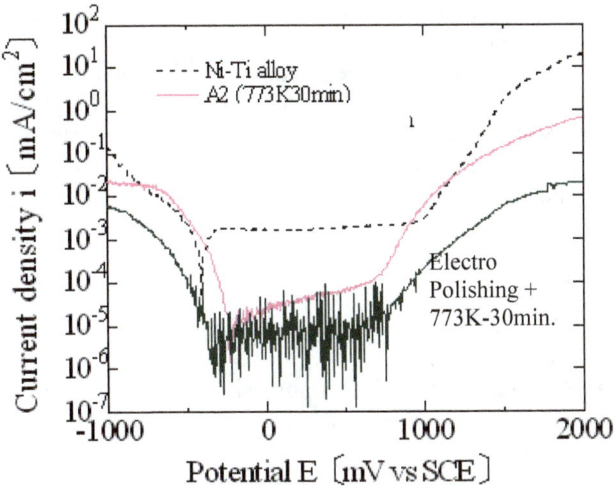

Figure 14: Polarization curves of complexly surface-modified Ni-Ti alloy obtained in lactic Ringer's solution of 310 K.

Ni-Ti alloy. From these figures, remarkable improvement in corrosion resistance was recognized in both the abovementioned surface-modified specimens compared with a simply air oxidized specimen. This remarkable improvement in corrosion resistance may be brought about through eliminating the Ni element in the extreme surface layer. Therefore, XPS analysis in the extreme surface layer has to be conducted in the near future. At the same time, more improvement in biocompatibility may also be accomplished judging from the abovementioned results.

4 Conclusions

The corrosion resistance of Ni-Ti alloys was improved through forming TiO_2 thin film by air oxidation and reactive sputtering. Surface morphology, crystal structure and composition of films were characterized by FE-SEM, XRD and XPS. Then, electro-chemical measurements in quasi-living body environment were conducted. In addition, the biocompatibility of air oxidized Ni-Ti alloy was evaluated by cell multiplication and morphological change of cell. The results obtained are summarized as follows.

A specimen with TiO_2 film whose thickness is more than 100 nm showed sufficient corrosion resistance in quasi-human body environment. The corrosion resistance in quasi-human body environment was influenced more by the film thickness and film defects than by the crystal structure of TiO_2.

In the case of cell after cultured 72 hrs in Ni ion concentration of 0.5 ppm, some morphological change, that is, linearization of cell was recognized. In contrast, in the case of cell after cultured 72 hrs on Ni-Ti alloy specimen surface with TiO_2 thin film, no remarkable change in cell morphology was generated. From these observed results, TiO_2 thin film formed by air oxidization also significantly contributes to improved biocompatibility of a Ti-Ni alloy.

The corrosion resistance of surface-modified Ni-Ti alloy may be significantly improved when air oxidization is conducted under the condition of suppressed partial pressure oxygen and the sequential combination of electro-polishing and air oxidation. In these cases, also remarkable improvement in biocompatibility of Ni-Ti alloy may be expected.

References

[1] Miyazaki, S., *Mechanical Properties of Shape Memory Alloys*, Yokendo, Chapters 1 & 2, 1993.

[2] Tanaka, K., Transformation thermomechanics in shape memory alloys. *Journal of the Society of Materials Science, Japan*, **48(2)**, pp. 1341–1349, 1999.

[3] Gil, F.J. & Planell, J.A., Shape memory alloys for medical applications. *Journal of Engineering in Medicine*, **212(6)**, pp. 473–488, 1998.

[4] Torrisi, L., The Ni-Ti superelastic alloy application to the dentistry field. *Bio-Medical Materials & Engineering*, **9(1)**, pp. 39–47, 1998.

[5] Eliades, T., Eliades, G., Athanasiou, A.E. & Bradley, T.G., Surface characterization of retrieved Ni-Ti orthodontic arch wires. *European Journal of Orthopedics*, **22(3)**, pp. 317–326, 2000.

[6] Ryhanen, J., Kallioinen, M., Serlo, W., Peramaki, P., Junila, J., Sandvik, P., Niemala, E. & Tuukkanen, J., Bone healing and mineralizaiton, implant corrosion, and trace metals after nickel-titanium shape memory metal intramedullary fixation. *Journal of Biomedical Materials Research*, **47(1)**, pp. 472–480, 1999.

[7] Wataha, J.C., Lookwood, P.E., Marek, M. & Ghazi, M., Ability of Ni-containing biomaterial alloys to activate monocytes and endothelial cells in vitro. *Journal of Biomedical Materials Research*, **45(3)**, pp. 251–257, 1999.

[8] Sugawara, H., Goto, H. & Komotori, J., Effect of thermal oxidation treatment on surface characteristic and corrosion resistance of Ni-Ti shape memory alloy. *Journal of the Society of Materials Science, Japan*, **55(10)**, pp. 965–970, 2006.

[9] Niu, Z., Jia, X. Zhang, W., Chen, W. & Qian, K.Y., Reactive sputtering TiO_2 films for surface coating of poly dimethylsiloxane. *Applied Surface Science*, **A252**, pp. 2259–2264, 2006.

[10] Morita, M., Hashimoto, T., Yamauchi, K., Suto, Y., Homma, T. & Kimura, Y., Evaluation of biocompatibility for Titanium-Nickel shape memory alloy in vivo and in vitro environment. *Materials Transactions*, **48(3)**, pp. 352–360, 2007.

[11] Homma, T., Kimura, Y. & Morita, M., Evaluation of compatibility of Nickel-Titanium shape memory alloy with femur surrounding tissue. *Transactions of Materials Research Society of Japan*, **28(3)**, pp. 853–856, 2003.

[12] Green, S.M., Grant, D.M. & Wood, J.V., XPS characterization of surface modified Ni-Ti shape memory alloy. *Materials Science and Engineering*, **A224**, pp. 21–26, 1997.

[13] Mizutani, M., Matsumura, Y., Komotori, J. & Tamauchi, H., Effects of thermal oxidation on biocompatibility of Ni-Ti shape memory alloy. *Transactions of the Japan Society of Mechanical Engineers, Ser. A*, **77(784)**, pp. 2144–2152, 2011.

[14] Fukushima, O., Yoneyama, T., Doi, H. & Hanawa, T., Corrosion resistance and surface characterization of electrolyzed Ti-Ni alloy. *Dental Materials Journal*, **25(1)**, pp. 151–160, 2006.

[15] Mizutani, M., Masuko, N., Honda, R., Komotori, J. & Kuritagawa, T., Creation of bioactive surface by nanosecond-pulsed laser irradiation of commercially pure titanium. *Journal of the Japan Society for Abrasive Technology*, **59(1)**, pp. 17–22, 2015.

Corrosion of mild steel and 316L austenitic stainless steel with different surface roughness in sodium chloride saline solutions

L. Abosrra, A. F. Ashour, S. C. Mitchell & M. Youseffi
University of Bradford, UK

Abstract

The corrosion behaviour of mild steel and 316L austenitic stainless steel was investigated in saline solution containing 1 and 3% NaCl. Specimens with surface roughness of 200, 600 grit emery paper and 1 μm diamond paste were investigated. The anodic polarization measurement technique was performed at a scan rate of 1 mV/s for a fixed period of 1 hour. Experimental results revealed that chloride ions have a significant effect on the corrosion behaviour of both steels as expected. As the surface roughness of 316L stainless steel increased, the breakdown potential (E_{break}), the free corrosion potential (E_{corr}) and the width of passivity decreased, hence corrosion rate increased. However, in the case of mild steel specimens, improving surface finish led to shifts in corrosion potential to more noble states and increased the corrosion rate. Metallographic examination of corroded specimens after electrochemical corrosion tests confirmed that the breakdown of the passive region was due to pitting corrosion.

Keywords: mild steel, 316L SS, anodic polarization, corrosion, surface roughness, saline solution.

1 Introduction

The study of corrosion properties of mild steel and stainless steel in aqueous solutions has received extensive attention. Mild steel is the most common structural material and is used in a wide range of environments. It is well known that when mild steel corrodes, anodic and cathodic areas develop over the corroded surface. Conventionally, these pits are known to change in shape and move across the surface, resulting in early corrosion that is approximately uniform [1]. However, when mild steel is exposed to saline solutions and marine environment, this type of corrosion is not observed. Anodic regions and micro-pits develop very quickly after early exposure and further presence of shallow broad pits occurs [2, 3]. When a high corrosion resistance is required, stainless steel is recommended. Type 316L stainless steel is austenitic and has been used in chemical and petrochemical industries and offshore structures for many decades. The excellent corrosion resistance of this stainless steel is attributed to the formation of a stable passive oxide layer, but nevertheless stainless steel is susceptible to localized corrosion by chloride ions [4]. In addition to choosing the right stainless steel grade for good corrosion resistance, it is equally important to specify the right surface condition of the materials used in many applications [5]. The surface condition affects the corrosion resistance to a certain extent, which implies that it is possible to meet certain requirements by specifying the proper finish rather than upgrading the chosen alloy. The effect of surface condition on corrosion resistance of 301, 304L and duplex stainless steel has been documented elsewhere [6–8]. Since the pitting potential actually defines a minimum condition under which pits can become stable, the aim is to lower the pitting potential. Metastable pitting occurs throughout the passive region of stainless steel and the potential at which the transition to stable pit growth occurs is an important parameter describing the stability of the metal [6, 9, 10].

In this study the effect of surface roughness on the corrosion resistance of mild steel and 316L austenitic stainless steel in the presence of different chloride concentration was investigated using the electrochemical technique. Corrosion parameters such as the breakdown potential, the free corrosion potential, and the corrosion rate were determined.

2 Experimental methods

2.1 Sample preparation

The chemical compositions are given in weight % for mild steel as: 0.84 Mn, 0.53 Cu, 0.24 Si, and 0.19 C and for 316L stainless steel as: 18 Cr, 14 Ni, 2.0 Mn, 0.75 Si, 0.03 C and 0.27 Mo. Corrosion tests were performed on cylindrical specimens with 9.5 mm diameter and 12.8 mm length with an exposed area of 4.5 cm^2. Specimen surfaces roughness was generated by wet grinding using silicon carbide papers of 200 and 600 grits and polished surface using 1 μm diamond paste.

2.2 Electrochemical measurements

Electrochemical tests were carried out in saline solutions containing 1 and 3% NaCl using the standard electrochemical corrosion cell containing the three electrodes, that is, saturated calomel electrode (SCE), counter electrode (graphite) and working electrode (the sample). All electrodes were immersed in a suitable glass vessel containing 600 ml of the saline electrolyte. The anodic and cathodic polarization curves were measured after 1 hr of working electrode immersion at ambient temperature (23± 2°C) in the investigated environment. Subsequently, the anodic polarization curves for both mild steel and 316L stainless steel were measured from potential values below and above the free corrosion potentials E_{corr} of ±250 mV and ±800 mV, respectively, with a scan rate of 1 mV/s.

Experiments were repeated 3–4 times for all specimens in all solutions. Corrosion rates were calculated for steel immersed in aqueous solutions using eqn. (1) [11].

$$CR = \frac{0.13 I_{corr} W}{DA} \quad (1)$$

where I_{corr} is the corrosion current density (μA/cm^2), W the atomic weight of steel, D the density of steel in mg/cm^3 and A is the exposed surface area of steel in cm^2.

3 Results

3.1 Electrochemical data analysis

3.1.1 Mild steel – potentiodynamic results

Fig. 1 gives the potentiodynamic polarization curves of mild steel with a surface roughness of 200 and 600 grit papers as well as 1 μm diamond paste polished surface tested in 1% NaCl solution. The corrosion behaviour was affected by the degree of surface roughness. Increase in roughness from polishing (like mirror) to 600- and 200-grit surface roughness shifted the free corrosion potential to more active values as shown in Figs. 1 and 3. Corrosion rate values were not expressing the same trend recorded for the free corrosion potential. The specimen of diamond polished surface showed a higher corrosion rate than that of the 600- and 200-grit surface finished specimens as shown in Fig. 4. Fig. 2 shows the corrosion behaviour of 1 μm diamond polished (like mirror), 600- and 200-grit surface

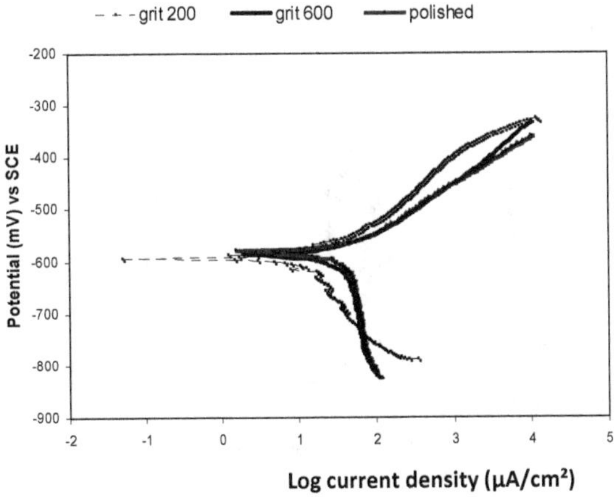

Figure 1: Polarization curves of mild steel at different surface roughness in 1% NaCl saline solution.

roughness mild steel specimens in 3% NaCl saline solution. Corrosion resistance was drastically reduced with increasing sodium chloride concentration up to 3%. Fig. 3 indicates that as the surface became rougher, the free corrosion potential moved to more active state. Corrosion rate measurements showed contradictory results and the conventional trend is that increasing the surface roughness decreases corrosion resistance. The specimens with diamond polished surfaces had the highest corrosion rate as illustrated in Fig. 4.

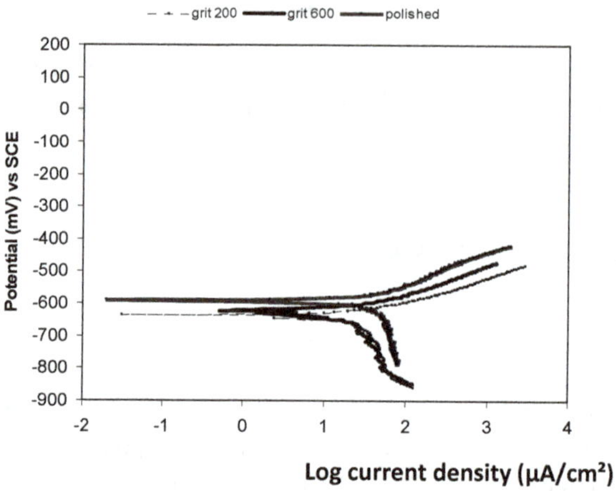

Figure 2: Polarization curves of mild steel at different surface roughness in 3% NaCl saline solution.

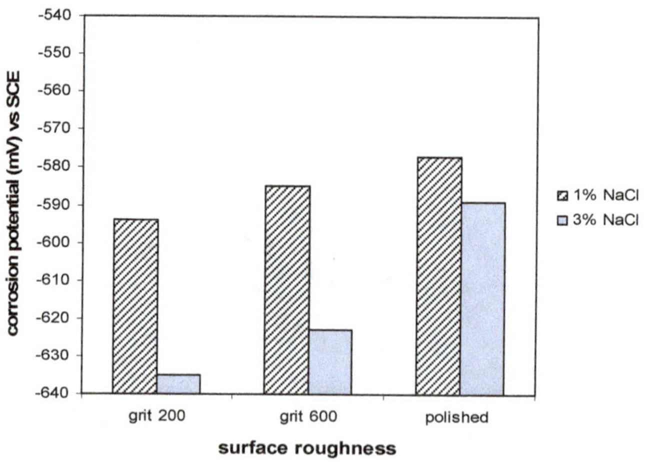

Figure 3: Corrosion potential vs. surface roughness in 1% and 3% NaCl saline solution.

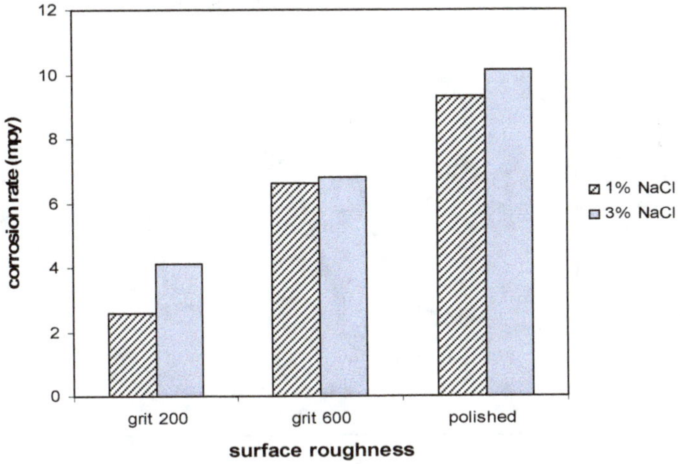

Figure 4: Corrosion rate vs. surface roughness in 1% and 3% NaCl saline solution.

3.1.2 Stainless steel (316L) – potentiodynamic results

Potentiodynamic polarization curves and the retrieved corrosion data for 316L stainless steel in 1% and 3% NaCl are shown in Figs. 5–9. The polarization curves shown in Figs. 5 and 6 indicate that the tendency of the stainless steel was to undergo, oxidation, passivation followed by breakdown, i.e. a typical characteristic behaviour. The polarization curves of specimens with surface finishes of 1 μm diamond polish, 600 and 200 grit papers tested in 1% NaCl saline solution are shown in Fig. 5. The passivity and hence the breakdown potential was affected by the surface roughness. The specimen with less roughness, i.e. 1 μm diamond finish surface had the highest breakdown potential of +317 mV, followed by 600- and 200-grit surface roughness specimens. The passivity of rougher surfaces of 200 and 600 grit failed at the less noble potentials of +286 mV and +296 mV, respectively. Fig. 6 shows the corrosion behaviour of specimens tested in 3% NaCl. The polarization curves of the investigated specimens showed the conventional trend of the effect of chlorides and surface roughness. As the chloride concentration and the surface roughness increased, the breakdown potential values decreased and the free corrosion potential moved to more active direction (less stable). The free corrosion potentials, breakdown potentials and corrosion rates in 1 and 3% NaCl were recorded and displayed in Figs. 7–9. Fig. 7 shows that for the same surface finish specimen, the free corrosion potential increased to more noble state as the surface roughness decreased. The chloride concentration had a significant effect and reduced the free corrosion potential as the NaCl content increased from 1 to 3%. Figs. 8 and 9 show the breakdown potential and the corrosion rate results, respectively.

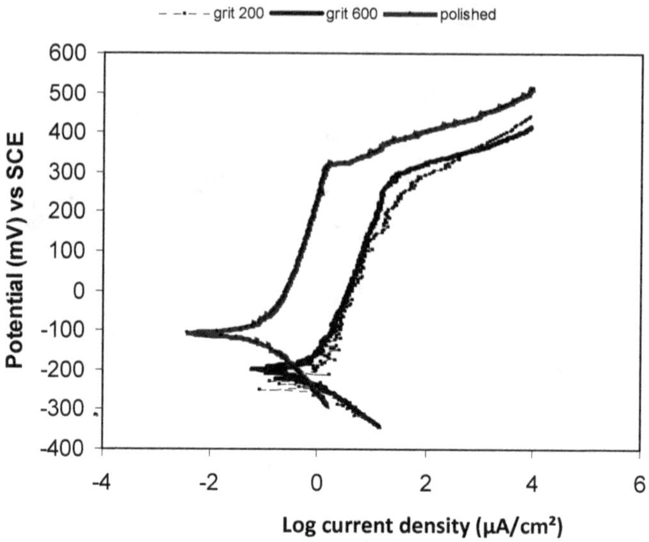

Figure 5: Polarization curves for different surface roughness of 316L stainless steel in 1% NaCl saline solution.

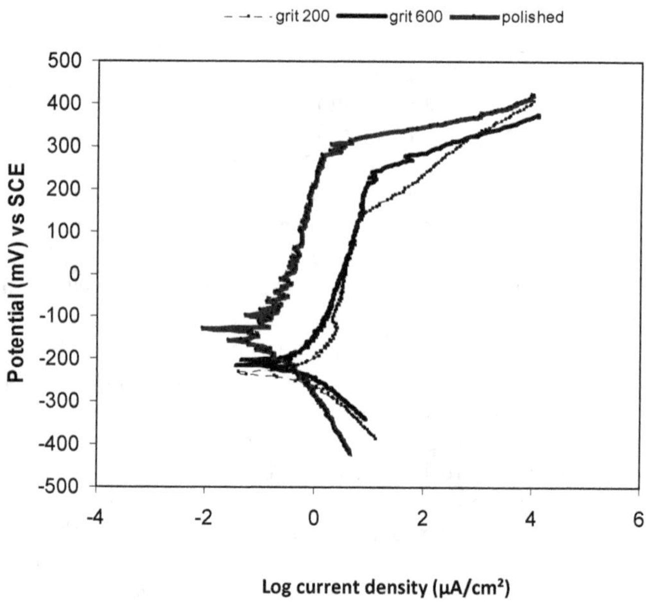

Figure 6: Polarization curves for different surface roughness of 316L stainless steel in 3% NaCl saline solution.

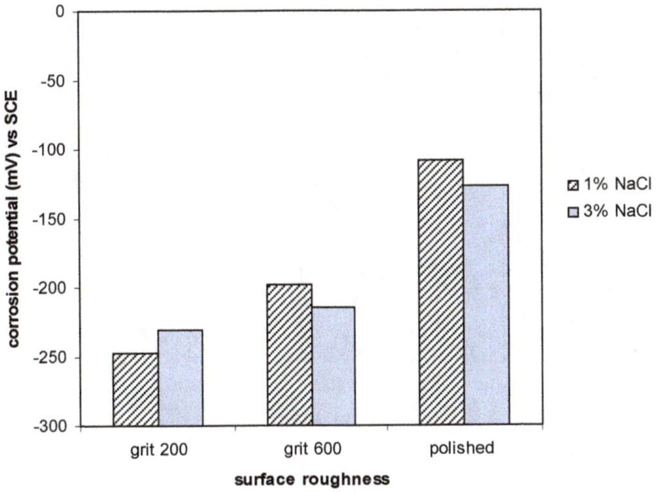

Figure 7: Corrosion potential of 316L stainless steel with various surface roughness in 1% and 3% NaCl saline solution.

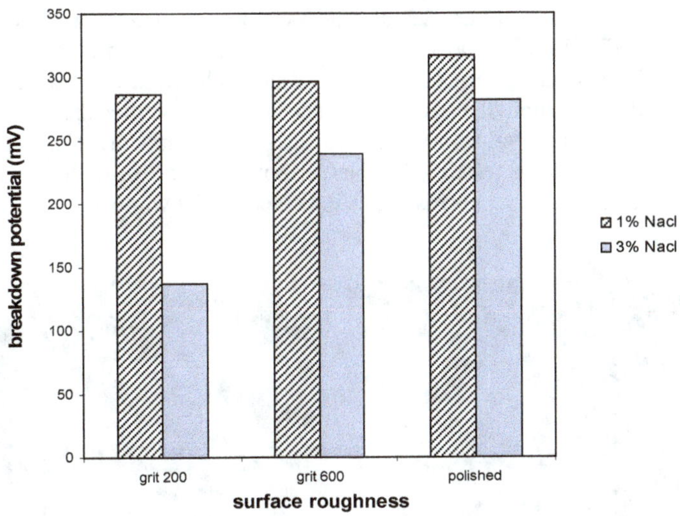

Figure 8: Breakdown potential of 316L stainless steel with various surface roughness in 1% and 3% NaCl saline solution.

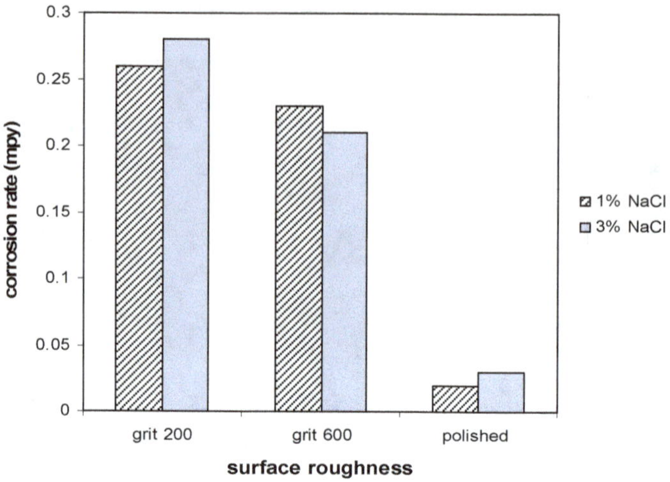

Figure 9: Corrosion rate of 316L stainless steel with various surface roughness in 1% and 3% NaCl saline solution.

3.2 Morphology results

Corroded specimens after testing were investigated at low and high magnifications for microscopy studies. Examples of mild steel specimen corrosion morphology are shown in Fig. 10. Fig. 10(a) shows the corrosion pits on a mild steel specimen of diamond polished surface tested in 3% NaCl. This figure shows clearly the severity of the corrosion damage and this confirms and justifies the reason for the highest corrosion rate values recorded for this specimen. Corrosion on 200 grit paper surface roughness tested in 3% NaCl, shown in Fig. 10(b), clearly reveals a cluster of less deep pitting corrosion, as indicated by the white arrows.

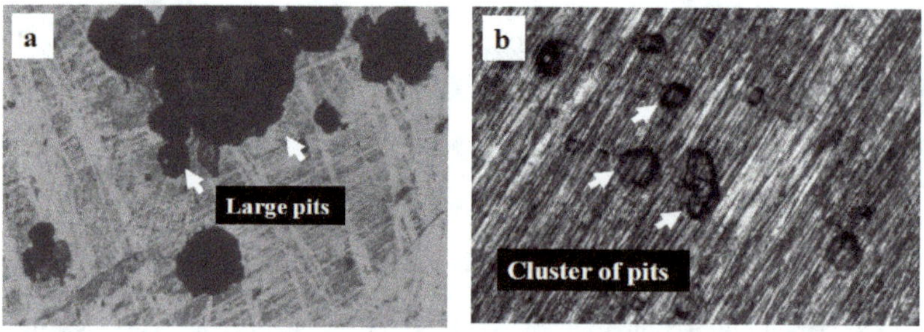

Figure 10: Optical photographs (X400) of corrosion pits on mild steel surfaces: (a) polished grit and (b) 200 grit in 3% NaCl saline solution.

Corrosion morphology on 316L stainless steel specimens was investigated using the scanning electron microscope (SEM). The results are shown in Fig. 11. Fig. 11(a) shows severe open corrosion pit on the surface of the 200-grit specimen. The pitting corrosion morphology transformed to another type called lacelike corrosion pits as the surface became smoother as can be seen in Fig. 11(b). These types of corrosion pits are due to the collapse of passive film and during pit propagation. Specimen of polished surface, Fig. 11 (c) showed high number of lacelike corrosion pit. Corrosion morphology shows that the lacelike corrosion propagated by the mechanism of diffusion till the final collapse where the pit became stable.

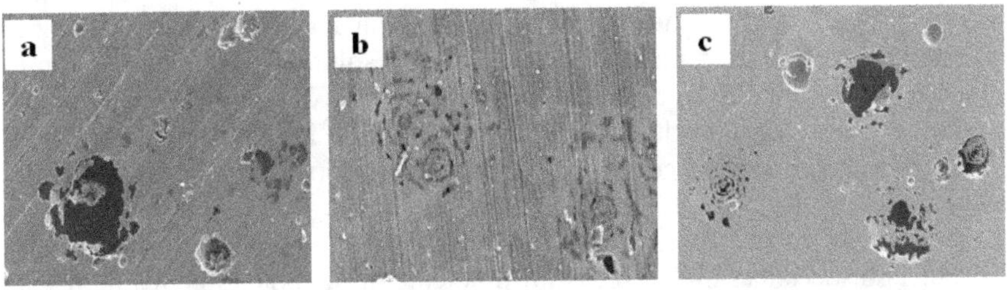

Figure 11: SEM images (X300): (a) 200-, (b) 600-grit surface roughness specimens and (c) diamond-polished specimen, respectively.

4 Discussion

Mild steel initially showed some weak passivation range and due to the presence of chloride in the vicinity of the steel surface, the passivity was easily destroyed. At the end of each experiment on mild steel samples, localized corrosion and red rust was visually observed indicating corrosion attack and products of various kinds due to the presence of aggressive chloride ions which are a powerful oxidizing agent and combine rapidly with the metal forming metal chloride. Polarization curves revealed weak passivation when sodium chloride increased from 1 to 3%. This is due to the absorption of water molecules and chloride ions on the steel surface causing breakdown of the oxide film. Corrosion mechanisms retrieved from these results are in agreement with the interpretation provided by others [12, 13].

Increasing the surface roughness shifted the corrosion potential to more noble direction; however, the corrosion current density decreased. This was attributed to a weak passive film which formed on the metal surface causing the early breakdown of passivity with more aggressive chloride ions. The interesting result obtained in this work was that mild steel with smooth polished surfaces showed the highest corrosion rate compared with 600- and 200-grit surface finished specimens. This can be attributed to the high rate of corrosion propagation after initiation. Pitting corrosion is controlled by the diffusion process and in this case once the pitting started, it propagated in a fast rate due to the continuous diffusion process and the formation of acid media at the bottom of the pits.

Corrosion resistance of stainless steel is attributed to the presence of chromium, forming its oxide film and protecting the steel from further corrosion in the passive region. However, the presence of certain ionic species such as chloride ions in the electrolyte can be very aggressive towards the corrosion behaviour. In this study, the anodic polarization curves in various chloride concentrations with different surface roughness clearly revealed that the free corrosion potentials and breakdown potential dropped from more noble to less noble values as sodium chloride increased from 1 to 3%. This is because chloride ions lower the activation energy required for corrosion reaction to occur. This interpretation is in agreement with the work reported by others [14–17]. The chloride formation did not stop the oxide film formation but reduced the passive region as seen in the polarization curves of figs. 5 and 6. There are other reasons for more aggressive behaviour reflected on reduction in passive region in saline solution which can be due to chloride ions with higher charge density and higher capacity to form soluble species. By entering into the lattice film the chloride introduces lattice defects which reduce the resistance of the oxide film to corrosion as seen from the corrosion rate values in Fig. 9.

The role of surface roughness on the breakdown potential was confirmed as smoother surfaces had higher breakdown potential. The potentiodynamic results in Figs. 5 and 6 showed that the width of passive region is highly dependent on the surface condition. When surface finish varied from smooth to rough, the passive film terminated at lower breakdown potential and the passive region became shorter due to the effect of surface topography which enhanced the presence of more aggressive corrosion media inside such rough surfaces. Pitting potential is more sensitive to surface roughness changes and the relative large increase in pitting potential by several tenth mV from the roughest to smoothest surface, suggested that both the nucleation and propagation of metastable pits depend on the steel surface condition [6]. This implies that for the metastable pit or pits to grow on smoother surface is more difficult than on rougher surface and the presence of chloride ions in solution results in the destruction of the passive film as observed elsewhere [9, 10]. The mechanism of pit growth is autocatalytic by anodic dissolution of the steel which leads to introduction of passive metal ions in solution that cause migration of chloride ions. In turn, metal chloride reacting with water causes the pH to decrease where the cathodic reaction is on the surface near the pit mouth. The results in this study confirmed that when the surface is smooth the pit will survive due more to the formation of lace cover on the pit mouth maintaining the diffusion process. The corrosion morphology for the diamond surface finish behaved in this manner and showed pitting corrosion with lacelike pits as seen in Fig. 11.

5 Conclusions

Corrosion rate measurements of mild steel specimens showed results contradictory to the conventional trend of the corrosion reduction as surface roughness increased. The specimens with diamond polished surfaces had the highest corrosion rate compared with 600- and 200-grit surface finished specimens.

The free corrosion potential, breakdown potential and corrosion rate of 316L stainless steel are chloride concentration dependent. As the chloride concentration increased, the free corrosion potential and the breakdown potential became less noble. The corrosion rate increased as the chloride concentration increased as expected.

Reduction in the corrosion resistance of the 316L stainless steel is attributed to the destruction of the passive film in the presence of chloride. The roughest surface of 200 grit showed early passivity breakdown at highest rate of corrosion and lower breakdown potential compared with the rest of the surfaces.

References

[1] Uhlig, H. & Revie, R., *Corrosion and Corrosion Control*, 3rd edition, John Wiley and Sons: New York, 1985.

[2] Forgeson, B., Southwell, C. & Alexander, A., Corrosion of metals in tropical environments, part 3-underwater corrosion of ten structural steels. *Corrosion*, **16**, pp. 105–114, 1960.

[3] Blekkenhorst, F., Ferrari, G., Wekken, C. & IJsseling, F., Development of high strength low alloy steels for marine applications, Part 1: results of long term exposure tests on commercially available and experimental steels. *British Corrosion Journal*, **21**, pp. 163–176, 1986.

[4] Congmin, X., Zhang, Y. & Cheng G., Pitting corrosion behavior of 316L stainless steel in the media of sulphate- reducing and iron – oxidizing bacteria. *Materials Characterization*, **245**, pp. 245–256, 2007.

[5] Mathiesen, T., Nielsen, T.S., Frantsen, J.E., Kold, J. & Boye-Moeller, A.R., Influence of various surface conditions on pitting corrosion resistance of stainless steel tubes type EN 1.4404. *NACE Corrosion 2006 Conference*, paper 06095, San Diego, CA, 2006.

[6] Sasaki, K. & Burstein, K.T., The generation of surface roughness during slurry erosion- corrosion and its effect on the pitting potential. *Corrosion Science*, **38**, pp. 2111–2120, 1996.

[7] Moayed, M.H., Laycock, N.J. & Newman, R.C., Dependence of the critical pitting temperature on surface roughness. *Corrosion Science*, **45**, pp. 1203–1216, 2003.

[8] Elhoud, A., Renton, N.C. & Deans, W.F., Effect of surface roughness on pitting corrosion of 25 Cr duplex stainless steel in chloride solution. *The 9th Libyan Corrosion Conference*, Tripoli-Libya, 2007.

[9] Hong, T. & Nagumo, M., Effect of surface roughness on early stages of pitting corrosion of type 301 stainless steel. *Corrosion Science*, **39**, pp. 1665–1672, 1997.

[10] Cruz, R.P., Nishikata, A. & Tsuru, T., Pitting corrosion mechanism of stainless steel under wet- dry exposure in chloride containing environments. *Corrosion Science*, **40**, pp. 125–139, 1998.

[11] Trethewey, K.T. & Chamberlain, J., *Corrosion for Science and Engineering*, 2nd ed., Pearson Education, UK, 1995.

[12] Jeffrey, R. & Melches, R., The changing topography of corroding mild steel surface in sew water. *Corrosion Science*, **49**, pp. 2270–2288, 2007.

[13] Kim, D.K., Muralidharan, S., Ha, T.H., Bae, J.H., Ha, Y.C., Lee, H.G. & Scandebury, J.D., Electrochemical studies on the alternating current corrosion of mild steel under cathodic protection in marine environments. *Electrochemical Acta*, **51**, pp. 5259–5267, 2006.

[14] Al-Fozan, S. & Umalik, A., Pitting behavior of type 316L stainless steel in Arabian Gulf seawater, Technical Report no. SWCC (RDC), 1992.

[15] Abd El- Meguid, E., Mohmoud, N. & Gouda, A., Pitting corrosion behavior of 316L in chloride containing solution. *British Corrosion Journal*, **33**, pp. 42–48, 1998.

[16] Chuan, M. & Tseng, W., Environmentally assisted cracking behavior of single and dual phase stainless steel in hot chloride solution. *Materials Chemistry and Physics*, **84**, pp. 162–170, 2004.

[17] Refaey, S.A.M., Taha, F. & Abd El- Malak, A.M., Corrosion and inhibition of 316L stainless steel in neutral medium. *International Journal of Electrochemical Science*, **1**, pp. 80–91, 2006.

Simultaneous effect of surface roughness and passivity on corrosion resistance of metals

A. S. Toloei, V. Stoilov & D. O. Northwood
University of Windsor, Canada

Abstract

Unidirectional surface roughness of varying magnitudes was created on both nickel and mild steel by grinding on SiC papers with grit sizes from G60 (roughest) to G1200 (smoothest) and the corrosion resistance in 0.5 M H_2SO_4 solution was determined using a potentiodynamic polarization technique. A different trend of corrosion rate versus roughness was seen for the active–passive metal (nickel) and non-active–passive metal (mild steel). For nickel, there was an increase in corrosion rate with increasing roughness, whereas for mild steel the corrosion rate decreased with increasing surface roughness. Furthermore, through a detailed examination of the surface before and after corrosion using techniques including profilometry, scanning electron microscopy (SEM) and energy dispersive spectroscopy (EDS), it was established that different corrosion mechanisms were operative for nickel and mild steel. In the case of nickel, the higher roughness provided a greater contact area between the corrosive medium and metal and there was trapping of the corrosive ions in the deep grooves. Both of these factors would lead to an increase in corrosion rate. Also, for the smoother nickel surfaces, it is easier to form a stable passive film. For mild steel, which does not form a passive film, corrosion rates are generally much higher than for nickel. For the rougher surfaces with the deeper grooves, the corrosion product, $FeSO_4$, can fill the grooves thereby acting as a barrier to further ingress of the corrosive ions to the un-corroded metal.

Keywords: corrosion resistance, passivity, surface roughness, potentiodynamic polarization.

1 Introduction

In metallic materials the surface roughness has a major impact on both general corrosion and the localized nucleation of metastable pitting. In most metals and their alloys, the general trend is that a higher surface roughness results in higher corrosion rate. Such trend was observed for copper, nickel, aluminium, stainless steel, magnesium and titanium alloys [1–6]. Zuo *et al.* [7] investigated the metastable pitting behaviour of 316L stainless steel with various degrees of surface roughness and showed that, as surface roughness increased, the nucleation rate of metastable pits as well as the overall corrosion rate increased. In contrast, a range of materials such as mild steel [8] and AE44 magnesium alloy [9] show the reverse trend namely, as surface roughness increases, the corrosion rate decreases. Alvarez *et al.* [9] showed that the corrosion rate in AE44 from polished coupons (with lower roughness) was notably greater than the corrosion rate from the semi-polished coupons (with higher roughness). Similar behaviour was observed by Abosrra *et al.* [8] on mild steel in saline solution containing 1 and 3% NaCl. In his work Abosrra *et al.* [8] showed that the improvement of the surface finish led to a shift in the corrosion potential to more noble states and to an increase of the corrosion rate. The obvious difference between these two groups of materials is that materials such as aluminium, copper, nickel, stainless steel passivate quickly by developing stable oxide films whereas AE44 magnesium alloy and mild steel, unlike the first group, do not form a stable passive film. For the material group without stable passive film (mild steel, magnesium alloys etc.) the phenomenon of the decrease of the corrosion with the increase of the roughness (contact area) is not well understood. The main goal of this work is to provide a better insight into the effect of the roughness and the main corrosion mechanisms of non-active–passive metals (such as mild steel) in comparison to active–passive metal (such as nickel).

2 Experimental procedure

The mild steel and nickel specimens were prepared by cutting mild steel and nickel sheet using wire electrical discharge machining (EDM) to pieces of $15 \times 15 \times 1\ mm^3$. The sample surfaces were abraded with 60, 120, 180, 240, 320, 400, 600, 800 and 1200 grade SiC papers in a manner that produced unidirectional roughness, washed with double-distilled water, degreased with acetone and dried in warm air. A smaller grit number represents a rougher finish. The letter G, in the used notation (G60 to G1200), stands for grit. After drying, the specimens were placed in a desiccator and then used for experiment. The 0.5 M H_2SO_4 solution was prepared by dilution of analytical grade H_2SO_4 with double distilled water.

All electrochemical measurements were conducted in a thermostatic conventional three-electrode cell using a Solarton SI-1287 (electrochemical interface) instrument. Mild steel and nickel were used as the working electrode (WE). Only 1 cm^2 areas of the samples were exposed to the electrolyte. A saturated calomel electrode (SCE) and a platinum wire were also used as the reference and auxiliary electrodes, respectively. All experiments were carried out in 100 mL of naturally aerated electrolyte maintained at 25°C. The scan rate of the experiments was 1.0 mV/s. In all experiments, the mild steel and nickel electrodes were allowed to reach their stable open-circuit potential (OCP), which occurred after 30 minutes. The surface morphology of the samples, both before and after corrosion, was studied using a JEOL 5800 SEM at an accelerating voltage of 20.0 kV. The EDS spectra were obtained both before and after corrosion to show the changes in the oxygen content in the surface layer. A profilometry technique was used to calculate different roughness parameters including R_a, R_q, R_z and R_t before and after corrosion using a Wyko Surface Profiling System NT-1100. The change in all parameters was calculated after corrosion and the relationship between each parameter and corrosion rate was determined.

3 Results

3.1 Potentiodynamic polarization technique

The three different curves for three different values of unidirectional roughness in Fig. 1 illustrate the potentiodynamic polarization test results for mild steel in 0.5 M H_2SO_4 solution at room temperature. The electrochemical parameters shown in Table 1, i.e. the open-circuit potential (OCP), the corrosion potential E_{corr}, the corrosion current density i_{corr} and the anodic β_a and cathodic β_c Tafel constants were collected from the plots using the extrapolation of the polarization curves. The corrosion current was determined by the intersection of the cathodic or the anodic Tafel line with the OCP (potential of zero current in the potentiodynamic curves or E_{corr}. This point determines the potential E_{corr} and current density i_{corr} for corrosion [10]. Polarization resistance R_p and corrosion rate CR were also determined using the equations [10]:

$$R_p = \frac{\beta_a \beta_c}{2.3 i_{corr} (\beta_a + \beta_c)} \tag{1}$$

$$CR = K i_{corr} \frac{EW}{\rho} \tag{2}$$

where K is a constant for converting units, i_{corr} the corrosion current density ($\mu A/cm^2$), ρ the alloy density (g/cm^3), EW the alloy equivalent weight (g/equivalent). The potentiodynamic polarization curves show that in the presence of surfaces with different roughness values, there is a decrease in anodic and cathodic current densities. The increase of the roughness (G1200 to G60) resulted in the curve shift toward lower corrosion current densities which meant lower corrosion rates. In addition, according to the data in Table 1, with the increase of the roughness values, the corrosion current density and corrosion rate decreased and polarization resistance increased.

On the other hand, the presence of unidirectional roughness only slightly modified the current densities of the anodic and cathodic curves and decreased corrosion. The same effect has already been observed in AE44 Mg alloy [9]. In both cases the metal has no ability to form a passive layer but in the case of other metals which form a passive layer, a reverse trend was observed [1–6]. As shown in Fig. 1, the polarization curves rise to parallel and it is clear that both cathodic and anodic branches show a lower current density indicating that the hydrogen evolution reaction is activation controlled [11, 12]. Thus, creating the unidirectional roughness does not affect the mechanism of this process and has been able to affect the corrosion performance. In order to be sure about the consistency of the results, all the tests were performed five times. As seen in

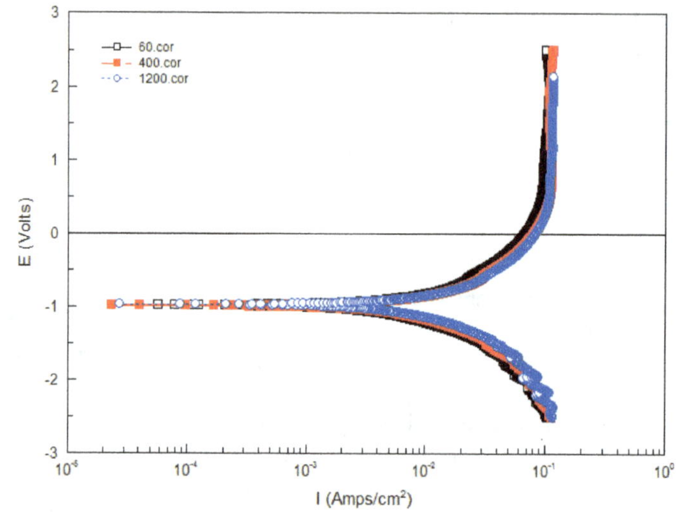

Figure 1: Polarization curves for mild steel samples.

Table 1: Corrosion testing results for mild steel.

Sample	E_{corr} (V)	i_{corr} (μA/cm^2)	β_a (V)	β_c (V)	R_p (Ω/cm^2)	CR (mil/year)
G60	–0.968	4141	0.295	0.570	20.46	1886
G120	–0.971	4373	0.290	0.507	18.37	1991
G180	–0.973	4432	0.305	0.457	17.93	2018
G240	–0.974	4724	0.291	0.513	17.05	2151
G320	–0.973	5027	0.306	0.543	16.93	2289
G400	–0.974	6113	0.349	0.690	16.49	2784
G600	–0.974	6255	0.358	0.688	16.35	2848
G800	–0.973	7239	0.388	0.628	14.44	3297
G1200	–0.973	7499	0.374	0.740	14.41	3415

Fig. 2(a), by decreasing the roughness from sample G60 to G1200, i_{corr} and consequently *CR* are increased. In contrast, as can be seen in Fig. 2(b), a different corrosion behaviour and a reverse trend was observed in the case of nickel in 0.5 M H_2SO_4 solution on surfaces with different unidirectional roughness [6]. As can be seen in previous work carried out by the authors, by increasing the roughness of nickel surface from sample G1300 to G60, the polarization curves shifted toward higher corrosion current densities which meant higher corrosion rates [13].

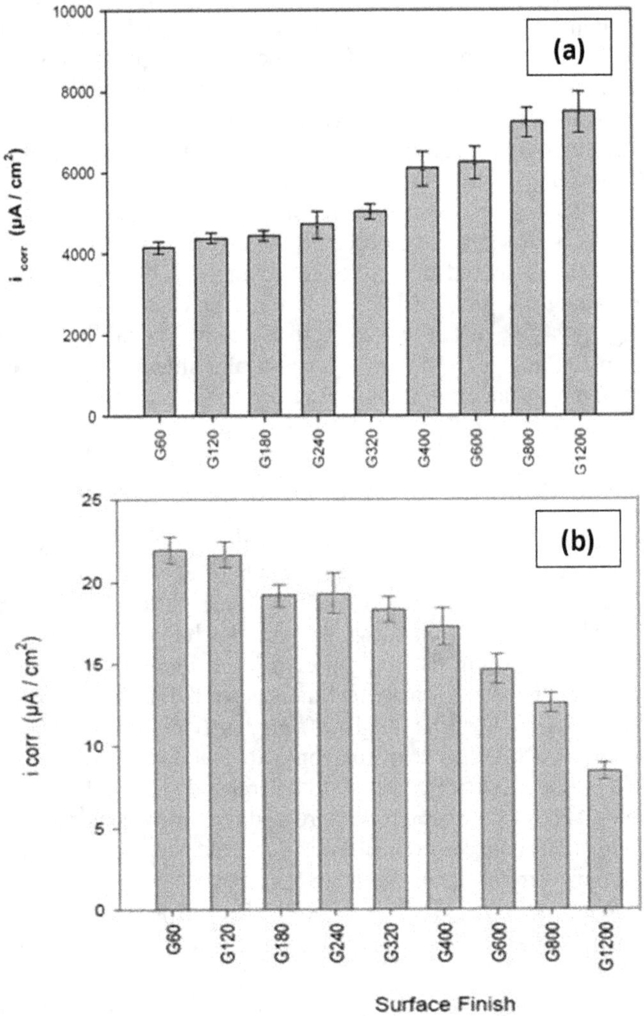

Figure 2: Dependence of i_{corr} on surface finish of (a) mild steel and (b) nickel.

3.2 Scanning electron microscopy

Figs. 3(a)–(d) illustrates the SEM micrographs of the surface morphologies of mechanically polished mild steel electrodes with unidirectional roughness before and after corrosion. Samples G1200 and G60 are shown as examples of smooth and rough surfaces. It can be seen that the surfaces of electrodes became rougher before corrosion as the grain size of the applied SiC paper increased (higher roughness value). Fig. 3(a) presents the smoothest mild steel surface (G1200) with nano-size unidirectional roughness whereas Fig. 3(b) presents the sample G1200 surface after corrosion in 0.5 M H_2SO_4 solution. Fig. 3(c) shows sample G60 as the sample with the roughest surface and Fig. 3(d) presents this sample surface after corrosion in 0.5 M H_2SO_4 solution. It is clear from the images that the surface of mild steel has changed in both cases. In the case of sample G1200 the metal has been corroded completely and the surface was strongly damaged but there are still some grooves on sample G60, but in a larger scale. Other samples also showed the same trend i.e. on rougher samples more grooves are still visible which shows less degradation on rougher samples. The results obtained from SEM showed good agreement with the results obtained from potentiodynamic polarization technique. In Figs. 3(e)–(h), a reverse trend is observed for the nickel samples in 0.5 M H_2SO_4 solution namely more degradation on rougher samples and corrosion seemed to be concentrated along the grooves [13]. This trend in change of corrosion versus surface roughness for nickel as an active–passive metal was observed by researchers in the case of stainless steel and aluminium [3, 5] which showed poorer corrosion resistance and trapping the corrosive ions on rougher surfaces of active–passive metals.

3.3 EDS analysis

Energy dispersive spectroscopy (EDS) was used to characterize the surface and measure the oxygen content of the surface both before and after corrosion testing. Table 2 presents oxygen concentration both before and after corrosion on surfaces of mild steel with different roughness. All samples had similar oxygen contents before corrosion testing but the oxygen content increased for all samples after corrosion. Looking at the oxygen content increase of the surface, all samples show a significant increase. For the roughest samples (G60–G120) the increase in oxygen content is a little bit less compared to that for the other samples. As demonstrated by the electrochemical results, G60 and G120 exhibited the lowest corrosion rates, and the corrosion rate decreased with increasing roughness.

All the chemical composition analysis for mild steel before the corrosion testing had almost similar results which were as expected. After corrosion in all samples, specially samples G180 to G1200, there is a significant increase in oxygen content which is related to a severe general corrosion on the surface. The same thing has happened for rougher surfaces (G60 and G120) but with lower amounts of oxygen and less degree of corrosion. The results are in good agreement with SEM and potentiodynamic polarization tests.

In the case of metals such as aluminium and nickel, the lower corrosion rate was related to the formation of a stable passive film on smoother surfaces as suggested by the authors [3, 13–16] but in the case of mild steel no stable passive film is formed and rougher surfaces showed less corrosion and lower oxygen content. Generally, more increase in the oxygen content was observed for mild steel indicating more oxides form on the surface.

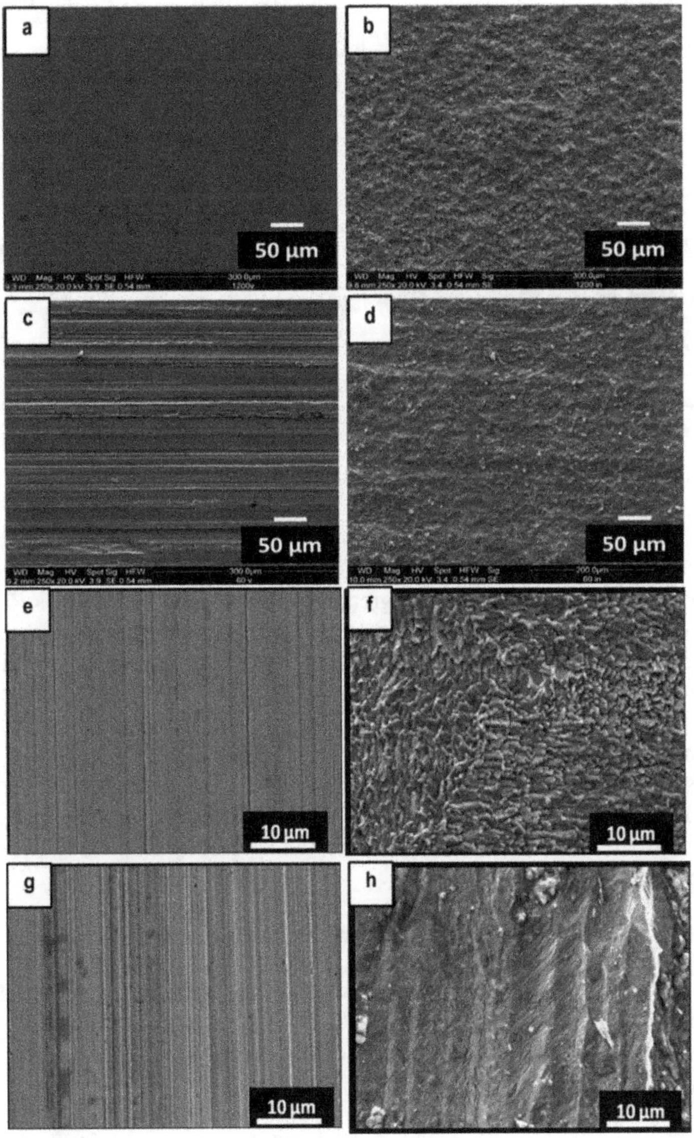

Figure 3: SEM micrograph of mild steel sample G1200 (a) before corrosion, (b) after corrosion, G60 (c) before corrosion, (d) after corrosion and nickel samples G1200 (e) before corrosion, (f) after corrosion and G60 (g) before corrosion, (h) after corrosion.

Table 2: EDS results for oxygen contents on the surface of the mild steel samples before and after corrosion.

Sample	Oxygen wt% before corrosion	Oxygen wt% after corrosion	Oxygen wt% difference
G60	2.51	24.63	22.12
G120	2.92	20.36	17.44
G180	2.93	31.61	28.68
G240	2.55	30.49	27.94
G320	2.38	33.67	31.29
G400	2.35	26.28	23.93
G600	2.23	31.95	29.72
G800	2.63	26.95	24.32
G1200	2.37	28.85	26.48

3.4 Roughness measurement

Profilometry (Wyko Surface Profiling System NT-1100) was used to characterize the surface and measure the average surface roughness R_a, root-mean-squared roughness R_q, the average of the ten greatest peak-to-valley separations R_z and the peak-to-valley difference calculated over the entire measured array R_t both before and after corrosion testing. In all parameters, the roughness value for mild steel has increased after corrosion which is due to the formation of corrosion products on the surface. As an example, R_a values before and after corrosion testing are listed in Table 3. The last column in this table is a list of SiC particle size on grinding papers. Nickel also showed an increase in roughness parameters after corrosion which is an indication of the formation of deeper grooves (on rougher surfaces where there is more corrosion) and passive layer (on all surfaces especially smoother surfaces) [6, 13].

3.5 Relationship between corrosion rate and roughness

Figs. 4(a), (b) and 5(a), (b) show the variation of the corrosion rate as determined by potentiodynamic polarization tests versus the average surface roughness R_a and the average ten largest peak-to-valley separations R_z for mild steel and nickel, respectively.

By increasing the surface roughness of mild steel, corrosion rate decreased. These results are in agreement with results of previously reported studies which investigated the appearance, oxygen concentration and roughness parameters in SEM, EDS and profilometry tests, respectively. This is a general trend in corrosion of metals with no ability to form a passive layer [8, 9]. This trend is exactly the opposite to the one in nickel (Fig. 4(b)) and aluminium with ability to form a protective passive film [3, 6]. A reverse trend was also reported by Li and Li [2] for Cu in a 3.5% NaCl solution which was expected. Looking at the general shape of the plot, it can be seen that after a limit, decreasing the surface roughness of mild steel dramatically increases the corrosion rate. The rate of change of corrosion rate with roughness decreases before this limit and, for the high roughness values, the corrosion rate appears to reach a plateau.

Table 3: Ra values before and after corrosion testing on mild steel.

Sample No	Roughness values		SiC particle size (μm)
	R_a (before corrosion testing) (nm)	R_a (after corrosion testing) (nm)	
G60	1446	1550	254
G120	916	1688	102
G180	819	1725	80
G240	401	1384	60
G320	272	1460	40
G400	217	1578	30
G600	185	1586	15
G800	171	1588	13
G1200	128	1254	9

4 Discussion

The result obtained in this work was that mild steel with smoothest polished surfaces (G1200) showed the highest corrosion rate compared to surfaces with higher roughness values. This can be attributed to the high rate of corrosion propagation after initiation. It is said that localized corrosion is controlled by the diffusion process and in this case once the corrosion started, it propagated at a fast rate due to the continuous diffusion process and the formation of acid media at the bottom of the grooves [8]. The results are in a good agreement with the observations of Alvarez on AE44 magnesium alloy which has similar characteristics as mild steel i.e. no ability to form a stable passive film [9]. These trends are opposite of the trends seen in the literature for aluminium, nickel, copper, titanium and stainless steel, which include the rough surfaces limiting diffusion out of the forming pits, more available active sites on the rough surfaces and fast formation of a stable oxide film on the smoother surfaces [3, 7].

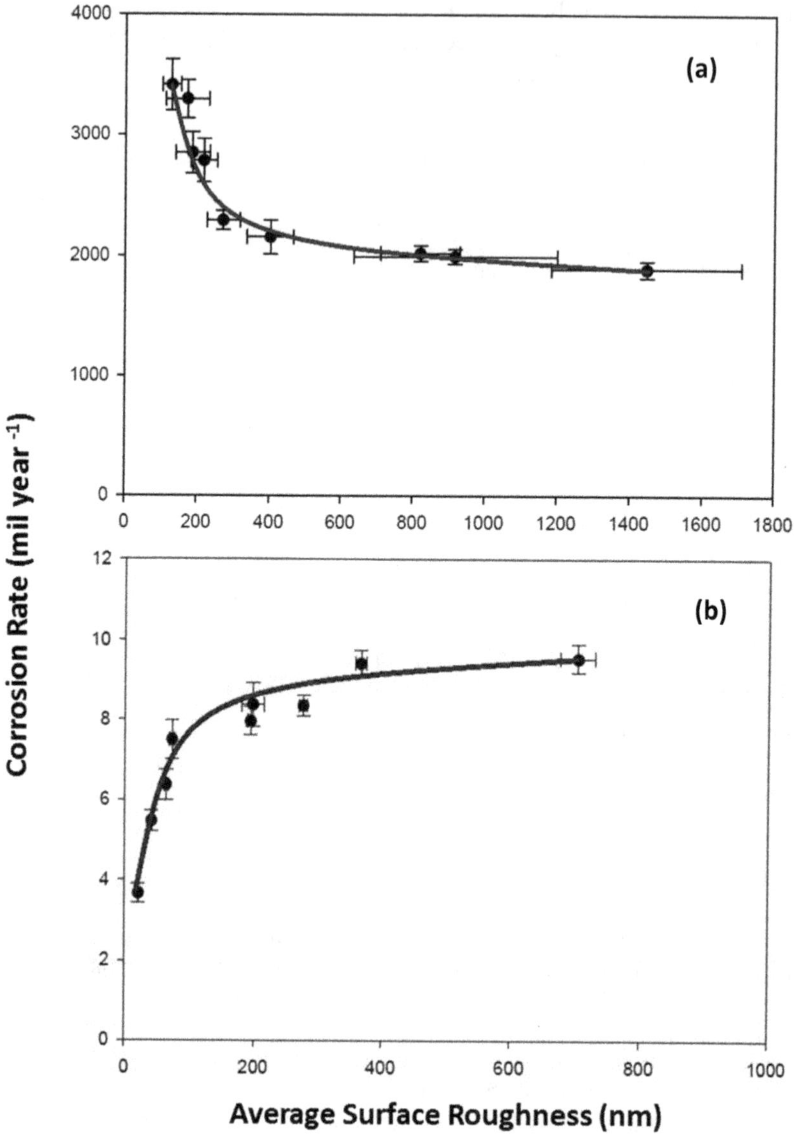

Figure 4: Dependence of corrosion rate on R_a for (a) mild steel and (b) nickel.

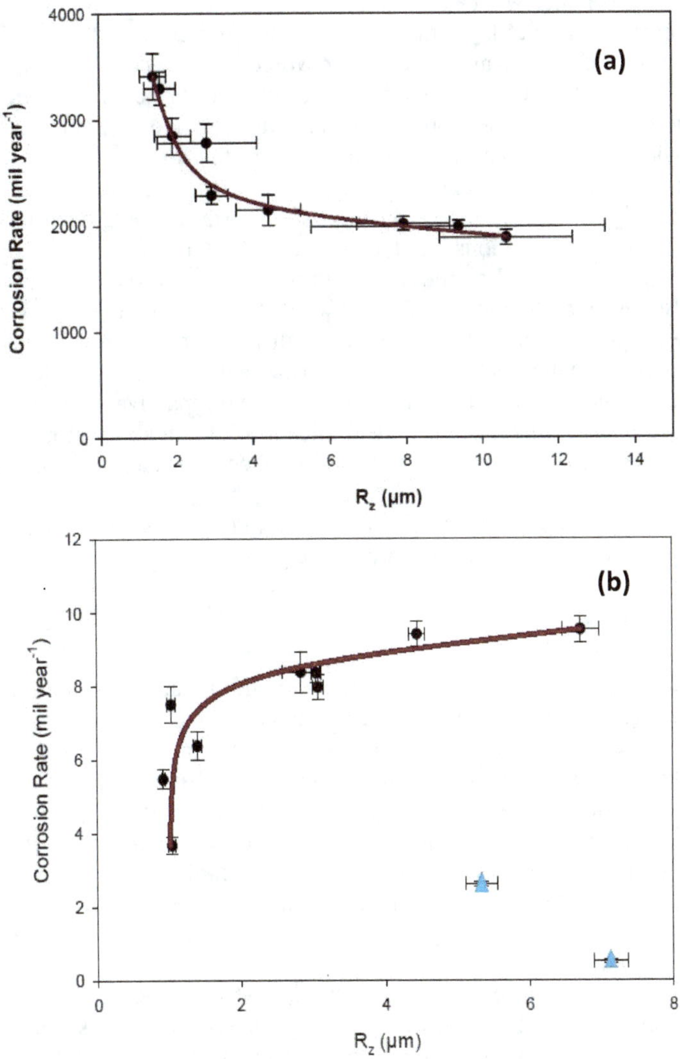

Figure 5: Dependence of corrosion rate on R_z for (a) mild steel and (b) nickel.

On mild steel and also magnesium [9], since there is no stable protective passive film on both smooth and rough surfaces, the active sites would be equally available on both surfaces. But there are some grooves which are left from polishing process that act like active sites. These grooves exist more on the surfaces with higher roughness values. Corrosion products and corrosive ions are trapped in such grooves which will result in more corrosion on rough surfaces. But this is not the whole story, because the rough surface might still possess some of the corrosion products, which is said to be less reactive than the bulk of metals. The same thing could also occur on mild steel or magnesium. Therefore, the possibility of corrosion will be decreased on rougher surfaces of mild steel because of the corrosion product layer that exists on the surface. But despite of having fewer places for corrosion nucleation, there would be more corrosion on smoother surfaces because there is no protection on the polished surface since the places for corrosion to occur would be more in contact with corrosive solution.

It is said that a rougher surface would prevent the diffusion of the corrosive ions out of the grooves and forming pits by trapping the ions [3], but simultaneously a rougher surface on mild steel surfaces could also prevent the diffusion of the corrosive ions into the grooves [7]. Thus, the limited diffusion of the aggressive ions into the grooves would reduce the corrosion on the rougher surface. But on smooth mild steel surface which has no corrosion product layer or a passive film, corrosion is dependent on the diffusion of the corrosive ions onto the surface because there would be nothing to prevent the corrosion [3]. In the case of metals with no ability to form a passive layer, no repassivation of the smooth surface will occur, unlike stainless steel, nickel or aluminium, but instead more corrosive ions will be in contact with the surface and help the occurrence of corrosion.

Thus, it is suggested that corrosion is dependent on both surface roughness and the ability of the material to form a protective passive film.

5 Conclusions

The surface roughness of mild steel as a metal with no ability to form a passive layer showed a significant effect on its corrosion resistance. A rougher surface has both a slight amount of corrosion product layer remaining, as well as a rougher topography that can prevent the corrosive ions from diffusing close to the surface to initiate corrosion but the smoother surface is corroded more quickly because the latter has nothing preventing corrosion in contrast to the rougher surface. For nickel, low surface roughness promotes the formation of a stable passive film and in rougher samples there is a greater contact area between the corrosive solution and metal together with trapping of corrosive ions in the deep grooves leading to processes such as pitting.

The potentiodynamic polarization technique was used to measure corrosion parameters related to samples with different roughness values. Less corrosion was observed on rougher samples of mild steel and by decreasing the roughness, corrosion rate increased. A reverse trend was observed for nickel.

SEM micrographs showed more degradation on smoother surface of mild steel but in the case of rougher surfaces some grooves are still visible in a larger scale. The micrograph for nickel showed that the corrosion pits were preferentially aligned along the grooves suggesting that the deep valleys on the ground surface were favourable sites for corrosion nucleation.

EDS analysis also measured oxygen content before and after corrosion and different elements were characterised from the peak analysis. Less increase in oxygen content was observed on rougher samples of mild steel. On nickel, however, the increases in oxygen content, especially on smoother surfaces, could be indicative of the formation of a stable passive film.

Profilometry was utilized to measure different roughness parameters to investigate corrosion. By increasing the roughness in mild steel, all roughness parameters had the same trend in corrosion rate (decreased and plateaued) and rougher samples showed less increase in roughness after corrosion. For nickel, by increasing the roughness, corrosion rate increased and plateaued.

References

[1] Cabrini, M., Cigada, A., Rondell, G. & Vicentini, B., Effect of different surface finishing and of hydroxyapatite coatings on passive and corrosion current of Ti6Al4V alloy in simulated physiological solution. *Biomaterials*, **18(11)**, pp. 783–787, 1997.

[2] Li, W. & Li, D.Y., Influence of surface morphology on corrosion and electronic behavior. *Acta Materialia*, **54(2)**, pp. 445–452, 2006.

[3] Suter, T., Müller, Y., Schmutz, P. & von Trzebiatowski, O., Microelectrochemical studies of pit initiation on high purity and ultra high purity aluminum. *Advanced Engineering Materials*, **7(5)**, pp. 339–348, 2005.

[4] Walter, R. & Kannan, M.B., Influence of surface roughness on the corrosion behaviour of magnesium alloy. *Materials & Design*, **32(4)**, pp. 2350–2354, 2011.

[5] Burstein, G. & Vines, S., Repetitive nucleation of corrosion pits on stainless steel and the effects of surface roughness. *Journal of The Electrochemical Society*, **148(12)**, pp. B504–B516, 2001.

[6] Toloei, A.S., Stoilov, V. & Northwood, D.O., The effect of different surface topographies on the corrosion behaviour of nickel. *WIT Transactions on Engineering Science*, **77**, pp. 193–204, 2013.

[7] Zuo, Y., Wang, H. & Xiong, J., The aspect ratio of surface grooves and metastable pitting of stainless steel. *Corrosion Science*, **44(1)**, pp. 25–35, 2002.

[8] Abosrra, L., Ashour, A.F., Mitchell, S.C. & Youseffi, M., Corrosion of mild steel and 316L austenitic stainless steel with different surface roughness in sodium chloride saline solutions. *WIT Transactions on Engineering Sciences*, **65**, pp. 161–172, 2009.

[9] Alvarez, R.B., Martin, H.J., Horstemeyer, M.F., Chandler, M.Q., Williams, N., Wang, P.T. & Ruiz, A., Corrosion relationships as a function of time and surface roughness on a structural ae44 magnesium alloy. *Corrosion Science*, **52(5)**, pp. 1635–1648, 2010.

[10] Roberge, P., *Corrosion Engineering Principles and Practice*, 1st ed., McGraw-Hill: New York, 2008.

[11] Slepski, P., Gerengi, H., Jazdzewska, A., Orlikowski, J. & Darowicki, K., Simultaneous impedance and volumetric studies and additionally potentiodynamic polarization measurements of molasses as a carbon steel corrosion inhibitor in 1M hydrochloric acid solution. *Construction and Building Materials*, **52**, pp. 482–487, 2014.

[12] Bouklah, K.O., Messali, M., Hammouti, B., Elidrissi, A. & Warad, I., A pyrrolidine phosphonate derivative as corrosion inhibitor for steel in H_2SO_4 solution. *Der Pharma Chemica*, **3(5)**, pp. 283–293, 2011.
[13] Toloei, A.S., Stoilov, V. & Northwood, D.O., An electrochemical impedance spectroscopy and potentiodynamic polarization study of the effect of unidirectional roughness on the corrosion of nickel. *International Journal of Computational Methods and Experimental Measurements*, **2(3)**, pp. 243–254, 2014.
[14] Toloei, A.S., Stoilov, V. & Northwood, D.O., A new approach to combating corrosion of metallic materials. *Applied Surface Science*, **284**, pp. 242–247, 2013.
[15] Toloei, A., Stoilov, V. & Northwood, D.O., The effect of creating different size surface patterns on corrosion properties of nickel. *Proc. of ASME International Mechanical Engineering Congress & Exposition* (IMECE2012), Houston, USA, 2012.
[16] Toloei, A., Stoilov, V. & Northwood, D.O., The relationship between surface roughness and corrosion. *Proc. of ASME International Mechanical Engineering Congress & Exposition*, San Diego, California, USA, 65498, 2013.

Plasma electrolytic oxidation coatings on Mg-alloys for improved wear and corrosion resistance

R. O. Hussein[1], X. Nie[2] & D. O. Northwood[2]
[1]Higher Colleges of Technology, Al Ain, United Arab Emirates
[2]University of Windsor, Canada

Abstract

Due to their high chemical reactivity, relatively low melting point and low hardness, magnesium and magnesium alloys have relatively poor corrosion and wear resistance. Since both corrosion and wear are surface phenomena, a number of surface engineering techniques have been used to improve corrosion and wear performance. Whilst some surface hardening/strengthening methods have led to improvements in wear properties, they have not, in themselves, significantly improved the corrosion performance. Plasma electrolytic oxidation (PEO) has the potential to produce hard, compact oxide coatings that are well adhered to the magnesium alloy substrate. Such coatings can provide both improved wear and corrosion resistance. In this contribution, it is shown how the nature of the PEO oxide coatings (thickness, microstructure, porosity, phase content, composition) can be changed by changing the PEO processing parameters (substrate alloy, electrolyte, current or voltage, processing time); this, in turn, effects the corrosion and wear performance. All PEO-coatings have a three-layer structure with a porous outer layer, an intermediate dense layer and a thin inner dense layer. From a corrosion aspect, the performance of coatings is determined by the time taken for corrosion to initiate since this is much shorter than the time taken for the coating to degrade. For PEO-coated Mg-alloys, this initiation time is primarily determined by the thickness, porosity and phase content of the inner dense layer at the coating/substrate interface. With respect to tribological properties, the coefficient of friction (COF) in dry sliding wear increases with increasing surface roughness of the PEO coatings. The wear rate is primarily determined by the thickness and hardness of the intermediate dense layer. Coatings containing less porosity and higher spinel-phase ($MgAl_2O_4$, Mg_2SiO_4) content are harder and more wear-resistant.

Keywords: magnesium alloys, PEO coatings, three-layer structure, corrosion, tribological properties.

1 Introduction

Magnesium is the lightest structural metal. In addition to its low density, a high specific strength, specific stiffness, good castability and weldability, good damping and good electro-magnetic shielding make magnesium alloys a desirable material for many applications [1]. However, their lower formability, relatively low creep resistance at elevated temperature and high chemical reactivity with associated low corrosion and wear resistance, limits the use of magnesium alloys in many industries [2].

It is generally recognized that magnesium has poor wear resistance and that this is related to a low hardness [3]. The main mechanism for improving the mechanical properties is precipitation hardening, e.g. by adding aluminium which forms a $Mg_{17}Al_{12}$ intermetallic phase which increases tensile strength and hardness. However almost all elements used so far for alloying increase magnesium's susceptibility to corrosion. The relatively poor wear performance of magnesium and its alloys is also related to magnesium's high reactivity and relatively low melting point. Surface treatment to form coatings which produce a protective ceramic, polymer or composite layers is considered as one of the most effective ways for corrosion and wear protection. Such coatings should be uniform, pore free, relatively thick, hard, well adhered, flexible to withstand an overload and insulating to minimize corrosion. Plasma electrolytic oxidation (PEO) fulfils most of these requirements and has been used for depositing a well adhered ceramic coating on light-weight metals (Mg, Al, Ti, Zr) that provides not only corrosion protection but also enhanced wear properties, hardness and toughness with better thermal stability and dielectric properties [4–7].

PEO is a process that transforms the magnesium metal surface into thick, ultra-hard ceramic oxides by a plasma discharge in an electrolytic bath [8, 9]. The PEO process for light-weight metals is strongly influenced by such parameters as electrolyte composition and concentration, current or voltage applied and substrate alloy [10]. Generally, these parameters have a direct influence on the phase transformations, crystallization and sintering.

Due to the existence of temperature gradients between the inner and outer layer of the coating, PEO coatings on light alloys have been found to have a three-layered structure [8]. The porous outer layer usually has poor mechanical properties. The intermediate layer is relatively dense and can provide good corrosion protection. The third (barrier) layer is a very thin interface layer (few hundred nm to 1 μm in thickness) is well adhered to the substrate and ultimately provides the best corrosion performance [8]. Mg alloy parts used for automotive and aerospace applications including helicopter gearbox covers, pistons and cylinders, are subject not only to corrosion but also to mechanical wear.

The electrochemical formation of surface oxide layers can occur through different mechanisms depending on the electrolyte, e.g. silicates, aluminates, phosphates. The PEO coatings are usually produced by AC or bipolar current mode, containing both anodic and cathodic components. An oxide layer formation is induced both by the ionic component of the current which is transmitted via surface discharges and by the anodizing current passing across the surface which is free of discharges [4]. The other components of the current cause secondary electrochemical processes which lead to liberation of electrode gases (H_2 and O_2) and anodic dissolution of the lightweight metal, where for magnesium alloys the main electrochemical reactions occurring at the metal/oxide and oxide/electrolyte interfaces are as follows [11]:

Metal-oxide interface:

$$Mg \rightarrow Mg^{2+} + 2e^- \quad (1)$$

$$4OH^- \rightarrow 2H_2O + O_2 \uparrow + 4e^- \quad (2)$$

Oxide-electrolyte interface:

$$Mg^{2+} + 2OH^- \rightarrow Mg(OH)_2 \quad (3)$$

$$Mg(OH)_2 \rightarrow MgO \downarrow + H_2O \quad (4)$$

$$Mg^{2+} + 2AlO_2^- \rightarrow MgAl_2O_4 \downarrow \quad (5)$$

$$2Mg^{2+} + SiO_3^{2-} + 2OH^- \rightarrow Mg_2SiO_4 \downarrow + 2H_2O \quad (6)$$

$$3Mg^{2+} + 2PO_4^{3-} \rightarrow Mg_3(PO_4)_2 \downarrow \quad (7)$$

Reaction (1) is the anodic dissolution and (2) is the oxygen evolution reaction. The cation released from the metal (reaction (1)) combines with the anion in the electrolyte to form compounds $Mg(OH)_2$, Mg_2AlO_4, Mg_2SiO_4 or $Mg_3(PO_4)_2$, depending on the different electrolyte by reactions (3), (5), (6) and (7). The unstable hydroxide $Mg(OH)_2$ dehydrates to MgO by the high temperature (reaction (4)), resulting from the plasma discharge.

Relatively few published papers can be found that combine both tribological and corrosion studies of PEO treated Mg alloys. In this contribution, the authors' own work is primarily used to illustrate the general relationships between PEO processing parameters and the subsequent tribological-corrosion performance of the coatings.

2 Corrosion and tribological properties of PEO-coated Mg-alloys

2.1 Corrosion resistance

Magnesium alloys exhibit very poor corrosion resistance due to their chemically active nature, especially internal galvanic corrosion [12], which can further cause severe pitting corrosion on the metal surface resulting in decreased mechanical stability and an unattractive appearance. For coated metals, including Mg-alloys, the corrosion performance of coatings is determined by the time taken to initiate corrosion in the metal substrate since this is much shorter than the time taken for the coating to degrade [13]. Generally, the corrosion resistance of PEO-coated magnesium alloys depends on many factors that are summarized schematically in Fig 1. The overall protective abilities of the PEO coatings are governed by:

- Coating structures, mainly coating compactness in terms of porosity and other defect levels.
- Relative thicknesses of the three layers, particularly the coating/substrate interface layer.
- Chemical composition of the coating.

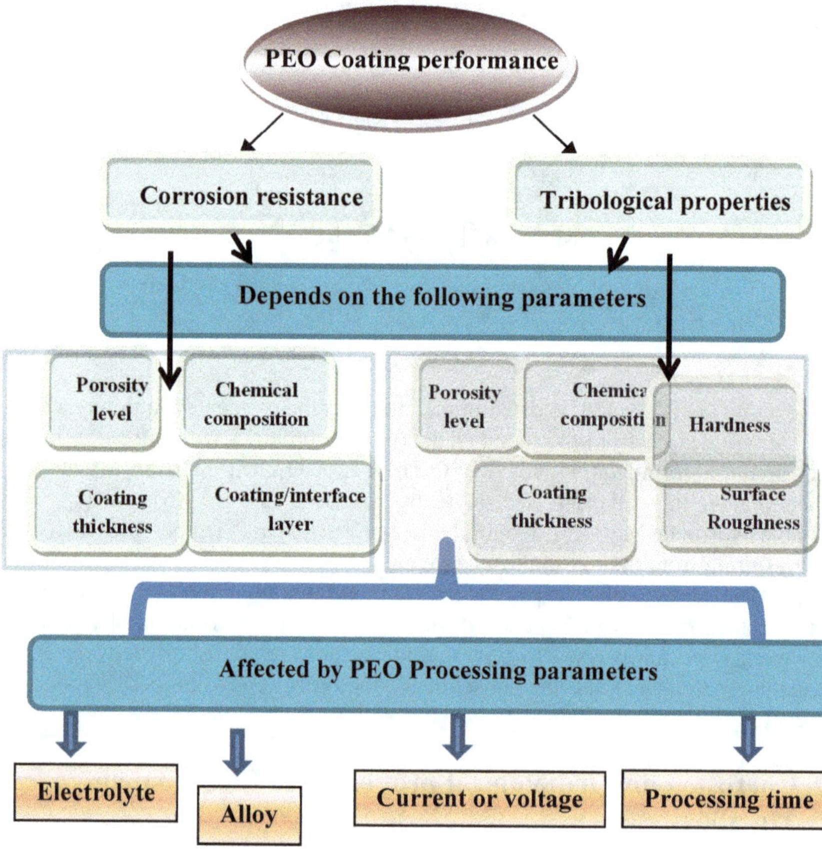

Figure 1: PEO processing parameters and performance.

Therefore, the more compact and thicker the coating layers, the more difficult it is for the corrosive anions or oxidants (Cl^{-1} ions) to penetrate to the base magnesium substrate. The coating/substrate barrier layer plays a key role in decreasing the substrate area exposed to the aggressive solution. The phases present in the oxide coatings are also important in the overall corrosion resistance. The spinel phases ($MgAl_2O_4$, Mg_2SiO_4) are more resistant to dissolution than MgO which can quite readily be converted to $Mg(OH)_2$ [14].

2.2 Mechanical/tribological properties

As a result of the PEO coating process, the magnesium alloy surface is converted into ceramic-like coating layers that are hard, well adhered to the substrate and wear resistant [15, 16]. The mechanical and tribological properties of PEO coatings depend on the characteristics schematically shown in Fig. 1:

- Hardness is strongly dependent on the coating thickness, the nature of the dominant phases present, their ratio and distribution and the density of porosity and micro-cracks in the coatings [9, 17, 18].
- The adhesion is influenced by the coating thickness. As the coating process progresses, significant changes take place in the inner layer structure as a result of diffusion processes [15] and adhesion is improved.
- The load that can be supported is affected by the coating thickness. Thicker coatings will give lower stresses at the coating-substrate interface and hence give better load support [9].
- The wear resistance of coatings mainly depends on hardness, coefficient of friction and roughness. Hence, porosity levels, chemical composition, thickness and structure of the coatings are key parameters for wear performance of Mg alloys [17]. As coating growth continues, the coating surface roughness and outer layer porosity level will increase with increasing processing time.

2.3 Correlation between tribological and corrosion properties

Since the coating hardness depends directly on the coating compactness, coatings produced with more porosity have lower hardness. Hence, a change of the process parameters, such as a current density and modes [5, 17] and electrolyte composition and concentration [4], are important in controlling porosity levels to improve coating hardness. Coatings produced in silicate solutions generally have a higher hardness than coatings produced in a phosphate solution under the same process conditions [18] since the oxide layers made in silicate solutions contain Mg_2SiO_4, which has a higher micro-hardness than MgO, $MgAlPO_5$ and $Mg_3(PO_4)_2$ that are produced in phosphate solutions [4]. The use of sodium aluminate ($NaAlO_2$) in the electrolyte for AJ62, AM60B and AZ91D Mg-alloys was investigated using different current modes [5, 8]. PEO coatings can increase the hardness of the magnesium alloys 5–9 times compared with the hardness of the base substrate [17]. However, the values of micro-hardness close to the substrate/coating interface generally have the highest value compared to other locations, which may be attributed to its higher compactness compared with the outer porous layer [19]. The wear resistance of the coated samples is a complex process that involves surface roughness and hardness which are controlled by phase contents and porosity levels.

Since PEO coatings contain unavoidable pores, although most of them are discontinuous and the number of through thickness pores is very small, the overall corrosion resistance of PEO coated magnesium alloy depends on the retarding effect of the porosity level of the oxide layer and on the corrosion resistance of the substrate/coating interface [8]. However, the more compact and thicker the coating and presence of phases, the slower and more difficult access of the corrosive species can be. The electrochemical behaviour of magnesium alloys and PEO coatings have been mostly studied using potentiodynamic polarization in 3.5% NaCl solution at 25°C and electrochemical impedance spectroscopy (EIS). The presence of a PEO coated layer on the magnesium alloys causes a decrease of the corrosion current density by up to four orders of magnitude [8]. There are situations where the structures for good corrosion and wear performance are not the same. An example would be for oil-lubricated wear where the presence of a porous outer layer in the PEO coating could act as an oil reservoir, improving wear performance [20].

3 Illustrative examples

In this section, it is shown how a coating can be formed with both good corrosion and wear properties by varying the PEO process parameters, in particular the electrolyte chemistry (concentration, composition) and current mode.

3.1 Experimental procedures

Commonly used Mg-alloys AZ91D (mass fraction: Al 9%, Mn 0.15%, Zn 0.8%, Si 0.1%, balance Mg), AM60B (mass fraction: Al 5.6–6.4%, Mn 0.26–0.4%, Zn $\leq$ 0.2%, balance Mg) and AJ62 (mass fraction: Al 6.1%, Mn 0.34%, Sr 2.1%, balance Mg) are used as example materials. The PEO coating system used to produce the oxide coatings consists basically of a container with alkaline electrolyte and an electrical source and is described in more detail in refs. [5] and [15]. A stainless steel plate in the bath acts as a counter-electrode (cathode) with the Mg-alloy coupons as the anode. The electrodes are connected to two pulsed DC power supplies operating under a constant current control function to generate different current waveforms.

3.2 Effect of electrolyte chemistry

3.2.1 Electrolyte concentration

PEO coatings were produced on AZ91D Mg-alloy using a base composition of 2 g/l KOH electrolyte and addition of two different aluminate $Na_2Al_2O_4$ concentrations 8 g/l (S1) and 15 g/l (S2). The PEO-coated materials show a higher coefficient of friction (COF), Fig. 2(a), but a lower wear rate, Fig. 2(b), than the uncoated alloy.

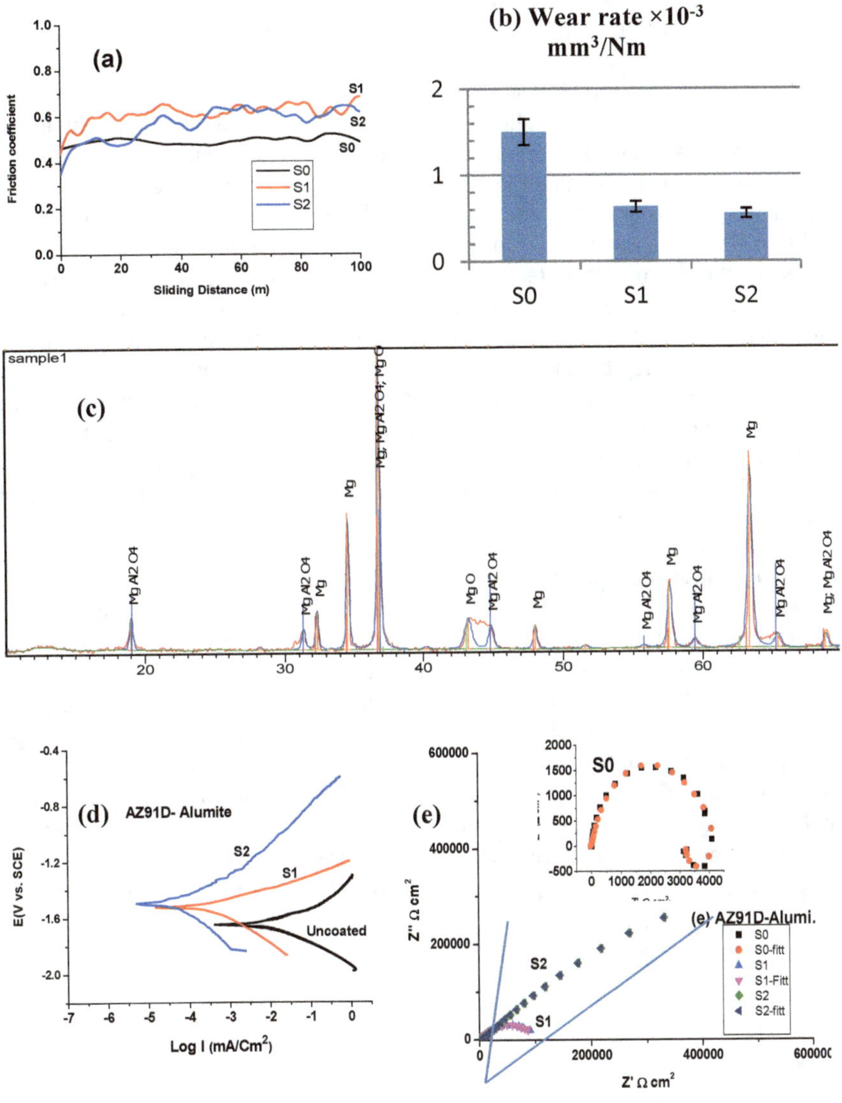

Figure 2: (a) Coefficients of friction vs. sliding distance for samples S1 and S2 and S0 uncoated Mg alloy substrate; (b) wear rates of the oxide coatings on AZ91D alloy for different electrolyte concentrations; (c) XRD pattern for S2 and (d) potentiodynamic corrosion resistance; and (e) EIS results.

As shown in Fig. 2(c), from the XRD pattern of the oxide coating made using a concentrated aluminate electrolyte, the coatings were mainly composed of MgAl2O4 and MgO phases. Based on the potentiodynamic polarization and EIS results (shown in Figs. 2(d) and (e)), it was found that all PEO coatings offered significant corrosion protection to the AD91D alloy for short immersion times (0.33 h) prior to corrosion testing.

Compared to uncoated alloy, the corrosion rate from potentiodynamic polarization was reduced by at least 50 and 300 times for S1 and S2, respectively. The EIS results demonstrated the same trends. As can be seen from Fig. 2(e), the impedance response of the uncoated Mg alloys (the enlarged graph) is very different from that of the coated specimens. The differences in impedance are related to the corrosion protection mechanisms for the alloys provided by the naturally formed oxide layer (few nm) which is easily corroded by the corrosive electrolyte. Sample S2, produced with a concentrated aluminate electrolyte, shows that both the tribological and corrosion properties can be enhanced at the same time. Increasing the aluminate concentration in the electrolyte from 8 to 15 g/l produces a coating that is denser and contains a higher amount of the spinel phase $MgAl_2O_4$ relative to MgO. This produces a more corrosion and wear resistant coating. The thickness, porosity and defect levels are decreased by using a concentrated aluminate-containing electrolyte.

3.2.2 Electrolyte composition

The electrolyte composition plays a key role in the PEO process and polyvalent metal anions, such as tungstate, are promising for the formation of coatings of different chemical compositions that enhance both corrosion and tribological performances. PEO coatings were produced on AZ91D Mg-alloy using a bipolar current mode and an electrolyte with a base composition of 8 g/l $Na_2Al_2O_4$ + 1 g/l KOH and additions of 0, 1, 2 and 6 g/l of Na_2WO_4 [21]. The Nyquist and the Bode (frequency dependencies of impedance modulus $|Z|$) diagrams from the EIS analysis of experimental and fitted curves of uncoated and PEO-coated materials for 0.33 h immersion time are given in Fig. 3. Table 1 summarizes all the fitted EIS results for the uncoated and PEO coated samples. The choice of equivalent

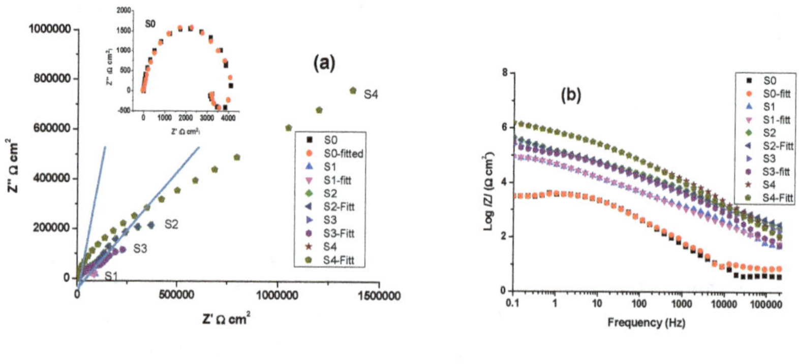

Figure 3: Plots of uncoated and PEO coated AZ91D Mg alloy after 0.33 h immersion time (a) Nyquist and (b) Bode plot-impedance.

Table 1: Tungstate composition, thickness and fitting results of EIS diagrams in 3.5%wt NaCl solution, for 0.33 h immersion time of the uncoated (S0) (equiv. circuit Rs+CPE1/(R1+ CPE2/R2+L)) and of PEO treated AZ91D Mg alloy using four different tungstate concentration (equiv. circuit Rs+ CPE1/(R1+ CPE2/R2)).

S	Na_2WO_4 g/l	Coating thickness μm	CPE1-Q μF/cm² s^{1-n}	CPE1-n	R1 kΩ·cm²	CPE2-Q μF/cm² s^{1-n}	CPE2-n	R2 kΩ·cm²	L H*
S0	----	0	18.66	0.73	−15.65	−2.498	0.33	18	0.25
S1	0.0	25.2 ± 2.5	3.499	0.621	0.049	1.235	1	270	
S2	1.0	30.0 ± 3.0	1.247	0.563	0.244	1.528	1	899	
S3	2.0	28.5 ± 2.9	1.38	0.584	0.324	0.001	0	710	
S4	6.0	43.4 ± 4.5	0.187	0.675	0.978	0.907	0.713	2269	

*Only applicable to uncoated material.

circuits employed for curve fitting was a balance between a reasonable fitting of the experimental values and a good description of the electrochemical system by keeping the number of circuit elements at a minimum [5].

The proposed equivalent circuit used describes the behaviour of a process characterized by two time constants, namely a high-frequency (HF) time constant (CPE1−R1) and a low-frequency (LF) time constant (CPE2−R2). The low frequency data are on the right side of the Nyquist plot and higher frequency data are on the left. CPE2 is the constant phase element for the double layer capacitance of the interface at, or near, the coating/substrate interface. R2 represents the polarization resistance which is the Faradic charge transfer resistance related to electrochemical reactions in the same coating/substrate interface region [5]. It is R2 that primarily controls the corrosion resistance of the coated alloy. The impedance of the CPE is given by the equation:

$$Z_{CPE} = 1/[Q(j\omega)^n] \tag{8}$$

where Z_{CPE} (Ω·cm²) is the impedance of the constant-phase-element; Q, a CPE constant ($\Omega^{-1}s^n cm^{-2}$), is a combination of properties related to both the surface and the electroactive species and is independent of frequency, j is the imaginary unit $\sqrt{-1}$, $\omega = 2\pi f$ the angular frequency (rad/s) of the sine wave, f the frequency in Hz; the value of n ranges between 0 and 1. Depending on the value of exponent n, the physical meaning of the CPE of the circuit can be related to pure resistance (n = 0), pure capacitance (n = 1), pure inductance (n = −1), or mass transport related impedance, i.e. Warburg impedance (n = 0.25–0.5). For the uncoated samples, the presence of Warburg element and its resistance R3 is evidence for a diffusion control process.

The tungstate addition produces changes in the microstructure and phase contents of the coatings. Increasing the tungstate content from 0 to 6 g/l in the electrolyte increases the thickness of the dense substrate/coating interface layer, enhances the corrosion resistance for immersion times (0.33 h) prior to corrosion, reduces the surface roughness which leads to a decrease in the COF in dry sliding wear (Fig. 4), and leads to the formation of the tungstate-containing phases WO_3 and $MgWO_4$ (see Fig. 5). All these factors, together with a reduction in porosity, leads to a harder coating with lower wear rates.

Fig 4(b) shows that sample S4 treated with high tungstate concentrations exhibits a lower coefficient of friction (COF) compared with that of samples treated without tungstate or with lower tungstate concentrations. However, the uncoated samples exhibit the minimum COF. The lowest wear rate was produced with the highest tungstate addition. However, all PEO coated samples show a lower wear rate than the uncoated alloy, Fig. 4(a). In the pin-on-disk tests, the COF is shown to be directly dependent on the surface roughness's parameters as shown in Fig. 6(b). The tungstate ions may also serve as a corrosion inhibitor and promote the formation of a passive film, which is essential for the deposition of PEO coatings on the Mg alloys.

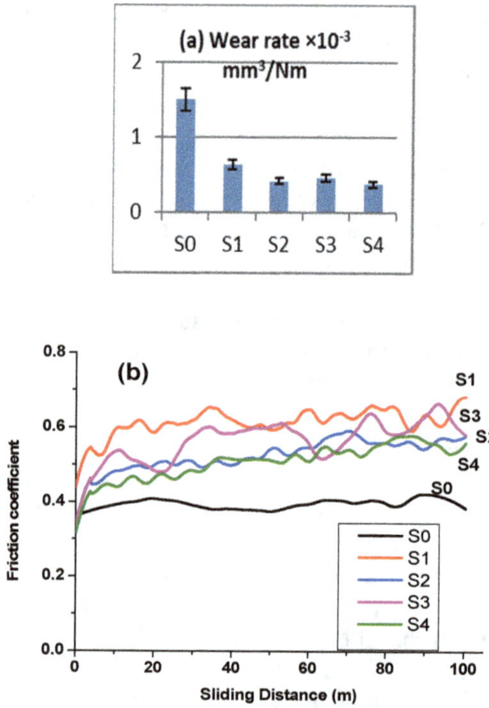

Figure 4: (a) Wear rates and (b) coefficient of friction vs. sliding distance of the uncoated Mg alloy substrate and the oxide coatings formed using different tungstate concentrations.

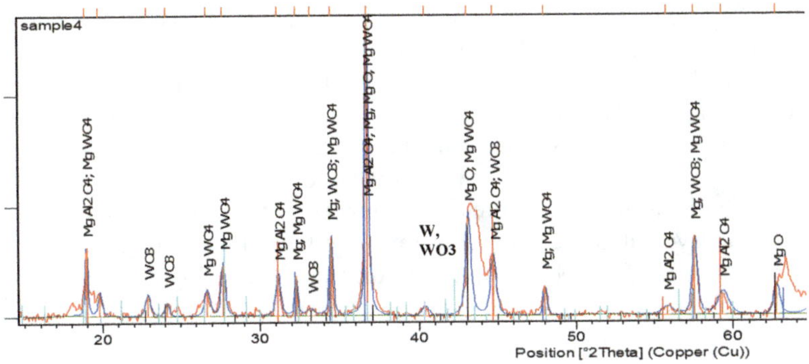

Figure 5: XRD patterns of oxide coatings formed in $Na_2Al_2O_4$ + 1 g/l KOH electrolyte with addition of 6.0 g/l Na_2WO_4 (sample S4).

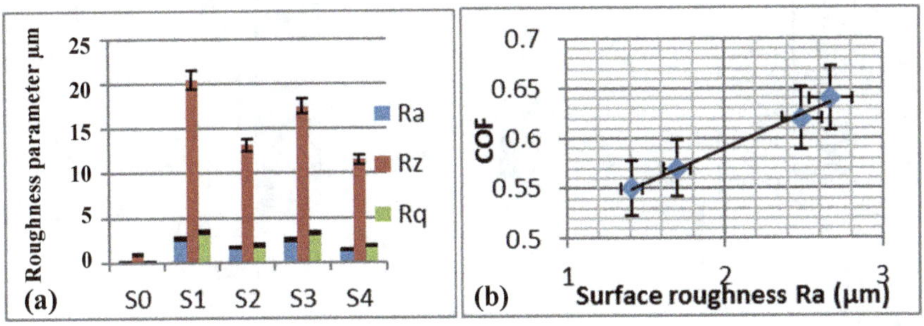

Figure 6: (a) Roughness parameters R_a, R_z and R_q of the oxide coatings formed using different tungstate concentrations and the S0 uncoated Mg alloy substrate. (b) Coefficients of friction vs. roughness parameter R_a of the oxide coatings on AZ91D-Mg alloy substrate.

3.3 Effect of current mode

The influence of current mode (unipolar U, bipolar B or hybrid H1 (unipolar followed by bipolar) and H2 (bipolar followed by unipolar)) on the corrosion and wear properties of PEO coatings formed on an AM60B magnesium alloy was investigated and are shown in Figs. 7 and 8 [5]. The wear resistance was evaluated using a pin-on-disk set-up under dry conditions and an applied load of 2 N. PEO coating decreases the wear rate, Fig. 7(a), but also increases the COF (Fig. 7(b)). This increase in COF is related to an increase in surface roughness, Fig. 7(c). Corrosion testing was performed in a 3.5 wt% NaCl solution using a potentiodynamic polarization technique. R_p, the polarization resistance, was used as the measure of corrosion resistance. The results are summarized in Fig. 8 and Table 2.

The reduction of the porosity and other defects, e.g. micro cracks, gas trapping, discontinuity between the coatings and the substrate, of the coatings, together with lower surface roughness that are produced by the B and H1 current modes compared to the U and H2 modes lead to improvement in both corrosion and wear resistance (including a decrease in COF), even though the coatings were thinner. Use of a bipolar current mode reduces the number of large pancake features on the surface, thus lowering the surface roughness and COF. In the pin-on-disk tests, the COF is shown to be directly dependent on the surface roughness, R_z (Fig. 7(d)). The corrosion performance is more closely related to the protective nature of the dense oxide layer at the coating/substrate interface. In this regard, the most protective oxide is formed by using the bipolar and hybrid 1 (H1) current modes (Fig. 8).

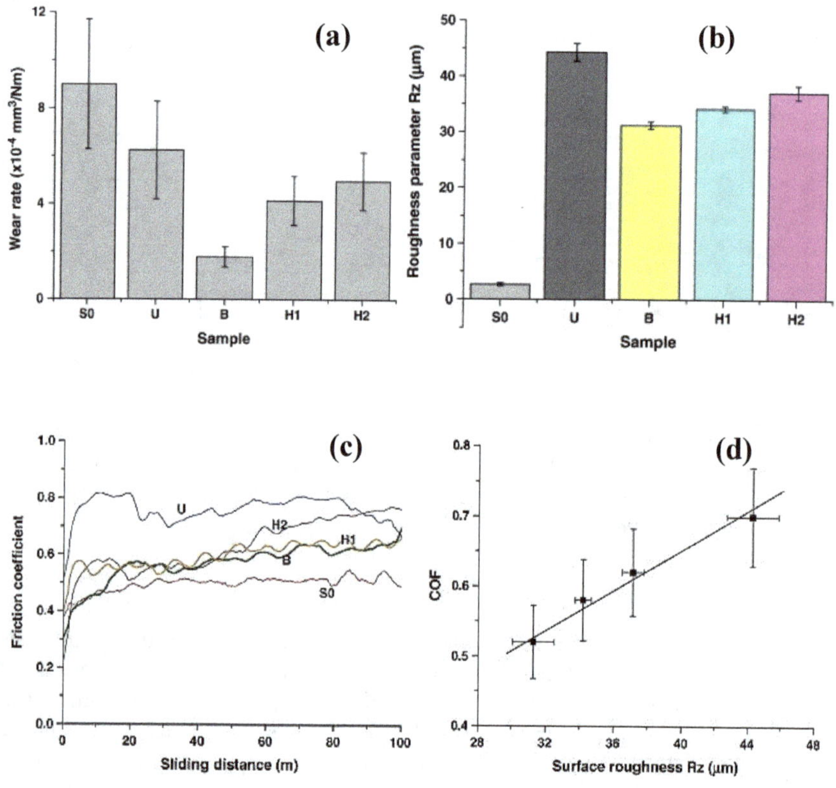

Figure 7: Wear rate (a), roughness parameter Rz (b), coefficients of friction vs. sliding distance (c) and coefficients of friction vs. roughness parameter Rz (d) for samples U, B, H1, and H2 and the S0 uncoated AM60B Mg alloy substrate and H2 and the S0 uncoated AM60B Mg alloy substrate.

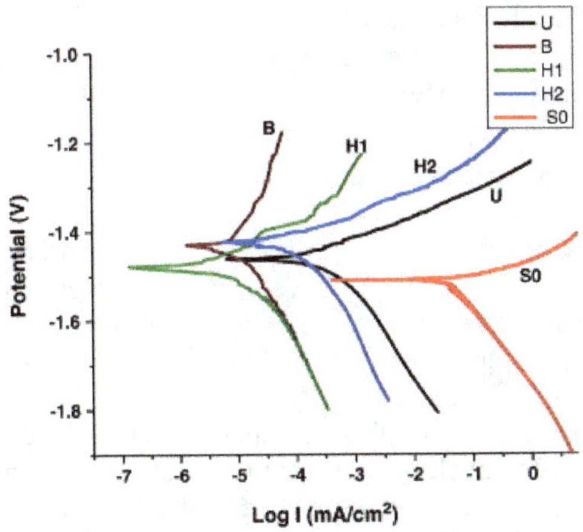

Figure 8: Potentiodynamic polarization curves of the uncoated (S0) and coated AM60B alloy samples using unipolar (U1), bipolar (B) and hybrid (H1 and H2) current modes.

Table 2: Characterization of coated samples and dry sliding wear results.

S	Coating thickness (μm)	Surface roughness R_z (μm)	Level of porosity and other defects	R_p $\Omega\cdot cm^2$	COF	Wear rate $\times 10^{-4}$ $mm^3/N\ m$
U	40 to 55	44.3 ± 1.6	Porous with many microcracks	5.8E07	0.72 to 0.8	6.24 ± 2.0
B	31 to 42	31.3 ± 0.7	Low level of porosity and microcracks	8.4E08	0.55 to 0.6	1.78 ± 0.4
H1	22 to 38	34.2 ± 0.5	Low level of porosity and microcracks	4.5E09	0.55 to 0.6	4.14 ± 1.0
H2	21 to 39	37.1 ± 1.2	Intermediate level of microcracks and porosity	1.9E08	0.65 to 0.7	4.96 ± 1.2
S0	NA	2.6 ± 0.2	NA	2.8E05	0.4 to 0.5	9.0 ± 2.7

4 Conclusions

The PEO process for coating Mg alloys is strongly influenced by such parameters as electrolyte composition and concentration as well as current mode (unipolar, bipolar and hybrid i.e. combination of both). Generally, these parameters have a direct influence on the discharging behaviour, porosity and microcrack levels, roughness, phase transformations, crystallization and sintering. This, then, affects the physical, mechanical and chemical properties of the coating. For PEO-coated materials, the primary enhancement of the corrosion and wear resistance comes from:

- the inner dense barrier layer, but also from the relatively dense intermediate layer.
- The denser coatings with lower surface roughness that are produced by the bipolar and hybrid (unipolar followed by bipolar) current modes lead to improved corrosion resistance and tribological performance compared to those produced using a unipolar or hybrid (bipolar followed by unipolar) current modes.
- Each coating has a different phase composition and surface morphology due to the effect of different electrolyte chemistry. It has been shown that increases in the $MgAl_2O_4$ spinel phase content and the tungstate-containing phases including WO_3 and $MgWO_4$, together with a reduction in porosity, lead to coatings with lower wear rates as well as increased corrosion resistance.

References

[1] Mordike, B.L. & Ebert, T., Magnesium properties-applications-potential. *Materials Science and Engineering*, **A302**, pp. 37–45, 2001.

[2] Hussein, R.O. & Northwood, D.O. Improving the performance of magnesium alloys for automotive applications. *WIT Transactions on the Built Environment*, **137**, pp. 531–544, 2014.

[3] Dobrzanski, L.A., Tanski, T., Cizek, L. & Domagala, J., Mechanical properties and wear resistance of magnesium casting alloys. *Journal of Achievements in Materials and Manufacturing Engineering*, **31(1)**, pp. 83–90, 2008.

[4] Hussein, R.O. & Northwood, D.O., Production of anti-corrosion coatings on light alloys (Al, Mg, Ti) by plasma electrolytic oxidation (PEO). *Developments in Corrosion Protection*, ed. M. Aliofkhazraei, InTech Open Publishers, 2013.

[5] Hussein, R.O., Zhang, P., Nie, X., Xia, Y. & Northwood, D.O., The effect of current mode and discharge type on the corrosion resistance of plasma electrolytic oxidation (PEO) coated magnesium alloy AJ62. *Surface & Coatings Technology*, **206**, pp. 1990–1997, 2011.

[6] Apelfeld, A.V., Borisov, A.M., Krit, B.L., Ludin, V.B., Polyansky, M.N., Romanovsky, E.A., Savushkina, S.V., Suminov, I.V., Tkachenko, N.V., Vinogradov, A.V. & Vostrikov, V.G., The study of plasma electrolytic oxidation coatings on Zr and Zr–1% Nb alloy at thermal cycling. *Surface & Coatings Technology*, **269**, pp. 279–285, 2015.

[7] Chen, Y., Nie, X., Northwood, D.O., Investigation of plasma electrolytic oxidation (PEO) coatings on a Zr-2.5Nb alloy using high temperature/pressure autoclave and tribological tests. *Surface & Coatings Technology*, **205**, pp. 1774–1782, 2010.

[8] Hussein, R.O., Nie, X. & Northwood, D.O., The application of plasma electrolytic oxidation (PEO) to the production of corrosion resistant coatings on magnesium alloys: a review. *Corrosion & Materials*, **38(1)**, pp. 54–65, 2013.

[9] Yerokhin, A.L., Nie, X., Leyland, A., Matthews, A. & Dowey, S.J., Plasma electrolysis for surface engineering, *Surface & Coatings Technology*, **122**, pp. 73–93, 1999.

[10] Hussein, R.O., Nie, X. & Northwood, D.O., Processing-structure-properties-performance relationships for Mg-Alloys coated using plasma electrolytic oxidation (PEO) method. *Materials Forum*, **37**, In press, 2014.

[11] Duan, H., Yan, C. & Wang, F., Effect of electrolyte additives on performance of plasma electrolytic oxidation films formed on magnesium alloy AZ91D. *Electrochimica Acta*, **52**, pp. 3785–3793, 2007.

[12] Roberge, P.R., *Corrosion Engineering Principles and Practice*, McGraw Hill: New York, 2008.

[13] George, G.A., Colwell, J.M., Trueman, T. & Will, G., Sensitive methods for studying the environmental performance of protective coatings. *Corrosion & Materials*, **39(2)**, pp. 46–53, 2014.

[14] Liu, L., Yang, P., S C., Guo H., An M., Microstructure and corrosion behavior of micro-arc oxidation film on magnesium alloy. *International Journal of Electrochemical Science*, **8**, pp. 6077–6084, 2013.

[15] Hussein, R.O., Nie, X. & Northwood, D.O., An investigation of ceramic coating growth mechanisms in plasma electrolytic oxidation (PEO) processing. *Electrochimica Acta*, **112**, pp. 111–119, 2013.

[16] Khan, R., Yerokhin, A.L., Pilkington, T., Leyland, A. & Matthews A., Residual stresses in plasma electrolytic oxidation coatings on Al alloy produced by pulsed unipolar current. *Surface & Coatings Technology*, **200**, pp. 1580–1586, 2005.

[17] Durdu, S., Bayramoglu, S., Demirtas, A., Usta, M. & Ücısık, A.H., Characterization of AZ31 Mg alloy coated by plasma electrolytic oxidation. *Vacuum*, **88**, pp. 130–133, 2013.

[18] Liang, J., Hu, L. & Hao, J., Characterization of microarc oxidation coatings formed on AM60B magnesium alloy in silicate and phosphate electrolytes, *Applied Surface Science*, **253(10)**, pp. 4490–4496, 2007.

[19] Li, Z., Yuan, Y. & Jing, X., Effect of current density on the structure, composition and corrosion resistance of plasma electrolytic oxidation coatings on Mg–Li alloy. *Journal of Alloys and Compounds*, **541**, pp. 380–391, 2012.

[20] Guo, L.J.W., Liang, J., Xue, Q. & Yan F., Tribological behavior of plasma electrolytic oxidation coating on magnesium alloy with oil lubrication at elevated temperatures. *Journal of Alloys and Compounds*, **481**, pp 903–909, 2009.

[21] Hussein, R.O., Nie, X. & Northwood, D.O., Plasma electrolytic oxidation (PEO) coatings on magnesium alloy AZ91D for improved corrosion and wear resistance. *CAMS2014 Conference,* Combined Australian Materials Societies, Sydney, Australia, paper V1.3, November 2014.

Corrosion resistance of Au/Ni thin film-coated stainless steel used as a polymer electrolyte fuel-cell separator

Y. Kimura & M. Hirano
Kogakuin University, Japan

Abstract

Recently, as a new clean and renewable energy, much attention has been paid to the polymer electrolyte fuel cell (PEFC) because of energy and environmental problems, such as depletion of petroleum and global warming. PEFC has the characteristics of a simple structure, is easy to maintain and has a low cost. One of the parts of the PEFC is called the 'separator'; metal separators have a major issue regarding compatibility with corrosion resistance and electrical conductivity, in addition to low cost. Using current technology, a stainless steel AISI316L with an Au/Ni skin film has here been selected. Therefore, there is a need to make a thinner skin film, especially an Au top film for suppressing the PEFC total cost. When thinking about skin film, defects in the Au top film increase gradually as the thickness of Au film decreases. As a result, extensive corrosion is generated due to defects in the Au/Ni skin film. There are few reports evaluating the correlation between the corrosion characteristics of stainless steel with an Au/Ni skin film under a PEFC environment and existing defects in the Au/Ni skin film. Therefore, in this study, the pinhole defect ratio especially in Au film was evaluated using the critical passivation current density (CPCD) method. Au/Ni double layered coating was formed on AISI316L stainless steel employing the sputtering method because of its high adhesion strength. As the quasi-PEFC environment, 1 M H_2SO_4 aqueous solution was selected and then the pinhole defect ratio was evaluated by the CPCD method. As a result, the pinhole defect area ratio of 80 nm thick Au film was obtained to be about 1.1% in 1 M H_2SO_4 aqueous solution.
Keywords: PEFC separator, AISI316L stainless steel, Au/Ni skin film, corrosion, pinhole defect area ratio evaluation, CPCD method.

1 Introduction

Recently, as a new clean and renewable energy, much attention has been paid to the polymer electrolyte fuel cell (PEFC) because of energy and environmental problems, such as depletion of petroleum and global warming. PEFC has the characteristics of a simple structure, is easy to maintain and has a low cost. One of the parts of PEFC is called "separator"; metal separators have a major issue regarding compatibility with corrosion resistance and electrical conductivity, in addition to low cost [1, 2]. Using current technology, stainless steel AISI316L with Au/Ni skin film has here been selected. Therefore, there is a need to form a thinner skin film, especially Au top film, for suppressing the PEFC total cost [3]. When thinking about a skin film, defects in Au top film increase gradually as the thickness of the Au film decreases. As a result, extensive corrosion is generated due to defects in Au/Ni skin film. There are few reports evaluating the correlation between corrosion characteristics of stainless steel with Au/Ni skin film under PEFC environment and existing defects in Au/Ni skin film. Therefore, in this study, the pinhole defect ratio, especially in Au film, was evaluated using the critical passivation current density (CPCD) method [4–7]. Also, the corrosion resistance of Au/Ni thin film-coated stainless steel under quasi-PEFC environment was investigated.

2 Experimental procedures

2.1 Pinhole defect area ratio in Au skin film

The substrate material used in this study was a pure Ni plate specimen with a size of $11 \times 11 \times 1$ mm^3. The surface of the substrate was polished until #2000 emery paper and then given mirror finish by buff polishing. For forming Au films with four different kinds of thickness, from 20 nm to 120 nm, a DC sputtering system (L-332S-FH, Canon ANELVA) was employed (Table 1) [8]. The pinhole defect ratio, especially in Au film, was evaluated using the CPCD method. This method is based on the critical passivation current density, just before the corrosion state shifts from active to passive with the potential sweep toward the noble direction, being proportional to the ratio of the pinhole defect area through the coating film.

First, to determine the optimal concentration of H_2SO_4 in an aqueous solution for obtaining larger current density data, solutions from 1 M to 5 M H_2SO_4 were employed and the anodic polarization measurement of substrate Ni specimen was conducted under the conditions shown in Table 2. When evaluating the pinhole defect area ratio of Au coating film, it is necessary to make sure the suppression of exfoliation of formed Au film. The corroded morphologies of the coating film surface were examined in detail by optical microscope, scanning electron microscope (SEM) and atomic force microscope (AFM).

Electrochemical measurement by three-electrode methods was conducted in H_2SO_4 aqueous solution after de-aeration treatment of more than one hour using N_2 gas.

Table 1: Details of specimen used for Au pinhole defect ratio evaluation.

Specimen	Remarks
Substrate: Pure Ni	#2000 emery paper polish and buff finish
Au coating	Au film thickness: 20, 50, 80, 120 nm made by sputtering

Table 2: Conditions of anodic polarization measurement for Au pinhole defect area ratio evaluation.

Potential sweep range	−500 mV ~ 500 mV (SCE)
Sweep rate	60 mV/min
Temperature	303 ± 0.5 K
Solution	1, 3, 5 M H_2SO_4

Then the defect area ratio R [%] of Au thin film was evaluated from the following equation

$$R = \frac{1}{f_s} \times \frac{i_{crit}\ \text{(coating/substrate)}}{i_{crit}\ \text{(substrate)}} \times 100\ (\%) \tag{1}$$

where, i_{crit} (coating/substrate) is the critical passivation current density of Au thin film-coated Ni specimen and i_{crit} (substrate) is that of non-coated substrate Ni specimen. The value of the shape factor of corrosion pit f_s is evaluated to be 2, because the morphology of corrosion pit is expected to be semi-elliptical.

2.2 Corrosion resistance of Au/Ni thin films coated stainless steel under quasi-PEFC environment

For evaluating the corrosion resistance of Au/Ni-coated stainless steel AISI316L under quasi-PEFC environment, AISI316L plate specimens with the chemical composition shown in Table 3 and a size of 11 × 11 × 2.5 mm^3 were selected as substrate material because of having good corrosion resistance, low hydrogen embrittlement susceptibility and actual results used as separator. The surface of the substrate was finished in the same manner as the Ni palate described in the previous section. A Au/Ni double layered coating film was formed under the conditions shown in Table 4. Then, for quasi-PEFC environment, 0.5 M H_2SO_4 aqueous solution at 353 K was employed and a static immersion test of 24 hrs was conducted (test conditions are described in Table 5) using the corrosion cell shown in Fig. 1.

Table 3: Chemical composition of material used in this study (wt. %).

Material	Ni	Fe	C	Cr	Mo
AISI316L	12.2	Bal	0.01	16.3	2.10

Table 4: Details of Au/Ni coated SUS316L specimen for H_2SO_4 corrosion test.

Specimen	Au film thickness
Au / Ni coated AISI316L (Ni thickness: 100 nm)	20, 50, 65, 80 nm

Table 5: Conditions of immersion corrosion test in H_2SO_4 aqueous solution.

Immersion time	24 H
Solution temperature	353±1 K
Solution	0.5 M H_2SO_4 aqueous solution

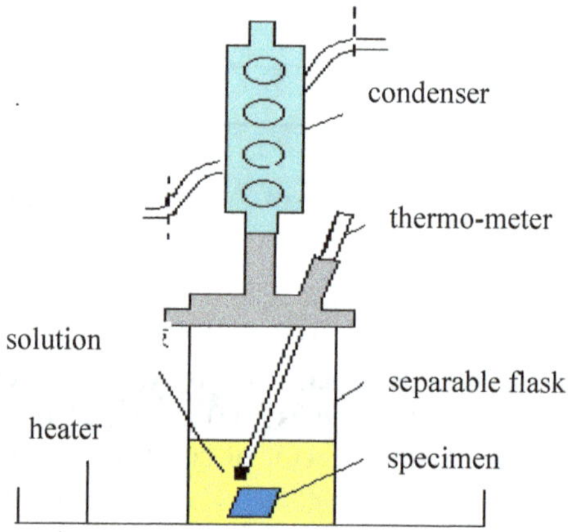

Figure 1: Setup of immersion corrosion test in H_2SO_4 aqueous solution.

2.3 Corrosion resistance of Au/Ni thin films coated stainless steel under quasi-PEFC environment by anodic polarization measurement

Electrochemical measurements were conducted using the three-electrode method (HZ-3000, Hokutodenkou, Japan) composed of the specimen (working electrode), Pt (counter electrode) and the saturated calomel electrode (SCE) (reference electrode). All tests were performed in 0.5 H_2SO_4 aqueous solution of 348 K. Potentio-dynamic corrosion tests were conducted with the sweep rate of 60 mV/min from –200 mV to 1500 mV (vs. SCE) considering the real potential range under actual usage (100–800 mV SCE) [9]. Through considering the separator's actual operating conditions, anodic polarization measurements were conducted under the conditions summarized in Table 6. In this case, specimens with 200 nm thick Ni intermediate film whose Au top film thickness was 50, 65 and 80 nm, were also prepared for understanding the effects of Ni intermediate film thickness upon corrosion resistance of Au/Ni/AISI316L specimen (Table 7).

Table 6: Conditions of anodic polarization measurement.

Potential sweep range	−200 mV–1500 mV (SCE)
Sweep rate	60 mV/min
Solution temperature	348 ± 0.5 K
Solution	0.5 M H_2SO_4 aqueous solution

Table 7: Details of Au/Ni/AISI316L specimen for anodic polarization measurement.

Ni intermediate film thickness	Au top film thickness
100 nm	50, 65, 80, 120 nm
200 nm	50, 65, 80 nm

3 Experimental results and discussions

3.1 Pinhole defect area ratio in Au skin film

First, to determine the optimal concentration of H_2SO_4 in aqueous solution for highly accurate measurements, aqueous solutions from 1 M to 5 M H_2SO_4 were employed and the anodic polarization measurement of substrate Ni specimen was conducted under the conditions described in Table 2. From the obtained results shown in Fig. 2, the 1 M H_2SO_4 aqueous solution was selected as the optimal concentration of H_2SO_4 aqueous solution for the CPCD method because of showing maximum current density within the potential range at which no anodic dissolution of Au was generated.

Then, the pinhole defect area ratio in the Au top skin film was evaluated employing the CPCD method [4–7]. The results obtained by anodic polarization measurements of specimens with different Au top skin film thickness on Ni substrate are shown in Fig. 3. The critical passivation current density, $i_{crit.}$, of the 20 nm thick Au film coated specimen was shown to be at almost the same level as that of the substrate pure Ni specimen. Other specimens with thicker Au coating showed relatively smaller $i_{crit.}$ as the thickness of the film became larger. A drastic suppression in critical passivation current density was recognized in specimens with Au film thickness between 50 nm and 80 nm.

In the case of 20 nm thick Au film coated specimen, almost all of the Au film was exfoliated after polarization measurement. Therefore, the Au top film defect area ratio R_{Au} in this case was not evaluated. Then, the defect area ratio R_{Au} with thicker Au film was evaluated using eqn (1) and is shown in Table 8. At the same time, an Au pinhole defect area ratio evaluation was also conducted by microscopic surface observation.

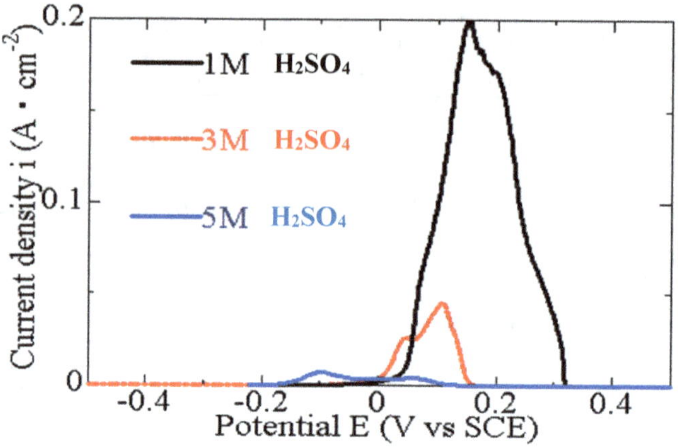

Figure 2: Results of anodic polarization measurements in aqueous solutions with different H_2SO_4 concentration.

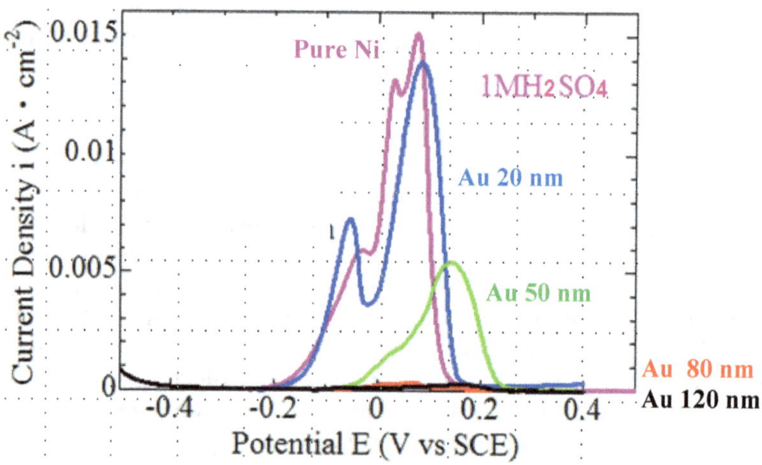

Figure 3: Results of anodic polarization measurements with different Au skin film thickness on Ni substrate.

Table 8: Results of Au film defect area ratio evaluation.

Au film thickness [nm]	Au pinhole defect ratio: R_{Au} [%] (CPCD method)	Au pinhole defect ratio [%] (Surface observation)
20	—	—
50	18. 0	6. 34
80	1. 12	0. 84
120	0. 854	1. 45

Also, in this Table 8, the Au pinhole defect ratio [%] evaluated by surface observation was summarized. From these results, a clear difference in pinhole defect area ratio values was recognized between data obtained by CPCD method and microscopic surface observation for 50 nm thick Au film. In the case of specimens with Au films thicker than 50 nm, there were no large differences between obtained values by both the CPCD method and microscopic surface observation. Therefore, consistency of evaluated results between two different evaluating methods was recognized. For making clear the reason of the difference in evaluated results obtained from 50 nm thick Au film coated specimen, its surface morphology after polarization measurement was investigated by SEM and shown in Fig. 4. In that figure, the surface morphology of the specimen with 50 nm thick Au film clearly shows to be different from that of the specimens with a thicker Au film. In the case of the specimen with 50 nm thick Au film, some trace of Au film swelling [10, 11] was recognized and shown by a circle in Fig. 4(a). In this case, the anodic dissolution of Ni substrate was expected to arise and be extended within the pinhole. For this reason, some excess current density may be detected in this case.

3.2 Corrosion resistance of Au/Ni thin film-coated stainless steel under quasi-PEFC environment

Next, for quasi-PEFC environment, a 0.5 M H_2SO_4 aqueous solution at 353 K was employed and a static immersion test of 24 hrs was conducted using the corrosion cell shown in Fig. 1 under the conditions shown in Table 5. From these tests, the effect of Au film thickness upon corrosion resistance of Au/Ni thin films coated AISI316L stainless steel under quasi-PEFC environment was evaluated. Then, the corrosion rate was calculated and shown in Table 9. As a result, both evaluated corrosion rates were suppressed. From these results, sufficient corrosion resistance of Au/Ni thin films coated AISI316L stainless steel under static quasi-PEFC environment of 0.5 M H_2SO_4 aqueous solution at 353 K was obtained within thin Au top films of thickness in the range of 20 nm and over.

For evaluating the corrosion resistance of Au/Ni thin films coated AISI316L stainless steel under a quasi-PEFC environment, a potentio-dynamic corrosion test was conducted with the sweep rate of 60 mV/min from –200 mV to 1500 mV (vs. SCE) considering the real potential range under actual usage (100–800 mV SCE) [9].

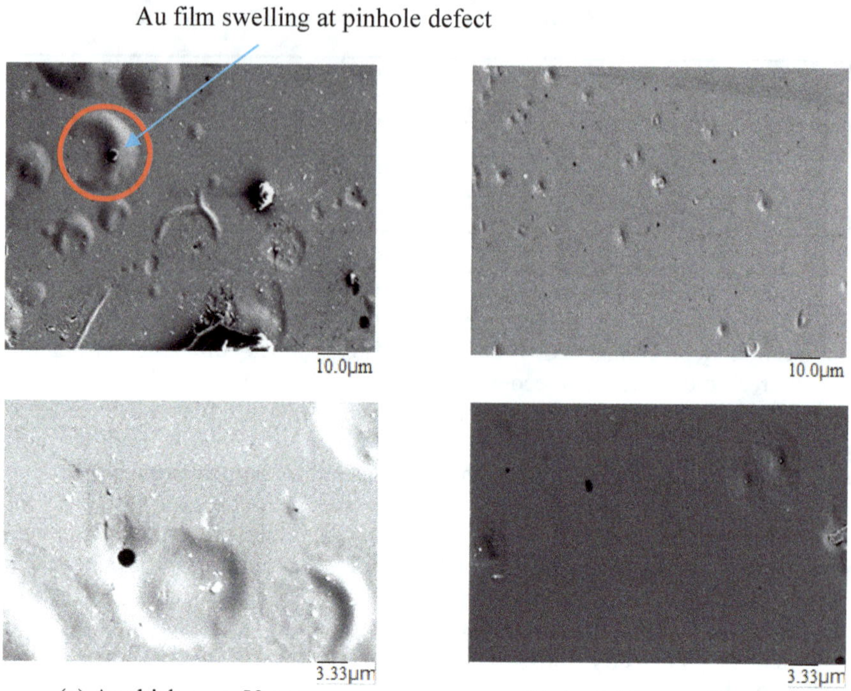

(a) Au thickness: 50 nm

(b) Au thickness: 80 nm

Figure 4: Surface morphology after polarization measurement.

Table 9: Calculated corrosion rate.

Au/Ni thickness (nm)	Corrosion rate (mg/dm^2 day)	Corrosion rate (mm/year)
20/100	9.30×10^{-3}	4.27×10^{-4}
50/100	9.38×10^{-3}	4.29×10^{-4}
65/100	9.72×10^{-3}	4.45×10^{-4}
80/100	15.8×10^{-3}	7.22×10^{-4}

3.3 Corrosion resistance of Au/Ni thin films coated stainless steel under quasi-PEFC environment evaluated by anodic polarization measurement

The obtained anodic polarization curves are shown in Figs. 5 and 6.

In Fig. 5, the results obtained from the Au/Ni/AISI316L specimen with 100 nm thick intermediate Ni film are indicated. In addition, those obtained from specimens whose intermediate Ni film thickness is 200 nm are shown in Fig. 6.

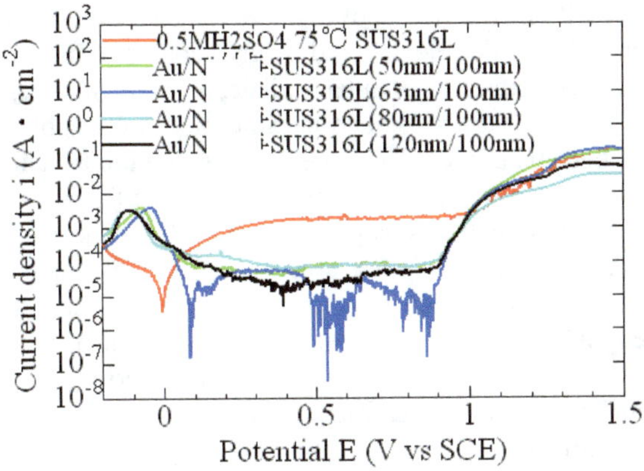

Figure 5: Results of polarization measurement obtained from Au/Ni/AISI316L specimen whose intermediate Ni film thickness is 100 nm.

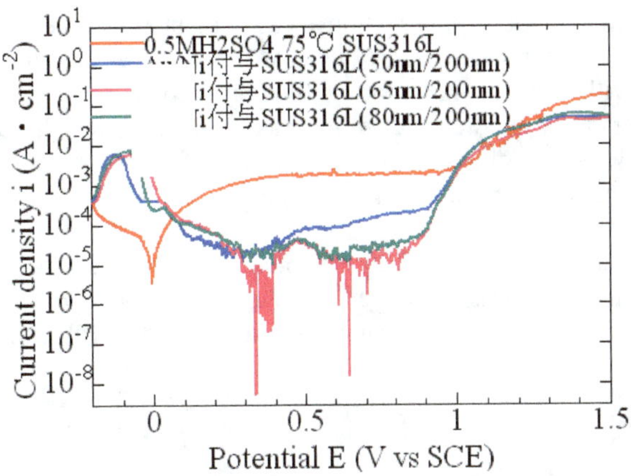

Figure 6: Results of polarization measurement obtained from Au/Ni/AISI316L specimen whose intermediate Ni film thickness is 200 nm.

The current density of Au/Ni thin films coated AISI316L stainless steel was suppressed down to about 1/100 level compared to that detected in bare substrate AISI316L stainless steel.

Also, under the potential range of actual usage (100–800 mV SCE), a double layer-coated AISI316L specimen indicated to stay still in passive state. The current density of Au/Ni thin films coated AISI316L stainless steel whose intermediate Ni film thickness was 200 nm showed furthermore suppressed passivation current density compared with that obtained from the same system specimen with the thinner intermediate Ni thickness of 100 nm. Thus, the improvement in corrosion resistance due to the increase of Au top film thickness was more significant in the case of specimens with a thicker intermediate Ni film.

From the results mentioned above, improvement of corrosion resistance was recognized in all the double layered Au/Ni coated AISI316L stainless steel specimens. The contribution of intermediate Ni film thickness increase to the improvement in corrosion resistance became more noticeable when the Au top film thickness became bigger. The adhesion strength improvement due to Ni film thickness increase may result in promotional effect of dense Au film formation together with Au film thickness increase effect itself. The best corrosion resistance in Au/Ni coated AISI316L stainless steel specimen was obtained when Au/Ni is 65 nm/200 nm. These results may be brought about due to optimization of adhesion strength [11] improvement and fill up of pinhole in Au top film by thicker Ni film formation. For this reason, an increase of intermediate Ni film thickness results in significant corrosion resistance improvement together with suppressing the top Au film thickness.

From these results, it was made clear that the CPCD method was applicable to Au top film defect area ratio evaluation of Au/Ni double layered coated AISI316L stainless steel. Through establishing adhesion strength improvement by intermediate Ni film Au/Ni double layered coated film has sufficient corrosion resistance under quasi-PEFC environment. At the same time, the outlook of specimens after anodic polarization measurement was investigated and typical results of these observations are indicated in Fig. 7. As shown in Fig. 7(a), a specimen with thin top Au coating film resulted in remarkable exfoliation of coated layers after anodic polarization measurement under the conditions shown in Table 6. However, this specimen indicated sufficient corrosion resistance under the conditions of 24 hrs static immersion test in 0.5 M H_2SO_4 aqueous solution of 353 K, as shown in Table 9.

In contrast, in the case of 120 nm thick Au top coated specimen, no damage was recognized in this specimen as shown in Fig. 7(b) even after anodic polarization measurement under the conditions listed in Table 6.

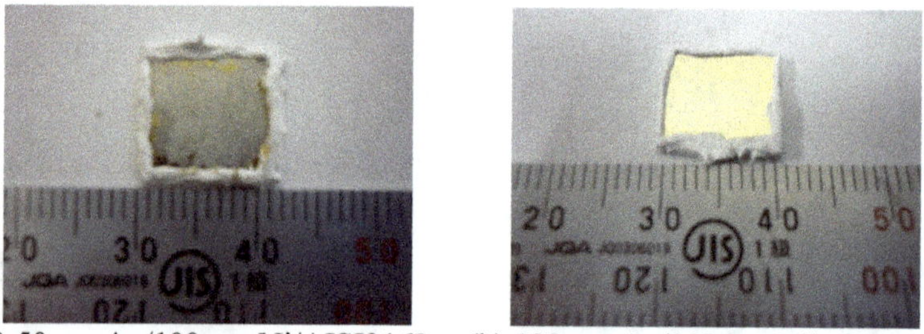

(a) 50 nm Au/100 nm Ni/AISI316L (b) 120 nm Au/100 nm Ni/AISI316L

Figure 7: Outlook of specimen surface after anodic polarization measurement.

4 Conclusions

There are few reports evaluating the correlation between corrosion characteristics of stainless steel with Au/Ni skin film under PEFC environment and existing defects in Au/Ni skin film. Therefore, in this study, the pinhole defect area ratio, especially in the Au top film, was evaluated using the CPCD method. Also, the corrosion resistance of Au/Ni thin films coated AISI316L stainless steel under quasi-PEFC environment was evaluated.

The results obtained are summarized as follows:

1. The CPCD method was applicable to Au top film defect evaluation of Au/Ni double coated AISI316L stainless steel. The pinhole defect area ratio in the Au thin film depends upon the thickness of film itself.
2. The consistency of the evaluated results of Au top film defect between CPCD method and surface morphology observation after polarization measurement was recognized.
3. Improvement of corrosion resistance was recognized in all the double layered Au/Ni coated AISI316L stainless steel. Current density of Au/Ni thin films coated AISI316L stainless steel suppressed down to about 1/100 level compared with that detected in bare substrate AISI316L stainless steel. Also, under the potential range of actual usage, double layered coated AISI316L specimen whose top Au film thickness is larger than 65 nm indicated to stay still in passive state.
4. Best corrosion resistance in Au/Ni double layered coated AISI316L stainless steel specimen was obtained when Au/Ni is 65 nm/200 nm under the condition of polarization measurement. However, a specimen whose top Au film thickness is less than 50 nm indicated sufficient corrosion resistance under the conditions of 24 hrs static immersion into 0.5 M H_2SO_4 aqueous solution of 353 K.

References

[1] Takagi, S., Development of new metallic separator material for PEFC "NANOCLAD". *Electric Furnace Steel*, **77(4)**, pp. 321–326, 2006.

[2] Ueda, M., *et al.* Development of Metallic Separator for Polymer Electrolyte Membrane Fuel Cell (PEFC) (Report 1), pp. 2–6, Kurimoto Technical Report, No. 55, 2006.

[3] Ueda, M., Mori, Y., Hashimoto, M., Yamamuro, S., Tanase, S., Aoi, Y., Iwasa, M. & Sakai, T., Development of carbon coated stainless separators and their polarization performance for polymer electrolyte cell. *Journal of Japan Institute of Metals and Materials*, **71(7)**, pp. 545–552, 2007.

[4] Sugimoto, K., *Proc. of the 95th Corrosion Symposium on Defect Evaluation of Corrosion-resistant Dry Coating Film*, Japan Society of Corrosion Engineering, pp. 1–11, 1993 (in Japanese).

[5] JSME standard S-010, Standard Method for Evaluating the Defects in the Coatings Made by Dry Processing, JSME, 1996.

[6] Nitta, S. & Kimura, Y., Evaluation of defects in ceramic coating films by various electrochemical techniques. *Transactions of Japan Society of Mechanical Engineers*, **61(589)**, pp. 1914–1920, 1995.

[7] Nitta, S. & Kimura, Y., Evaluation of defects in TiN thin films by various electrochemical techniques, *Advances in Surface Engineering, vol. II:*

Process Technology, eds. P.K. Datta & J.S. Burnell-Gray, The Royal Society of Chemistry, pp. 39–47, 1997.

[8] The Surface Finishing Society of Japan, Basics and Applications of PVD and CVD Coatings, Tokyo, Makishoten, 1994.

[9] Chiba, H., Komazaki, S., Watanabe, M. & Kohno, Y., Influence of mechanical and structural factors on durability of PEFC with gel electrolyte membranes, *Electrochemistry, The Electrochemical Society of Japan*, **75(2)**, pp. 238–242, 2007.

[10] Nitta, S. & Kimura, Y., Morphologies and quantitative evaluation by electrochemical methods of the defects in ceramic coating films. *Journal of Material Testing Research Association of Japan*, **42(1)**, pp. 4–14, 1997.

[11] Ohata, M. & Kimura, Y., Correlation between localized corrosion morphology and the state of interface of coated TiN thin film. *WIT Transactions on Engineering Sciences*, **33**, pp. 75–83, 2001.

Field-based prediction of localized anodic dissolution events taking place on ZnAl alloy coatings in the presence of 5% NaCl solution

S. G. R. Brown & N. C. Barnard
School of Engineering, Swansea University, UK

Abstract

An extension to a previously developed numerical model is presented that predicts the sites of localized anodic dissolution on exposed surfaces of ZnAl alloy coating often employed in the protection of steel-strip substrate material. Long-established thermodynamic concepts and galvanic-coupling mechanisms are used to identify sites susceptible to local metallic dissolution. Estimation of metal loss over time is linked to the electrode potentials predicted at the exposed alloy surface. Both alloy composition and the concentration of multiple species in the NaCl solution are considered in determining these electrode potentials. The temporal evolution of the Zn-Al-Fe system – represented using a cellular automaton framework – is predicted via field-based calculations on steady-state voltage and non-steady state diffusion/migration fields. Concentration perturbations in the electrolyte are captured and resultant potential fields are generated using a straight-forward finite difference technique on an irregular-structured computational grid. The influence of the microstructural features in the ZnAl alloy coatings is assessed in terms of metal loss, current density fields and pitted-depth. Simulated results are validated at the meso-scale in comparisons made to experimental observations in accelerated testing of these alloys in concentrated aqueous NaCl solutions. Simulations have been performed that quantitatively assess the localized corrosion at the surface of ZnAl alloy coatings and cut-edge scenarios where Zn, Al and Fe are exposed. The model is extended to predict the effect of altering the processing conditions, i.e. cooling rate and coating thickness.

Keywords: Zn-Al coatings, galvanic corrosion, de-alloying, finite difference.

1 Introduction

In an effort to protect steel from the deleterious effects of corrosion, it has been a long-standing practice to use zinc and zinc-bearing alloys as coatings. The use of Galfan-type coatings to protect sheet is well-established and forms the substrate for much of the organically coated cladding and guttering of modern buildings. Galfan-type coatings are typically applied via hot-dip routes and the coating thickness, alloy composition and bath chemistry are the focus of much attention [1–3]. A further attempt to model the micro-scale corrosion effects experienced by ZnAl alloy coatings applied to a 0.47 mm thickness mild steel substrate is presented here.

Whilst the performance of the zinc-aluminium coatings themselves is of interest, the performance of these coating when both coating and steel substrate is exposed, i.e. at a cut-edge, is of primary concern since the coating surface is protected by an organic coating layer. During cooling of the coating from the ZnAl hot-dip bath, at ~4.5wt.% Al, primary zinc dendrites solidify within a predominantly lamellar ZnAl (~5wt.% Al) eutectic. During subsequent accelerated aqueous corrosion testing it is the zinc from the dendritic phase that undergoes anodic dissolution via the equation

$$Zn_{(s)} \rightarrow Zn^{2+}_{(aq)} + 2e^- \tag{1}$$

In the case of the surface exposure of the metallic coating layer, it is the eutectic phase that acts exclusively as the cathodic surface, whereas in the case of cut-edge exposure, the coupled steel acts cathodically with respect to zinc. In aerated solutions at near-neutral pH, these represent the sites for the supporting oxygen reduction reaction:

$$O_2 + 2H_2O + 4e^- \rightarrow 4OH^-_{(aq)} \tag{2}$$

During the processes described by eqns (1) and (2), dissolution from the coating layer gives rise to the deposition of a hydroxide layer; impeding the cathodic oxygen reduction reaction [4]. This reaction involves the rapid hydrolysis of water via the equations

$$Zn^{2+} + H_2O \rightarrow ZnOH^+ + H^+ \tag{3}$$

$$ZnOH^+ + H_2O \rightarrow Zn(OH)_2 + H^+ \tag{4}$$

In order to predict the localized degradation experienced by these steel coatings in the surface and cut-edge situations, eqns (1)–(4) need to be adequately described by a model. Here, the evolution and movement of anions and cations, in addition to the effect of electrical potential on the kinetics of these processes, is included.

The model described here has been developed from a metallurgical standpoint, rather than an electrochemical one and originated as a cellular automata (CA) finite difference model. This approach dealt with the evolution of a representative concentration and electrical potential throughout the electrolyte only [5]. The model was able to predict morphological features such as localized corrosion pits and capping but was limited to qualitative simulation. However, the CA method has found alternative applications in the growth of corrosion pits where use of CA and Monte Carlo (MC) has been studied [6], including an associated probability for re-passivation and film formation. The CA approach has also been employed in simulating the simple corrosion process taking place at an electrode [7] and most recently in simulating the initiation and growth of multiple pits [8].

Multiple concentration fields are used here in an attempt to capture the dominant effects brought about by eqns (1) and (2) on the spatially discrete anodic and cathodic areas formed during exposure to 5% NaCl solution. Multi-ion electrolyte simulation has also been documented [9] using a nested radial basis function (RBF) approach to predict concentration profiles around a rotating disk. Here, the evolution of $[M^{n+}]$, $[O_2]$, $[OH^-]$ and $[H^+]$ fields around a planar interface is predicted, governed by corrosion rates determined using data obtained from the rotating disc technique.

Previously [3], quantitative predictions regarding the corrosion experienced at the cut-edge with respect to the cooling rate applied to the Galfan-type coating have been shown to be in agreement with experimental findings [1]. However, the trend in corrosion rate at the surface of the ZnAl coating layer was inconclusive. Refinement of the current model has resulted in the observed trend being replicated in simulations and an overview of the results is given here. As an extension to the work, modelling the cut-edge corrosion behaviour has been extended to predict the corrosion rate experienced by steel that has been coated at different weights in slightly differing compositions [3]. The result of these alterations in processing conditions is a change in primary phase fraction (β-Zn) and dendrite morphology.

The experimental work carried out by Elvins *et al.* [1] and Penney *et al.* [3] has been performed using *in situ* corrosion studies based on the scanning vibrating electrode technique (SVET) [10]. The SVET measures the voltage drop over an amplitude of ~30 μm giving a measure of the vertical current density, J_y ($A \cdot m^{-2}$), 100 μm above a corroding surface. An estimation of the material loss experience can then be made from the current density recorded at discrete time intervals.

2 The model

The algorithm used in the model is described in more detail below. There are effectively four main sections to the algorithm:

1. Set up: geometry and initial values on the computational mesh.
2. Determination of electrical potential throughout the solid electrolyte.
3. Dissolution/deposition.
4. Diffusion/migration of multiple species in electrolyte.

2.1 Initialization

The model uses an irregular structured cell-centred finite difference computational mesh. Grid spacings are typically 1–5 μm for cells in the coating layer and rise to

10-25 μm in steel cells close to the centre of the substrate and electrolyte cells far from the solid-liquid interface. The microstructure of the coating layer is entirely synthetic; however, the geometry of the dendritic phase and volume fraction of this primary phase is based upon metallographic observation.

Once a computational mesh has been set up, all variables are initialized and the simulation begins. Each site on the mesh is described by several variables. For electrolyte sites and partially corroded sites these include concentration, C_n, (individual fields for M^{n+}, dissolved O^2, OH^-, and H^+), volume fraction of corrosion product, *p*, volume fraction of solid material in a cell (to allow cells to corrode gradually), *frac*, and electrical potential, ϕ. Un-corroded solid cells are defined by electrical potential, ϕ, and whether the site is primary Zn, eutectic, inert (polymeric material) or steel substrate. In all cases below, the side faces of the computational mesh possess an insulation boundary condition.

2.2 Electrical potential

The first stage in calculating the electrical potential field, ϕ is to subdivide all elements into 4 groups; electrolyte only elements, electrolyte interface elements, solid interface elements and solid only elements.

Interface elements are any elements where electrolyte and solid materials are in contact, they may be solid elements in contact with electrolyte or vice versa. They will also include all those elements that are partially corroded and contain both solid and electrolyte components (i.e. $0 < frac < 1$). The electrode potential, ϕ_0 (V) at all electrolyte interface elements is calculated (weighted via areas in contact for any given element) using the equation

$$\varphi_0 = E_0 - \frac{RT}{nF}\ln\left(\frac{[Ox]}{[Red]}\right) \tag{5}$$

where E_0 is the standard electrode potential (vs. standard hydrogen electrode) in volts, *R* the molar gas constant ($J \cdot K^{-1} \cdot mol^{-1}$), *n* the electronic charge, *T* the temperature (K) and [*Ox*] and [*Red*] are the concentrations of the oxidized and reduced form of the species ($mol \cdot dm^{-3}$); *F* is Faraday's constant ($C \cdot mol^{-1}$).

These calculated electrolyte interface electrical potentials are then used as Dirichlet boundary conditions to determine the electrical potential field throughout the solid elements (solid interface elements and solid only elements) via the Laplace equation. For solid interface elements (which may be partially electrolyte), the electrical conductivity ($S \cdot m^{-1}$) is calculated via

$$\kappa^* = \left(\kappa_{solid}\right)^{frac}\left(\kappa_{electrolyte}\right)^{1-frac} \tag{6}$$

A precursor step to solving the electrical potentials within the electrolyte is carried out, whereby electro-neutrality is enforced at the solid-electrolyte interface via the adjustment of all solid-containing interface cells. These potentials are held as Dirichlet boundary conditions and the final electrical potential field throughout the electrolyte is calculated using

$$\nabla(\kappa_{electrolyte} \nabla \phi) = 0 \tag{7}$$

2.3 Dissolution/deposition

The amount of dissolution that may occur at interface elements containing solid over the current time increment, Δt (s), is calculated via the equation

$$i = i_0 \left[\exp\left(-\frac{\alpha n F \eta}{RT} \right) - \exp \frac{(1-\alpha) n F \eta}{RT} \right] \tag{8}$$

where i_0 is the exchange current density ($A \cdot m^{-2}$), α is the transfer coefficient and η is the overpotential (V); (ϕ_0-E_0) is predicted using eqn (5); the total dissolution current is defined as $i_{dissolution}$.

Parameters necessary to determine the corrosion rate of these cells are derived from rotating disk experimentation carried out in 5% NaCl. Polarization curves obtained for pure zinc and commercially pure aluminium are shown in Fig. 1. In the case of dissolution of the eutectic phase, two rates are calculated for zinc and aluminium and a harmonic mean is applied to describe the rate of degradation, weighted by volume fractions present in the eutectic phase. However, the amount of corrosion that will occur will depend on the maximum cathodic support that is available, controlled by the diffusion of oxygen to the surface. The availability of oxygen during the current time step, Δt, is determined by the availability of oxygen within one diffusion length (m), of the interface. This total is defined as $i_{capacity}$. If $i_{capacity} < i_{dissolution}$ then the amount of corrosion to occur during the current time increment is scaled down uniformly for each element so that $i_{capacity} = i_{dissolution}$.

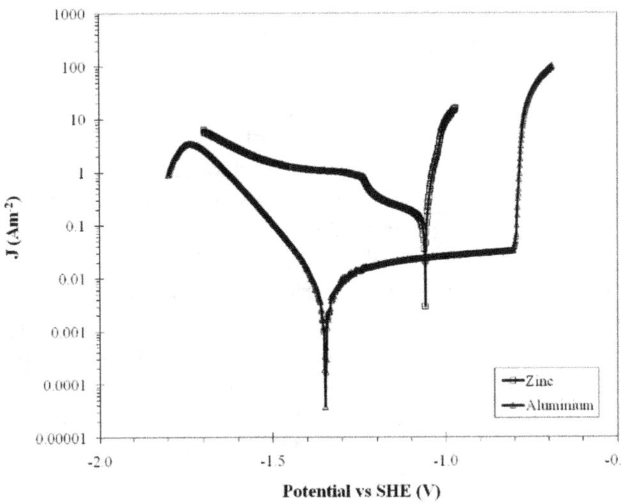

Figure 1: Polarization curves in 0.86M NaCl of a zinc and aluminium electrode.

Each corroding interface site will now have its amount of solid material, *frac*, reduced. Concentration contributions of M^{n+}, O_2 and OH^- from each solid interface cell are distributed geometrically to adjacent electrolyte interface cells.

It is assumed that 5–10% of the volume of material that is lost during dissolution over the time step is re-deposited at the surface of the material as corrosion product. Electrolyte interface sites are selected for corrosion product deposition according to voltage and concentration tolerances and here further ionic concentration contributions arise in M^{n+} and H^+.

2.4 Diffusion

Fickian diffusion (not including interaction terms but including migration effects where appropriate) is carried out for four species within the electrolyte, $[M^{n+}]$, $[O_2]$, $[OH^-]$ and $[H^+]$, according to the standard equation

$$\frac{\partial c_i}{\partial t} = -\nabla\left(D_i \nabla c_i\right) + z_i F \nabla(u_i c_i \nabla \phi) \quad i = 1,4 \tag{9}$$

where D_i is the diffusion coefficient ($m^2 \cdot s^{-1}$) in 0.86M NaCl solution at 298 K, C_i is the molar concentration ($mol \cdot dm^{-3}$) and u is the ionic mobility ($m^2 s^{-1} V^{-1}$) of species i. Side faces of the computational mesh are insulated whereas the top and bottom faces are both fixed at the initial concentration value. For corroding cells at the interface that contain corrosion products, an effective diffusion coefficient, D_{eff}, must be calculated. In the absence of accurate data, this is done using a quadratic form [11] so that the effect of corrosion product at the surface is to increasingly prevent diffusion occurring through that cell. Finally, once the diffusion step is completed for all four species, both [OH-] and [H+] are adjusted to enforce the condition $[OH^-].[H^+] = 10^{-14}$ at all elements containing electrolyte.

3 Results and discussion

During previous investigations, the predicted trend at the cut-edge of Zn – 4.5wt.% Al coating layers that have been cooled at different rates replicated that from SVET experimentation [3]. However, the predicted trend reported in Brown and Barnard [11] did not match that observed. To this end, changes have been made to the solid model that not only better represent the microstructures of the coatings, but also surface depressions around eutectic grain boundaries have been better represented. Again, the cooling rates described here relate to the power output of the coolers that are used after the hot-dip bath and so are expressed as a percentage of the total capability rather than an absolute cooling rate usually associated with microstructural solidification, that is, 55%, 80% and 100%. All cooling rates simulated correspond to 25 μm coating thickness and primary-Zn fraction of 20%.

A further examination of the extended model was performed in the cut-edge situation. This involved simulating the localized effect brought about by galvanic corrosion when the weight of the Galfan-type coating applied is altered. Changes in the coating weight and associated changes in the coating composition modelled are outlined in Table 1.

Table 1: Properties of Galfan-type coatings applied at different weights [3].

Coating Weight (gm^{-2})	Coating thickness (μm)	Al content (%)	β-Zn (%)
150	16.5	4.5	32.6
255	21.5	4.8	22.0
300	39.0	4.9	16.8

3.1 Anodic dissolution

The anodic dissolution modelled using eqns (1) and (8) can be seen to give rise to perturbations in metal cation concentrations from zinc-rich phases present in both surface and cut-edge cases shown in Fig. 2 below. Despite the fact that the eutectic material is permitted to act anodically and cathodically in the case of the cut-edge corrosive case (Fig. 2(a)), perturbations predominantly arise from regions occupied by zinc dendrites owing to the weighting applied in eqn (8) bringing the dissolution rate of the ZnAl eutectic close to that of pure aluminium.

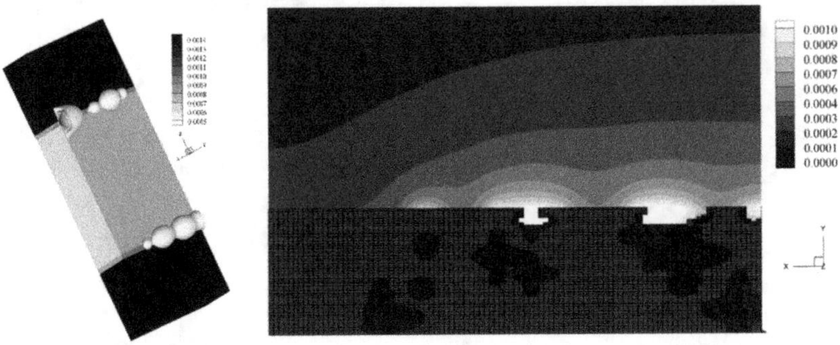

Figure 2: M^{n+} [M] Concentration fields predicted around solid model of 150 gm^{-2} alloy coating after 9.6 hours exposure: (a) Iso-contours with bulk electrolyte removed, and (b) 2D section of ZnAl coating.

The distribution of cation concentration across the cut-edge is in agreement with studies of a Fe/Zn galvanic couple [10]. Diffusional effects are in evidence in Fig. 2(b) where the highest iso-contour of cation concentration can be seen to exist within a pit formed by dissolution from a much smaller dendrite. A reduced zinc loss can be seen to be offset by the more tortuous route from the pit, impeded further still by closer proximity of cells containing corrosion product.

3.2 Cathodic oxygen reduction

The result of the supporting cathodic reaction, eqn (2), on the dissolved oxygen field can be seen for cut-edge case and surface corrosion in Figs. 3(a) and (b), respectively. The lowest values of oxygen can be seen in both to be lowest at sites where entire solid removal has taken place, i.e. the longest diffusion path. This has been distinguished in Fig. 3(b) at A, whereas at B it is the reduction of oxygen that has been the cause of local decrease in oxygen. It can be seen that the cathodic reaction creates a more diffuse field at the cut-edge owing to the presence of the steel substrate. It is only at these sites that the overpotential, η, in eqn (8) will be affected as cathodic and anodic processes are identified on a metallurgical basis.

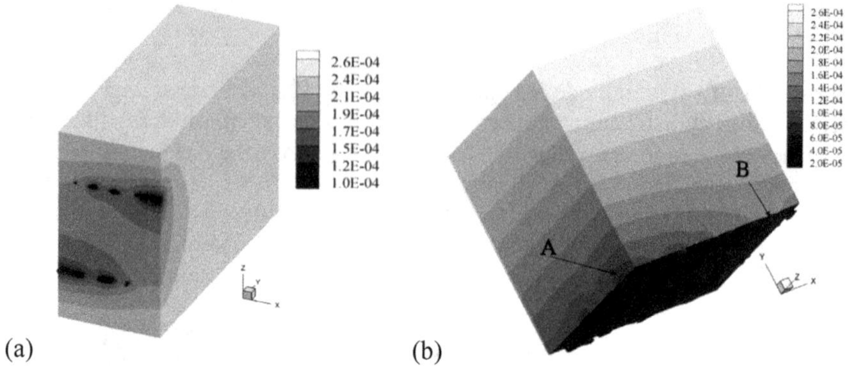

Figure 3: Dissolved oxygen (mol dm^{-3}) in 5% NaCl solution after (a) 12 hours contact with 255 gm^{-2} alloy coating at cut-edge, and (b) 19 hours in contact with surface of 80% cooling output.

3.3 Cooling rate vs. surface corrosion

In a previous attempt to model the surface degradation of Galfan-type coatings, it was assumed, as is the case in cut-edge simulations, that the morphology of the dendritic phase was the dominant factor in the corrosion rate. However, the trend predicted did not correspond to that in experimentation. It is proposed however, the influence of the depressions around the eutectic cells present dictates the localized corrosion effects that ensue. The cell size has been observed to decrease with an increase in cooling rate [1], giving the greatest opportunity for the dendrites to be present at the surface for the faster cooled coating layers. Fig. 4 shows the solid model including the steel substrate at A, primary zinc and eutectic cell depressions with slices showing the electrical potentials into the electrolyte at 30 μm intervals. The effect of the surface depressions is evident, where the incidents of anodic attack do not, as was the case in Brown and Barnard [11], occur evenly over the surface but at positions along depressions. The anodic potentials still correspond to those sites where dendrites make it to the surface. Interestingly, from Fig. 4 the anodic effects present at the points indicated at B do not consistently impact the voltage field in the electrolyte at regions remote from the corroding surface (where an SVET would operate). The authors are thus reluctant to express the corrosion rate as a function of anode lifetime or dendrite removal predicted at 100 μm as in much experimental reporting.

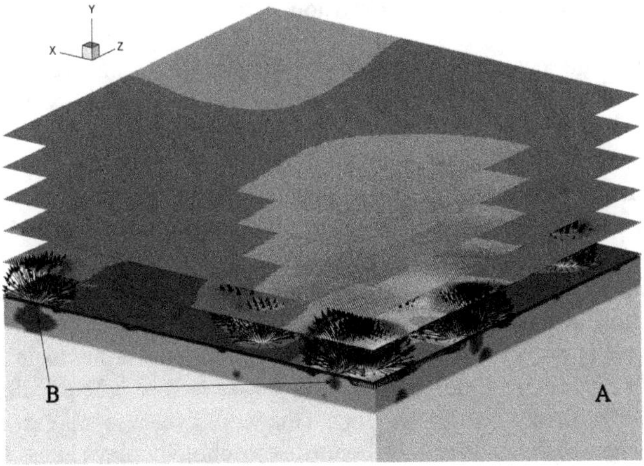

Figure 4: Planes showing anodic (light) to cathodic electrical potentials from the solid electrolyte interface to 150 μm above; current density vectors also shown.

Fig. 5 shows the predicted vertical current densities 100 μm above the solid model. It can be seen from Fig. 4 that this is a degraded representation but nevertheless a source of validation. A coupled approach to SVET experimentation and numerical prediction also considers this [12]. Fig. 5 shows a marked decrease in the frequency of anodic instances from Brown and Barnard [11] and it is apparent that the reduction in eutectic size corresponds to an increase in corrosion rate. The intensity of the anodic events is now more consistent within simulations and observation.

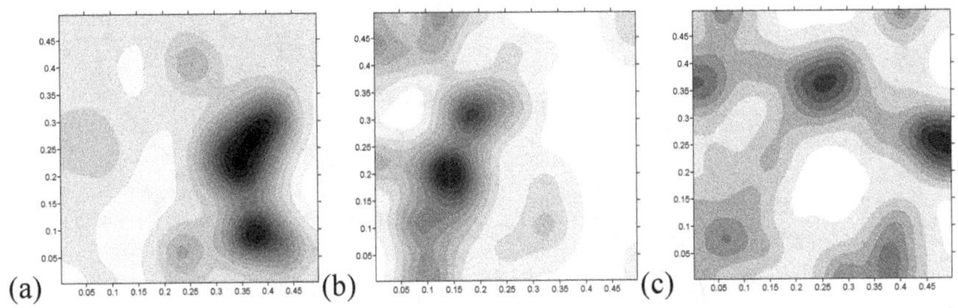

Figure 5: Vertical current density maps 100 μm above ZnAl coating after 12 hours exposure: white $-3.5\ A{\cdot}m^{-2}$ to black $+3.5\ A{\cdot}m^{-2}$. (a) 55%, (b) 80% and (c) 100% cooler output power (length scale mm).

In addition to the increased corrosion rate, decreasing the eutectic cell size coincides with an increase in maximum pit depth: rising from 14.6 μm to 19.2 μm over a 24-hour period – a clear indication of the relative barrier properties.

3.4 Coating weight vs. cut-edge corrosion

In simulating the cut-edge corrosion at differing coating weights, coating thickness, dendrite morphology and volume fraction of the primary phase will all have a significant influence. Fig. 6(a) shows that the material loss during the early stages of accelerated corrosion testing can be almost indistinguishable. After 2 hours simulated exposure to 5% NaCl solution, differences in the coating layers can be seen to take effect. It is the 150 gm^{-2} coating weight that exhibits the greatest acceleration in mass loss rate. As the steel substrate thickness in all cases is equivalent, the greatest anodic activity would be expected at this coating weight since the cathode to anode area is greatest in this case. This is further compounded by the increase in primary phase fraction. In Fig. 6(a), at certain exposure times, the degree of corrosion of the 255 gm^{-2} all can be seen to be equivalent to both 150 gm^{-2} and 300 gm^{-2} coating weights. This would seem to suggest that at the initial solid-electrolyte interface, the dendrite morphology plays a less significant role than volume fraction and coating thickness, but dendrite mass (smaller in lesser coating weights) becomes more significant later on exacerbating the situation in the 150 gm^{-2} coating later in the simulation. Also evident in Fig. 6(a) is the adjustment of the time step ensuring that no solid fraction greater than that present at the interface can transform in 1 time step.

Fig. 6(b) shows the total material removed from the solid model over a simulated 24-hour period. It can be seen that the greatest material loss is experienced by the 150 gm^{-2} coating, and decreases as the coating weight is increased. This is in good agreement with experimentation, although as can be seen in Fig. 6(b) the simulated mass loss is an underestimation, however the experimental findings themselves are calculated from current density maps [3].

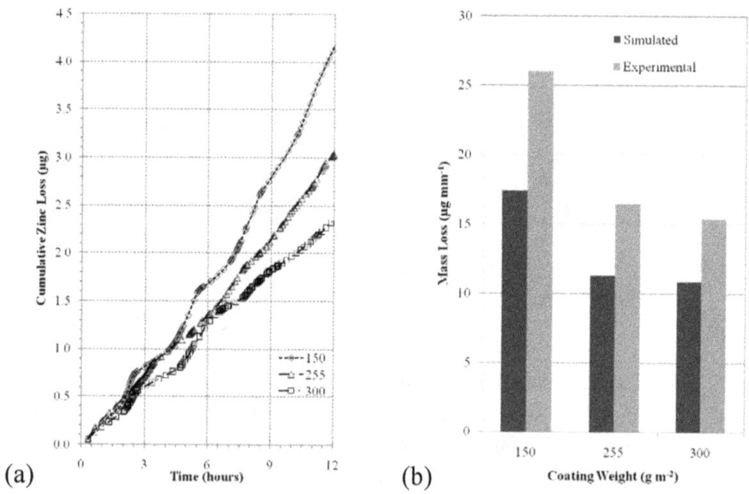

Figure 6: Material loss from the exposed steel cut-edge coated at different weights of ZnAl; (a) cumulative loss over 12 hours, and (b) mass loss during 24 hours' exposure compared with SVET testing [3].

Also in agreement is the changing trend as, unlike in cooling rate where a linear trend is observed, the decrease in mass loss as coating weight is increased is not linear and the difference between 255 gm^{-2} and 300 gm^{-2} is to a much lesser extent than is the case between 150 gm^{-2} and 255 gm^{-2}.

In terms of validating the model the most accessible data is that from SVET testing. Fig. 7 shows the vertical current density 100 μm above the initial solid-electrolyte interface after 12 hours simulated exposure. The form of the current density maps compare favourably to those reported [3]. The intensity of anodic events corresponding to the coating layer can be seen to decrease as the coating weight increases and it would be expected that heavier coating weights would experience lower delamination of applied organic coating layers. Curiously, the emergence of one coating layer offering the bulk of cathodic protection to the steel at heavier coating weights is also predicted by the model, which cannot be attributed to processing conditions since the coating layers are formed using the same parameters.

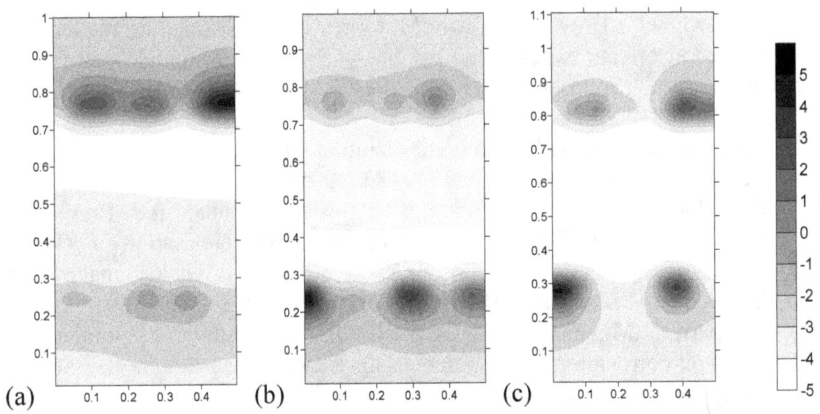

Figure 7: Vertical current density maps, J_y (A·m^{-2}) 100 μm above ZnAl-coated steel after 12 hours exposure; (a) 150 gm^{-2}, (b) 255 gm^{-2} and (c) 300 gm^{-2} (length scale mm).

4 Conclusions

Extensions to a previously described algorithm have resulted in the capability for the prediction of corrosion at both the surface and cut-edge of steels coated with Galfan-type zinc-aluminium coatings. Using well-established electrochemical relationships and straight-forward finite difference techniques, the relationship between corrosion performance and coating weight of ZnAl coatings at an exposed cut-edge has been successfully simulated.

References

[1] Elvins, J., Spittle, J.A. & Worsley, D.A., Microstructural changes in zinc aluminium alloy galvanising as a function of processing parameters and their influence on corrosion. *Corrosion Science*, **47(11)**, pp. 2740–2759, 2005.

[2] Elvins, J., Spittle, J.A. & Worsley, D.A., The effect of low level magnesium additions on the microstructure and corrosion resistance of zinc aluminum alloy galvanized steels. *Proc. of the 7th International Conference on Zinc and Zinc Alloy Coated Steel Sheet*, Osaka, Japan: Iron and Steel Institute of Japan, pp. 653–658, 2007.

[3] Penney, D.J., Sullivan, J.H. & Worsley, D.A., Investigation into the effects of metallic coating thickness on the corrosion properties of Zn–Al alloy galvanising coatings. *Corrosion Science*, **49(3)**, pp. 1321–1339, 2007.

[4] Worsley, D.A., McMurray, H.N. & Belghazi, A., Determination of localised corrosion mechanisms using a scanning vibrating reference electrode technique. *Chemical Communications*, **24**, pp. 2369–2370, 1997.

[5] Barnard, N.C., Brown, S.G.R. & McMurray, H.N., Modelling the localized corrosion effects experienced by zinc–4.5wt.% aluminium steel coatings in 5% NaCl solution. *WIT Transactions on Engineering Sciences*, **48**, pp. 99–108, 2005.

[6] Malki, B. & Baroux, B., Computer simulation of the corrosion pit growth. *Corrosion Science*, **47(1)**, pp. 171–182, 2005.

[7] Córdoba-Torres, P., Nogueira, R.P., de Miranda, L., Brenig, L., Wallenborn, J. & Fairén, V., Cellular automaton simulation of a simple corrosion mechanism: mesoscopic heterogeneity versus macroscopic homogeneity. *Electrochimica Acta*, **46(19)**, pp. 2975–2989, 2001.

[8] Pidaparti, R.M., Fang, L. & Palakal, M.J., Computational simulation of multi-pit corrosion process in materials. *Computational Materials Science*, **41(3)**, pp. 255–265, 2008.

[9] La Rocca A. & Power H., Free mesh radial basis function collocation approach for the numerical solution of system of multi-ion electrolytes. *International Journal for Numerical Methods in Engineering*, **64(13)**, pp. 1699–1734, 2005.

[10] Souto, R.M., González-García, Y., Bastos, A.C. & Simões, A.M., Investigating corrosion processes in the micrometric range: A SVET study of the galvanic corrosion of zinc coupled with iron. *Corrosion Science*, **49(12)**, pp. 4568–4580, 2007.

[11] Brown, S.G.R. & Barnard, N.C., Micro-scale computer modelling of the relationship between metallurgical microstructure and localized corrosion effects. *WIT Transactions on Engineering Sciences*, **54**, pp. 3–12, 2007.

[12] Thébault, F., Vuillemin, B., Oltra, R., Ogle, K. & Allely C., Investigation of self-healing mechanism on galvanized steels cut edges by coupling SVET and numerical modelling. *Electrochimica Acta*, **53(16)**, pp. 5226–5234, 2008.

Improving corrosion performance by surface patterning

M. Bigdeli Karimi, V. Stoilov & D. O. Northwood
University of Windsor, Canada

Abstract

Based on the idea that hydrophobic (low or non-wettable) surfaces can decrease the contact area between a corrosive solution and a surface, thereby potentially rendering the material more corrosion resistant, the effect of surface patterning on the corrosion behaviour of nickel was investigated. The surface patterning consisted of an array of holes of various diameters (D) and inter-hole spacings (L) that were produced by a laser ablation process. The corrosion behaviour of the patterned surfaces was studied using a potentiodynamic polarization method in a 0.5M H_2SO_4 electrolyte and compared with that of a polished reference sample. Following the potentiodynamic polarization corrosion test cycle, the corroded surfaces were examined using scanning electron microscopy (SEM) for morphological features and white light interferometry (WLI) to determine the surface roughness. The changes in surface morphology were related to the corrosion behaviour. A relationship was found between D, L and the corrosion current density (I_{corr}), whereby the higher the $(D/L)^2$ ratio, the higher the I_{corr} value. The corrosion potential (E_{corr}) of all surface patterned samples was lower (less noble) than that of the reference sample in all tests.
Keywords: corrosion resistance, surface patterning, hydrophobicity, laser ablation.

1 Introduction

Wettability of solid surfaces is an important property and depends both on the surface chemistry and on the surface topology. Recently, hydrophobic and superhydrophobic surfaces (water contact angles larger than 150°) have received a lot of attention, due to their important applications ranging from self-cleaning materials and in-flight icephobic treatment [1] to microfluidic devices. The archetype superhydrophobic surface is that of the lotus leaf, on which a water droplet apparently forms a sphere, reducing the area of contact. This behavior, known as the lotus-leaf-effect, is found to be a result of the hierarchically patterned structure. Effectively the fluid cannot penetrate the air gaps between the patterned pillars on the surface, and therefore it forms a heterogeneously wetted contact surface of alternating liquid/solid contacts and air pockets. As a result the overall solid liquid contact area is significantly reduced [2–5].

In the present work, surface patterning is employed to create heterogeneous wetting on metallic surfaces. It is shown that heterogamous wetting in metallic surfaces leads to reduction in corrosion rates and increase in corrosion resistance.

2 Experimental procedures

2.1 Specimen preparation

Pure nickel (99.7 Wt.%) was selected as a model metal. Samples of 1.5 cm × 1.5 cm size were polished to a standard finish with a roughness that did not exceed 50 nm. Laser ablation with a single pulse copper bromide (CuBr) metal vapour laser was used to create special surface textures. During laser ablation, nitrogen (N_2) was blown to protect the surfaces from oxidation and clean any debris. The pulse duration was selected at 30 ns. Different hole/pattern sizes were achieved with different laser output power within the range of 20–80 W. Fig. 1 shows a schematic of the applied pattern. The diameter of the holes and the inter-hole spacing were varied to obtain different textures. Table 1 presents the selected hole size and the inter-hole spacing. The distances between the holes are labeled L1, L2 and L3, where as the D1, D2, D3 etc are the hole diameters.

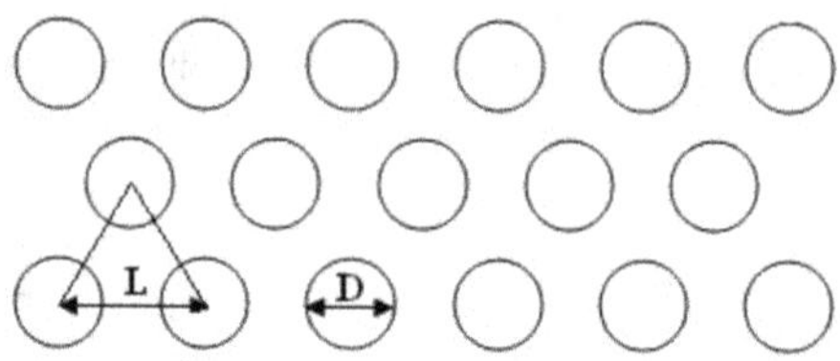

Figure 1: Schematic presentation of the proposed surface texture; D relates to the hole diameter and L defines the inter-hole spacing.

Table 1: The hole sizes and inter-hole spacing.

Hole diameter (D) µm	**Inter-hole spacing L_1, µm**	**Inter-hole spacing L_2, µm**	**Inter-hole spacing L_3, µm**
5	5	10	20
10	10	20	30
20	20	30	40

For easy identification of the samples, the following sample labeling system has been adopted: DxLy, where x is the diameter of the hole in μm and the y is the inter-hole spacing in μm.

2.2 Corrosion tests

The corrosion resistance of the samples was determined using a potentiodynamic polarization corrosion test, wherein the samples were immersed into a 0.5M H_2SO_4 solution at room temperature (24°C). A conventional three-electrode system was used in which a standard calomel electrode (SCE) served as the reference electrode and a platinum electrode as a cathode electrode. In the tests, the applied potentials were in the range of –0.7 to 1.5 V (with respect to SCE) with a scan rate of 1.0 mV/s. Corrosion rates were calculated in terms of the corrosion current density, I_{corr}, by using linear polarization resistance (LPR) curves. The relationship between I_{corr} and the polarization resistance, R_p, is obtained from the Stearn–Geary equation [6]:

$$R_p = \frac{\Delta E}{\Delta i_{app}} = \frac{\beta_a \beta_c}{2.3 I_{corr} (\beta_a + \beta_c)} \tag{1}$$

where β_a and β_c are the Tafel slopes of the anodic and cathodic reactions, respectively, I_{corr} is the corrosion current density and the ratio ($\Delta E/\Delta i_{app}$) is the polarization resistance. After the corrosion test, the sample surfaces were examined using JEOL 5800 scanning electron microscopy (SEM), and energy-dispersive X-ray spectrometry (EDS).

3 Results and discussion

3.1 Corrosion tests

The electrochemical characteristics of the patterned Ni samples were investigated by potentiodynamic techniques described in section 2.2. Polarization curves for all patterned samples and the reference sample were analyzed and the corresponding corrosion current density, I_{corr}, and potential, E_{corr}, were obtained. Fig. 2 shows the variation of the I_{corr} for different hole diameters (1/D) in samples with the same ratio of patterned to non-patterned area $(D/L)^2$. Clearly samples with D = 10 μm (1/D = 0.1 μm^{-1}) have the worst corrosion properties compared to any other samples with other pattern density, that is, $(D/L)^2 = 0.25$ and $(D/L)^2 = 1.0$. On the other hand, samples with D = 20 μm (1/D = 0.05 μm^{-1}) show the lowest I_{corr}, and therefore better corrosion resistance at either pattern density. In addition, the corrosion resistance observed in samples with D = 20 μm exceeds the performance of the reference sample (1/D = 0.0 μm^{-1}), which is a significant improvement of the corrosion properties of the metal surface. The observed trends in Fig. 2 suggest that a decrease of the hole size (D) μm towards nano-scale size could decrease the corrosion current density, I_{corr}, even further. Fig. 3 presents the E_{corr} values for samples with a pattern density of $(D/L)^2 = 1$ and $(D/L)^2 = 0.25$. The corrosion potential (E_{corr}) of all surface patterned samples was lower (less noble) than that of the reference sample (1/D = 0.0 μm^{-1}) in all tests.

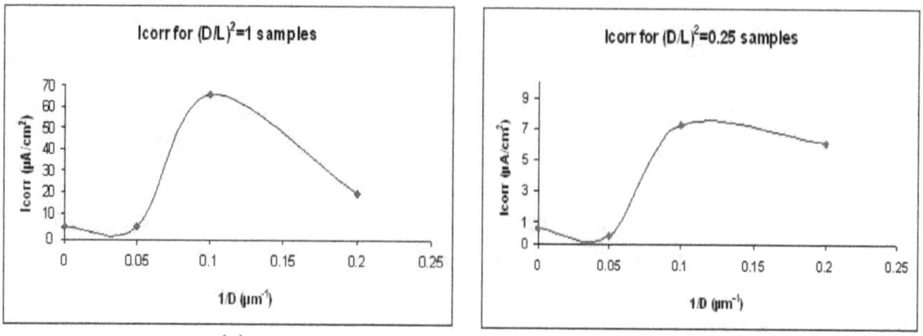

(a) (b)

Figure 2: I_{corr} values versus inverse hole diameters (1/D) for all samples with $(D/L)^2 = 1$ and $(D/L)^2 = 0.25$ after the corrosion test.

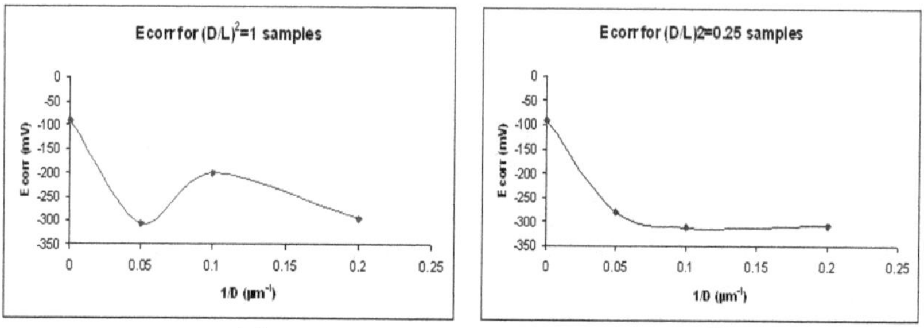

(a) (b)

Figure 3: E_{corr} values versus inverse hole diameters (1/D) for all samples with $(D/L)^2 = 1$ and $(D/L)^2 = 0.25$ after the corrosion test.

3.2 SEM images after corrosion tests

The patterned samples were examined by SEM before and after the above corrosion tests (see Figs. 4–6). The original REF sample showed some severe, localized corrosion. It is evident that those areas were formed by coalescence of small pits.

Sample D10L30 had a severely corroded appearance (Figs. 5a and 5b). According to I_{corr} of this sample, there is a good agreement between corrosion current density and the surface condition.

Sample D20L40 exhibited the lowest amount of patterning degradation. Comparing the original sample before and after the corrosion test there was no obvious change of the surface pattern (Figs. 6(a) and 6(b)). The measured I_{corr} for this sample was the lowest of all samples including reference and patterned. No surface damaged was detected also in samples D20L20 and D20L40. Clearly the observed surface damage in all samples was in good agreement with the trends predicted by the measured I_{corr} values.

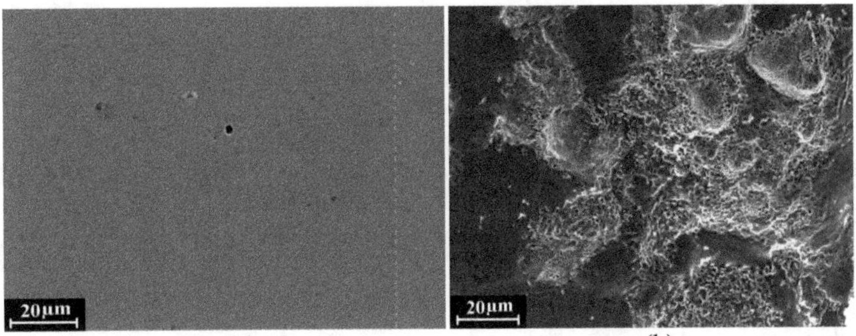

(a) (b)

Figure 4: SEM images of the reference sample (REF) (a) before and (b) after corrosion testing.

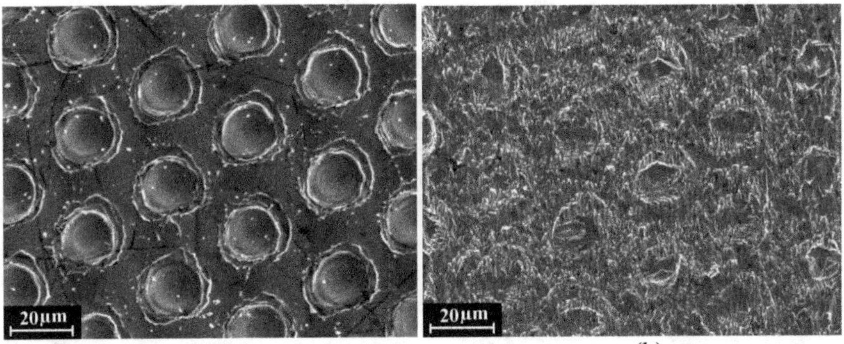

(a) (b)

Figure 5: SEM images of the D10L30 (a) before and (b) after corrosion testing.

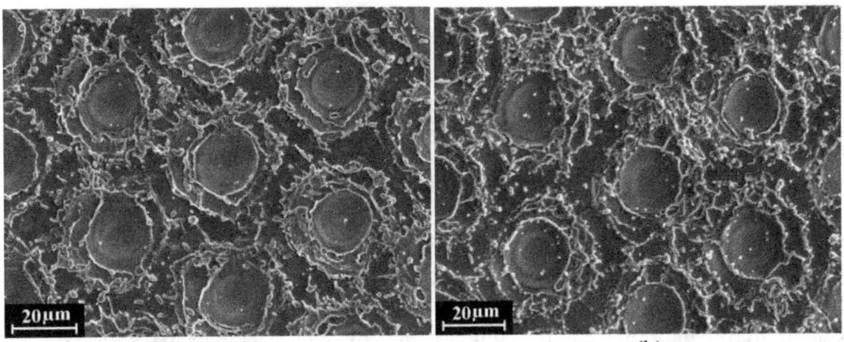

(a) (b)

Figure 6: SEM images of the D20L40 (a) before and (b) after corrosion testing.

3.3 EDS analysis of the patterned samples

In order to investigate the significant improvement of the corrosion resistance in some samples, energy-dispersive X-ray spectrometry (EDS) was used. The main goal was to evaluate the change of the O concentration on the surface of the samples and on the inside of the patterned holes. Fig. 7(a) refers to sample REF before the corrosion test. In the spectrum, three elements, nickel (Ni), oxygen (O) and carbon (C) are observed. The intensity of the most intense peaks (Ni: $L_\alpha = 0.851$ and O: K_α=0.523 keV) was used to calculate the ratio of Ni/O (see Fig. 7(a)).

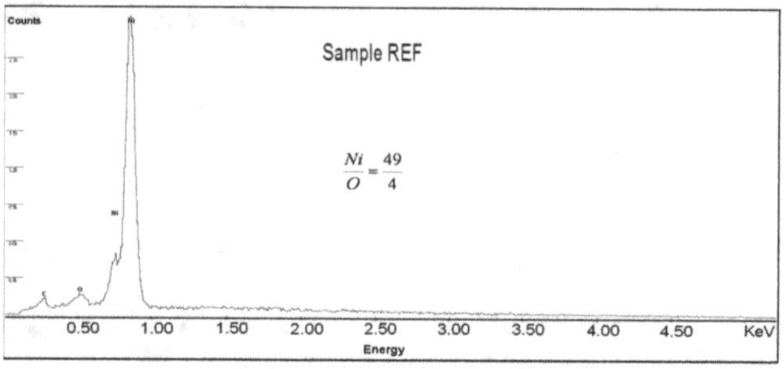

(a)

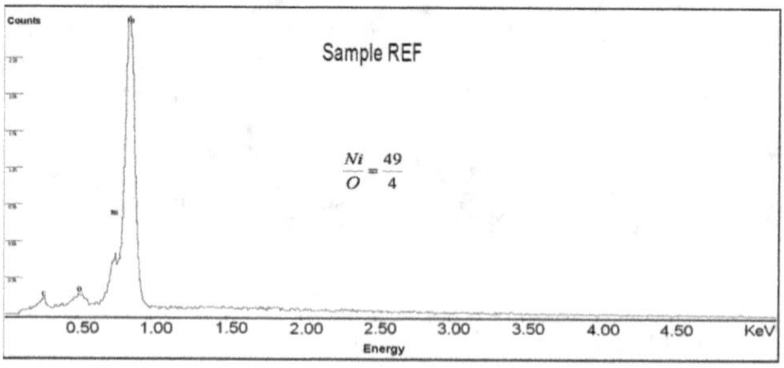

(b)

Figure 7: EDS analysis of the reference sample REF (a) before and (b) after corrosion testing.

Comparing sample REF before the test (Fig. 7(a)) and after the test (Fig. 7(b)) shows that there is virtually no change in the O concentration on the surface. The Ni/O ratio remained at 49/5, indicating that a passive oxide layer was not formed. Similar EDS spectra were obtained for all corrosion damaged samples. For instance, the EDS spectra of sample D5L20 on the surface (Fig. 8(a)) and in the hole (Fig. 8(b)) after corrosion testing do not show any significant difference and the Ni/O ratio is the same as for the polished REF sample. However, the EDS spectra for the patterned samples with the improved corrosion properties (D20L20, D20L30, and D20L40) exhibit different behaviour. The Ni/O ratios for sample D20L40 before and after the corrosion test are shown in Figs. 9(a), 9(b) and

Figs. 10(a), 10(b). The concentration of the O on the surface before and after the corrosion test slightly changed, which is consistent with the other corrosion damaged samples.

However, the bottom of the patterned holes shows significant increase in O concentration (Figs. 8(b) and 10(b)). This clearly indicates that Ni oxides were formed but were not dissolved by the electrolyte. A possible explanation could be that the fluid did not actually reach the bottom of the patterned hole. In other words, in the samples with better corrosion resistance, the contact between the electrolyte and the metal surface is heterogeneous wetting – alternating solid/liquid zones and air pockets.

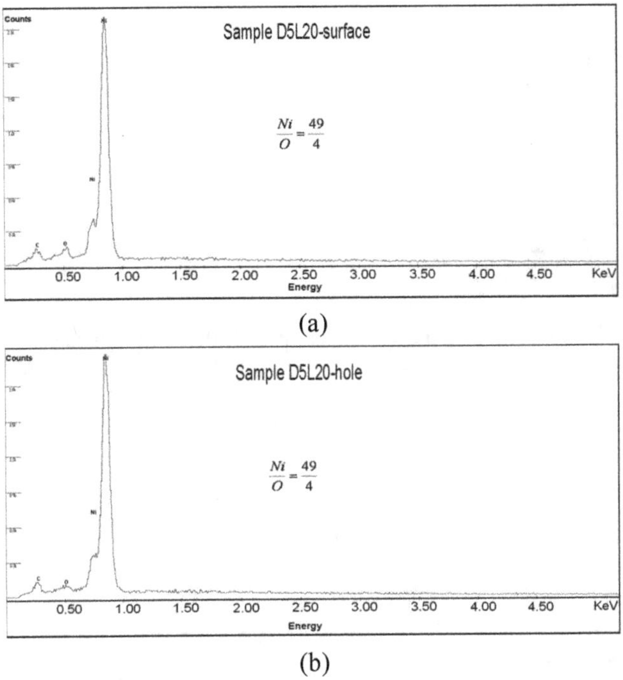

Figure 8: EDS analysis of the sample D5L20 after the corrosion test (a) surface and (b) hole.

4 Conclusions

Surface patterns with different pattern density were successfully created on Ni by laser ablation. The patterns had a distinct effect on the corrosion properties of the metallic surfaces. A specific group of patterns led to significant decrease of the corrosion current density and the corresponding corrosion rate. For this group of patterns, it has been shown that a possible reason for the decrease in corrosion rate is the detected heterogeneous wetting on the patterned surface.

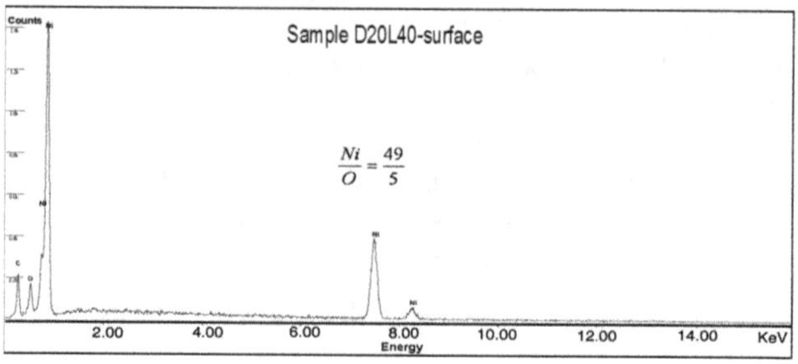

(a)

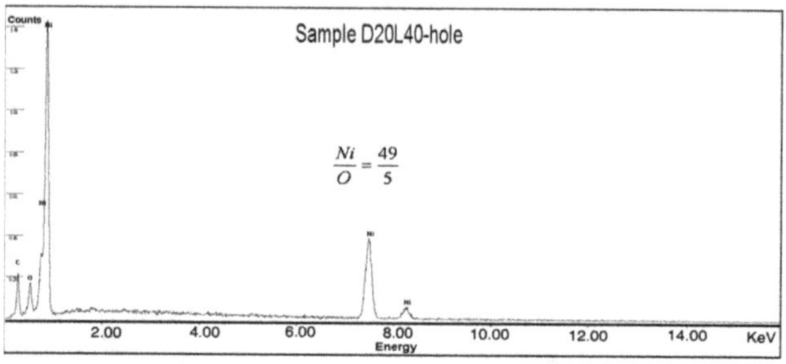

(b)

Figure 9: EDS analysis of the virgin sample D20L40 (a) surface and (b) hole.

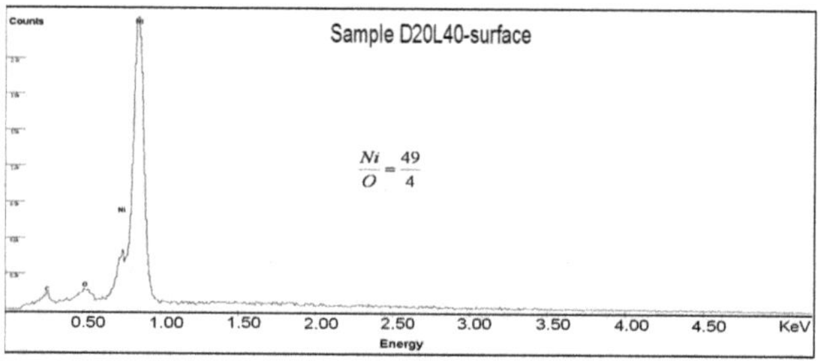

(a)

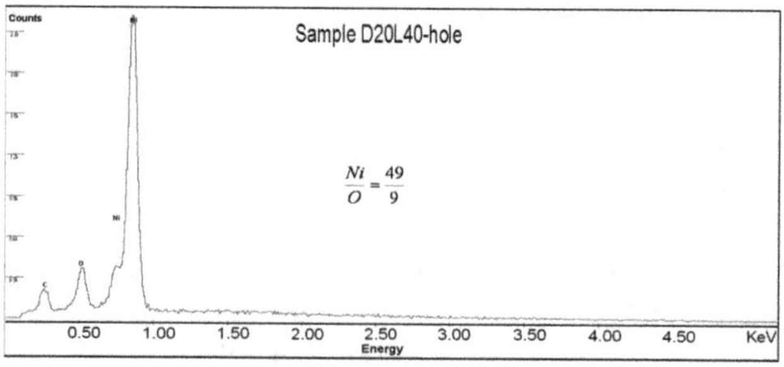

(b)

Figure 10: EDS analysis of the sample D20L40 after the corrosion test (a) surface and (b) hole.

References

[1] Varanasi, K.K., Deng, T., Smith, J. D., Hsu, M. & Bhate, N., Frost formation and ice adhesion on superhydrophobic surfaces. *Applied Physics Letters*, **97(23)**, 10.1063/1.3524513, 2010.

[2] Shafiei, M. & Alpas, A.T., Nanocrystalline nickel films with lotus leaf texture for superhydrophobic and low friction surfaces. *Applied Surface Science*, **256**, pp. 710–719, 2009.

[3] Barthlott, W. & Neinhuis, C., Purity of the sacred lotus, or escape from contamination in biological surfaces. *Planta*, **202**, pp. 1–8, 1997.

[4] Lee, S.J., Huang, C.H., Lai, J.J. & Chen, Y.P., Corrosion-resistant component for PEM fuel cells. *Journal of Power Sources*, **131**, pp. 162–168, 2004.

[5] Liu, T., Yin, Y. & Dong, L., New application of the underwater super-hydrophobic surface in the corrosion protection. *Advanced Materials Research*, **79–82**, pp. 1115–1118, 2009.

[6] Fontana, M.G. & Greene, N.D., *Corrosion Engineering*, 2nd edition, McGraw Hill, 1978.

Self-healing of steel corrosion in a model alkaline medium: electrochemical response and surface analysis

D. A. Koleva[1], J. Hu[2], H. Kolev[3] & K. Van Breugel[1]
[1]Delft University of Technology, The Netherlands
[2]South China University of Technology, PR China
[3]Bulgarian Academy of Sciences, Institute of Catalysis, Bulgaria

Abstract

The application of polymeric nano-particles for corrosion control in reinforced cement-based systems was studied. As an initial investigation, steel electrodes (St37) were tested in modified cement extract (CE), which is a model pore solution of pH 12.9 and chemical composition: 201 mg/l Ca; 3.85 mg/l K; 1.33 mg/l Na; 4 mg/l Al, Fe < 1 mg/l. Corrosion initiating factor was 5% NaCl. PEO_{113}-b-PS_{780} vesicles at 0.0024 wt.% in the CE were employed. These are nano sacs of 220 nm, enclosing a volume with a thin membrane, thus able to effectively entrap water-soluble compounds. Ion binding (bulk liquid medium), steel surface adsorption and/or participation in oxide layers formation were the expected mechanisms of particles' involvement. Both "empty" and "Ca-containing" vesicles were studied to evaluate "barrier" effects only vs. self-healing mechanisms. Increased steel corrosion resistance in the presence of both vesicle types was observed, however, it was more pronounced when Ca-containing ones were involved. Cyclic voltammetry and X-ray photoelectron spectroscopy analysis substantiate the self-healing phenomena, through direct evidence of pitting formation and propagation and consecutive healing (and no further pit formation) when Ca-containing vesicles are present.
Keywords: corrosion, cement extract, vesicles, self-healing.

1 Introduction

The alkaline nature of concrete (pH = 12.6–13.5) generally provides a physicochemical barrier against corrosion of the embedded steel and assures the formation of a stable passive layer on the steel surface [1]. However, corrosion is initiated due to carbonation (pH drop < 9) or chloride contamination (pH drop < 5) [2–8], resulting in corrosion propagation, accelerated damage and rapid failure [9]. Various protection methods and techniques have been investigated and are successfully applied, e.g. protective coatings/sealers [10, 11]; polymer coatings and cathodic protection (CP) (the latter combination leading to increased current demand/de-bonding [12] and possible hydrogen embrittlement or bond-strength degradation [13–18]); concrete re-alkalization [19]; corrosion inhibitors (quickly exhausted and with none or even negative effects on the bulk cement-based matrix [19–21]). It is well known that each method/technique has its limitations and side effects. Moreover, there is no technique that considers both steel reinforcement and cement-based bulk matrix, aiming at improved properties of both materials and superior global performance of reinforced concrete as a system. To this end, in the frame of a novel approach and (initially) within the IOP Self-Healing project SHM08743, executed in TU Delft and financially supported by Agentschap, NL, tailor-made polymeric nano-materials were studied. The aim was to increase steel corrosion resistance via self-healing of corrosion damage, along with improved microstructural characteristics.

As far as the cement-based bulk matrix properties were concerned, the suggested hybrid core-shell nano-particles already proved their positive effect, for example: in the presence of polyethylene oxide $(PEO)_{113}$-b- polystyrene $(PS)_{218}$ micelles (50 nm size) of only 0.025 wt.% per cement weight, the porosity of the micelle-modified matrix was reduced two times, whereas water permeability was reduced by six orders of magnitude; a more uniform distribution of calcium-silicate-hydrate (C-S-H) and enhanced mechanical properties, but undisturbed water balance was evident from nano-indentation tests [22–25]. For establishing the effects on corrosion resistance, the aforementioned micelles, as well as stabilized (kinetically frozen) PEO_{113}-b-PS_{760} vesicles (220 nm size), "empty" and Ca-containing ones, were studied [26–28]. The aim of these studies was to investigate the possibility for maintained positive effects on the cement-based microstructure, along with increased corrosion resistance and self-healing (or repair) of corrosion damage, resulting from Ca-core release. Initial tests were performed in model pore solutions, prior to studies in reinforced mortar and concrete. This paper briefly presents some main outcomes from the investigation on corrosion resistance of St37 steel electrodes in cement extract, modified with the above-defined vesicles [28]. Further, this work substantiates improved corrosion resistance and evidence of self-healing effects through correlating electrochemical performance and steel surface analysis.

2 Materials

2.1 Polymeric vesicles

Vesicles were prepared via dialysis from PEO_{113}-b-PS_{760} block-polymer [24–28]. In order to distinguish between "barrier" effects and "self-healing" due to Ca-release, both "empty" and Ca-containing vesicles were used. Their hydrodynamic radius was approximately 220–250 nm, identified by dynamic light scattering (DLS) measurement and transmission electron microscopy (TEM) (Fig. 1).

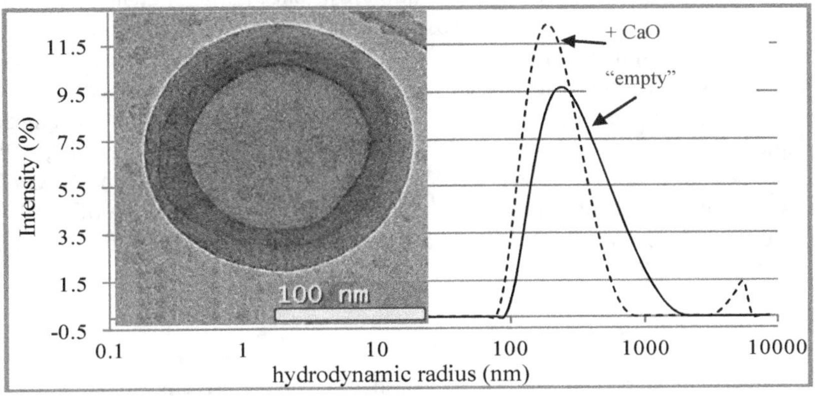

Figure 1: DLS and TEM of vesicles.

2.2 Cement extract

As a model, pore solution was prepared by mixing ordinary Portland cement CEM I 42.5N and tap water in 1:1 ratio; after 24 h rotation, the suspension was filtrated and this solution was used as a model medium. The pH of the cement extract (CE) is 12.6–2.9, the chemical composition, derived via inductive coupled plasma spectrometry (ICP) was: Ca – 201 mg/l; K – 3.85 mg/l; Na – 1.33 mg/l; Al – 4 mg/l and Fe < 1 mg/l; NaCl was corrosion-inducing factor, added in a concentration of 10 g/l to the CE. The aqueous solution of PEO_{113}-b-PS_{760} vesicles was added to CE, thus the vesicles concentration was 0.0024 wt. % in chloride-free and chloride-containing CE.

2.3 Steel electrodes

Low-carbon steel St37 with surface area of 4 cm^2 was used. All electrodes were equally treated (grinding, polishing, acetone cleaning and water rinsing) prior to treatment in the relevant solutions and electrochemical measurements. Three replicates per condition were tested. Sample designation is as follows: group CE – steel in CE only; CEn – steel in chloride-containing CE; CEV and CEVC stand for steel in CE modified with "empty" and "Ca-containing" vesicles, whereas CEVn and CEVCn stand for the respective corroding cells.

3 Methods

3.1 Electrochemical measurements

Autolab-Potentiostat PGSTAT 30, combined with FRA2 module using GPES and FRA interface, was used. Linear polarization resistance (LPR), potentio-dynamic polarization (PDP), impedance spectroscopy (EIS) and cyclic voltammetry (CVA) were employed, using a common 3-electrode cell arrangement (saturated calomel electrode, SCE, as reference electrode). EIS was performed at open circuit potential (OCP) in the frequency range of 50 kHz to 10 mHz by superimposing an ac voltage of 10 mV; LPR used a polarization window of ±20 mV vs. OCP; PDP was in the range of –0.2 V to +1.0 V vs. OCP at a scan rate 0.5 mV/s; CVA scans were in the potential region of −1.4 V to 0.6 V at a scan rate of 5 mV/s (all electrochemical measurements are as generally employed for the systems under study [22–27]).

3.2 Surface analysis

Morphology and composition of the product layers were investigated using scanning electron microscopy (SEM), coupled with energy dispersive X-ray (EDX) analysis as a qualitative and semi quantitative technique, using ESEM Philips XL 30. X-ray photoelectron spectroscopy (XPS) was carried out in the UHV chamber of an electron spectrometer ESCALAB-MkII VG Scientific with a base pressure of about 1×10^{-10} mbar (during the measurement 1×10^{-9} mbar). The photoelectron spectra were obtained using unmonochromatized Al Kα (hν = 1486.6 eV) radiation. Passing through 6 mm slit (entrance/exit) of a hemispherical analyser, electrons with energy 20 eV are detected by a channeltron. The instrumental resolution measured as the full width at a half maximum (fwhm) of the Ag3d5/2, photoelectron peak is about 1 eV. The energy scale is corrected to the C1s (peak maximum at 285 eV) for electrostatic sample charging. The fitting of the recorded XPS spectra was performed, using a symmetrical Gauss–Lorentzian curve fitting after Shirley-type subtraction of the background. Except scanning the steel surface after 7 days, i.e. product layers formed on the steel electrodes in each relevant condition, four-times consecutive mechanical polishing of equal pressure and time duration was performed in depth of the layer in the vacuum chamber of the spectrometer. This technique was chosen to avoid influence on the chemistry of the formed layers, which would otherwise be relevant if Ar sputtering was used. The XPS measurements thus resulted in qualifying and quantifying the composition of the product layers in depth, i.e. for each specimen, four in depth scan results were recorded.

4 Results and discussion

4.1 Electrochemical measurements

4.1.1 OCP evolution and polarization resistance

A general perception is that the evolution of OCP determines the time to corrosion initiation. For reinforced concrete and alkaline medium, as the here used CE, the steel surface is considered to be passive if OCP is equal or more anodic than –270 mV SCE. When chlorides (or other corrosion-inducing factors) are present, passivity is lost and localised corrosion initiated – OCP readings shift to more cathodic (more negative) values in this case. Fig. 2 depicts OCP evolution for the control (non-corroding) cases (dotted notation), Fig. 3 shows these values for the corroding cases. Additionally, Figs. 2 and 3 depict the derived (via linear regression) polarization resistance values (R_p) (R_p in global terms corresponds to corrosion resistance). As can be observed, OCP values fall in the region of passivity and shift towards more noble potentials at the end of the test. These correspond to the gradual increase of R_p values, the highest being for the steel treated in "empty" vesicles containing CE. Since there is no corrosion accelerator in the medium for the control samples, only gradual passivation is expected rather than any pronounced effect of the Ca-rich inner "core" of the Ca-containing vesicles. However, adsorption of both vesicle types on the steel surface and their incorporation in the passive layer could result in possible alterations in compactness and/or speed of layer formation. Obviously, barrier effects in terms of diffusion limitations from the adsorbed "empty" vesicles are relevant for specimen CEV, since OCP values are more cathodic, whereas R_p values are the highest. This does not necessarily mean a better corrosion resistant layer, but possibly limited oxygen access and hence locally impeded growth of a more inhomogeneous layer. In contrast, the CE and CEVC specimens exhibit similar R_p values.

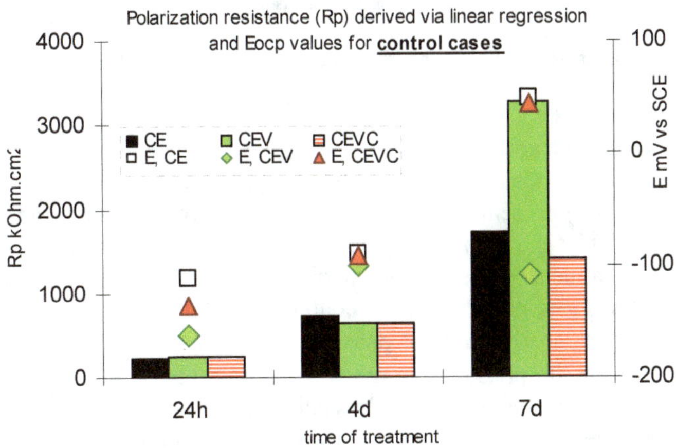

Figure 2: OCP and Rp values for control cases.

Apparently, the Ca-containing vesicles either act as "nucleation sites" on the steel surface, or adsorb on active spots or both and therefore do not result in diffusion limitations as in the case of CEV specimens.

For the corroding cases (Fig. 3), a shift to more active (cathodic) OCP values is as expected and as observed for specimens CEn (steel treated in vesicles-free solutions). This corresponds to the decreasing R_p values with time of treatment. In contrast, specimens CEVn and CEVCn depict OCP values in the passive range at the end of the test (more anodic than –200 mV) and higher R_p values. The influence of Ca-containing vesicles is here already well pronounced: after initially lower values, and even lowest after 4 days treatment, the R_p values for specimens CEVCn depict a significant increase at the end of the test. Since the steel electrodes, treated in CE containing "empty" vesicles do not show this trend (specimens CEVn), the only plausible explanation is that the Ca-containing vesicles exert a "self-repair" or 'self-healing" of the already induced corrosion damage on later stages of treatment. These observations are actually confirmed by EIS, PDP and CVA tests and supported by surface analysis.

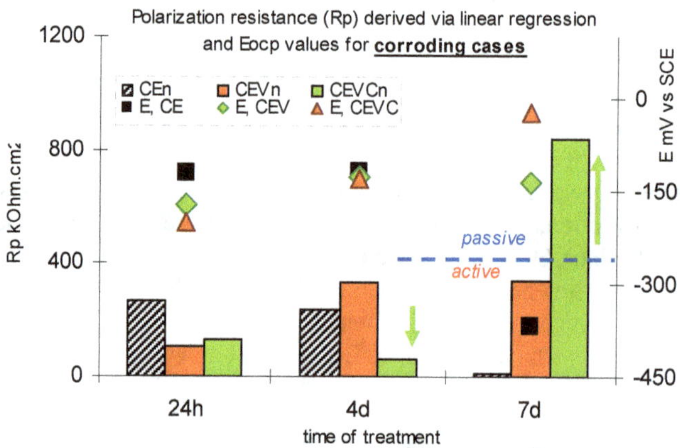

Figure 3: OCP and Rp values for corroding cases.

4.1.2 EIS, PDP and CVA results

The EIS response is not subject to elaborated discussion in this paper, but is presented as a direct visualisation of the electrochemical behaviour of the steel electrodes. Fig. 4(a) presents the response for the control specimens after 7 days. Figs. 4(b)–(d) depict the response of all corroding cases after 24 h (b), 4-day (c) and 7-day (d) treatment. As seen from the plots, the magnitude of impedance $|Z|$ is higher for the steel treated in vesicles containing solutions at all time intervals, accompanied by a mostly constant phase angle of approx. 80°, whereas for the vesicles-free cases the phase angle drops to approx. 65° at the end of the test. The EIS results clearly show the "self-repair" of the product layer on the steel surface when Ca-containing vesicles are involved: specimens CEVCn depict a response with two time constants after 24 h treatment, meaning a pronounced

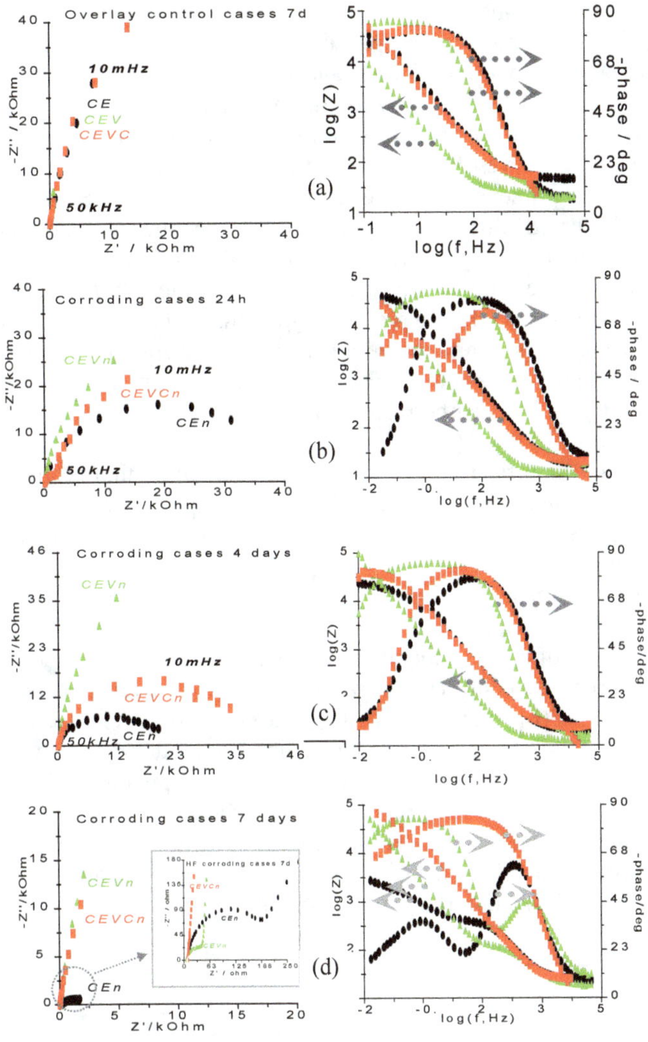

Figure 4: 7-day EIS response for CE, CEV and CEVC (a) and CEn, CEVn, CEVCn after 24 h (b), 4 days (c) and 7 days (d).

contribution of product layer alterations in parallel to charge transfer resistance phenomena. Later on, the magnitude of $|Z|$ slightly drops, whereas the phase angle increases, which can only be attributed to modified product layer morphology and composition, rather than increase in charge transfer. Furthermore, after 7 days' treatment, both magnitude of $|Z|$ and phase angle increase, attributed to improved and more corrosion resistant steel surface (one time constant only at 7 days). The only barrier, rather than "self-healing" effect in specimens CEVn ("empty" vesicles involved) is seen from the gradual decrease of impedance and phase angle and pronounced contribution of a second time constant at the end of the test. These observations are in line with the electrochemical behaviour with external polarization – PDP and CVA response. Fig. 5(a) depicts PDP curves for all corroding cases as an overlay of 4-day and 7-day response; Fig. 5(b) presents the response of specimens CEVCn only for all time intervals.

After 4 days of treatment, the corrosion current density for specimens CEn and CEVn are similar (curves 1 and 2, Fig. 5(a)), the anodic current with external polarization for specimens CEVn is even higher than that for CE. This proves the previously discussed diffusion limitations in specimen CEVn and barrier effects of the "empty" vesicles, which do not result in a stable and homogeneous product layer, since anodic currents are not limited but increasing with polarization. Specimens CEVCn present slightly higher corrosion current density, but a significant anodic limitation with external polarization is relevant (curve 3, Fig. 5(a)). The latter observation is in line with the previously discussed attempts to "repair" initially induced corrosion damage. After 7-day treatment, the CEn specimens (curve 1a, Fig. 5(a)) exhibit the highest anodic and corrosion currents (as expected) and significant shift of corrosion potentials to more cathodic values. In contrast, specimens CEVCn (curve 3a, Fig. 5(a)) present a drop of corrosion and anodic current densities, a shift to more noble corrosion potentials and maintained anodic currents limitations.

After 7-day treatment, specimens CEVn present similar to the 4-day behaviour of active dissolution with external polarization (curves 2a and 2, Fig. 5(a)). The self-healing effect of the "Ca-containing" vesicles, present in the medium of CE+ NaCl are well seen if a comparison is made of the response for specimens CEVCn with time – Fig. 5(b): after initially low currents (curve 1, 3 h), corrosion is initiated (curve 2, 24 h); the process is maintained until after 4 days, when slightly higher corrosion current is recorded, most likely corresponding to product layer transformations and Ca-content release and/or nucleation sites effect due to the "Ca-containing" vesicles on the steel surface. The result with prolonged treatment is product layer "repair", evident by a significant drop of corrosion and anodic currents (curve 4, 7 d) and further anodic limitations.

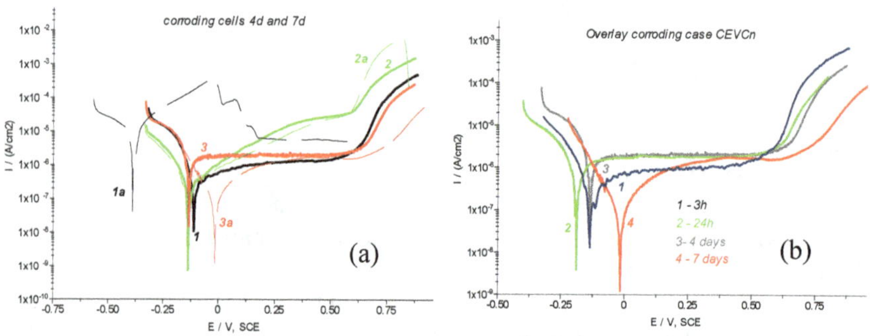

Figure 5: (a) PDP curves for specimens: CEn at 4 d and 7 d (curves 1 and 1a); CEVn at 4 d and 7 d (curves 2 and 2a); CEVCn at 4 d and 7 d (curves 3 and 3a); (b) PDP response for specimen CEVCn at 3 h, 24 h, 4 d and 7 d (curves 1, 2, 3 and 4 respectively).

CVA tests strongly support all above observations. Fig. 6 presents the 1st, 10th and 20th scans, recorded after treatment in the relevant solutions. As seen from the plots, pitting initiates for specimens CEn and CEVn (at approx. –325 mV) with the 1st scan, whereas no pitting was observed for specimens CEVCn, the latter exhibiting the lowest current densities in the reverse scan and most noble re-passivation potential (Fig. 6). After 10 scans, all specimens, including CEVCn show pitting formation at around –500 mV (Fig. 6) and shift towards higher currents in both anodic and cathodic sweeps. Finally, within 20 scans (Fig. 6), the self-healing effect of the Ca-containing vesicles is evident from the recorded significantly low currents for specimens CEVCn in both anodic and cathodic sweeps, accompanied by the lack of pitting formation. In contrast, for specimens CEn and CEVn, pitting propagation is still relevant, accompanied by already largely increased currents in both anodic and cathodic sweeps.

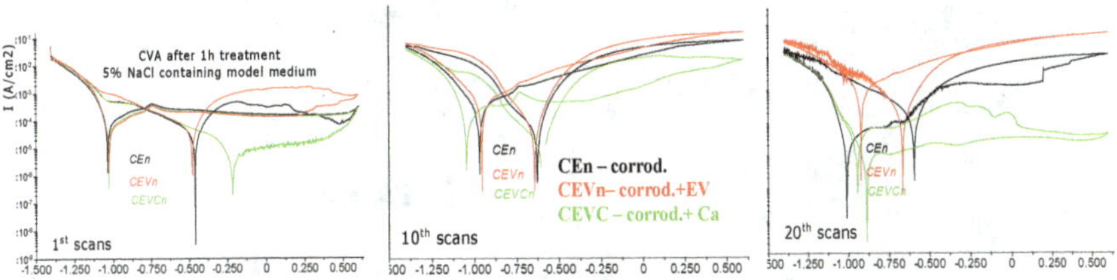

Figure 6: CVA scans for specimens CEn, CEVn and CEVCn.

4.2 Surface analysis

The previously discussed possible characteristics, with respect to electrochemical response, of the product layers formed on the steel surface were actually as observed via ESEM, EDX and XPS analysis. Fig. 7 presents ESEM micrographs and the relevant EDX analysis for the corroding cases (b–d); for a comparison, the control case (CE) is depicted in Fig. 7(a) (the layers at higher magnification are given as insets in the EDX patterns).

As it can be seen in Fig. 7, the morphology and composition of the product layer for corroding specimen CEVCn (Fig. 7(d)) resembles the one observed for the control specimen CE (Fig. 7(a)) i.e. a more compact layer, depicting mainly calcium carbonate was observed. It should be noted that for all specimens, $CaCO_3$ was detected by both ESEM/EDX and XPS (Fig. 8) (the XPS and ESEM/EDX investigations were performed after removing the specimens from the model solutions and drying on air). Since the specimens were in immersed condition, carbonation of the solutions during treatment is minimal. Moreover, if a substantial concentration of CO_2 in the solutions was present, $FeCO_3$ will form, which would be either detected by XPS and ESEM, or would rather rapidly transform when the specimens are taken on air to more stable iron oxides ($FeCO_3$ or "green rust" is not stable in oxygen containing environment). Therefore, the detected surface $CaCO_3$ is a result of the reaction of Ca-containing compounds (as CaO and $Ca(OH)_2$ previously deposited on the steel surface during treatment) with atm. CO_2 upon drying of the specimens on air. Consequently, the presence and

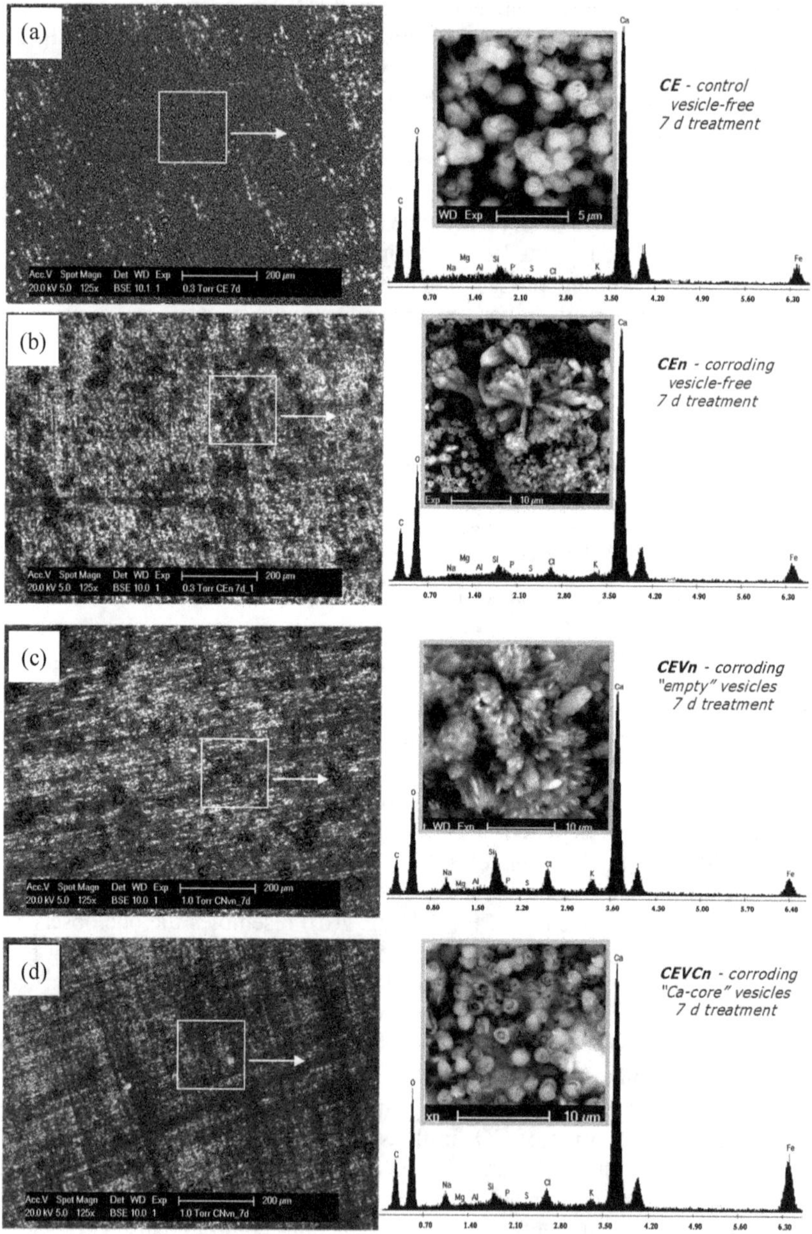

Figure 7: ESEM micrographs and the relevant EDX analysis for control (a) and corroding cases CEn (b), CEVn (c) and CEVCn (d).

amounts of $CaCO_3$ on the steel surface can serve as an indirect indication for the presence of adhered Ca-containing compounds during treatment in immersed condition. To this end, the control cases and specimens CEVCn present mainly carbonate-rich product layers, whereas for specimens CEn and CEVn (Figs. 7(b), (c)) except $CaCO_3$, chloride-containing iron oxides/hydroxides are relevant. This is in line with the lower corrosion resistance, detected via electrochemical measurements. In specimens CEn and CEVn localised corrosion is basically observed, evidenced by the deposition of voluminous, needle-shape corrosion products (Figs. 7(b), (c)).

An elaborated discussion on the otherwise extensive XPS analysis is outside the scope of this paper and will be separately reported. What needs to be mentioned are only the main observations and features, supporting the above results and discussion.

As previously defined, XPS was performed on the surface of the as-treated specimens (after 7 days) and in depth of the corresponding product layers. Fig. 8 presents the ionizations of Fe, O and C on the "as-treated" surface (scans 1 and 2) and in depth of the products (scans 1' and 2'). With respect to iron oxides and hydroxides, all specimens exhibit a mixture of these with a larger portion of the latter in the outermost sections of the layers (shape and features of the scans 1 and 2 in Fe2p, pronounced peak at 531.6 eV in O1s, corresponding to OH^-) and a larger portion of the former (oxides) in depth of the layers (hematite/magnetite based layers as evident from the characteristic features of Fe2p, 1′ and 2′ and the peaks at lower binding energy ~ 529 eV in the O1s for 1′ and 2′). As also observed, metallic iron appears in depth of the layers as well (~706 eV), which is as expected within product layer removal. The presence of adsorbed in the product layers vesicles for specimen CEVCn is confirmed by the characteristic binding energies in the C1s ionizations i.e. 283 eV for PS, 286.5 eV for PEO and 534 eV for PEO

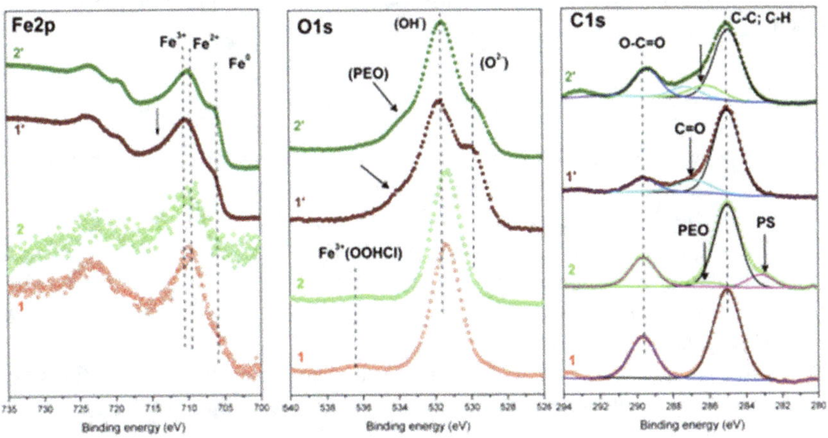

Figure 8: XPS iron (Fe2p), oxygen (O1s) and carbon (C1s) ionization for specimens CEn (curves 1 and 1′) and CEVCn (curves 2 and 2′).

in the O1s ionizations. Iron oxy/hydroxy chlorides are detected only in the outermost layers for (mainly) specimens CEn (peak at 536.1 eV in O1s), which is actually supported by the morphological observations on the surface of CEn (Fig. 7(b)). In other words, morphological and chemical alterations in the product layers, formed on the steel surface in the presence of Ca-containing vesicles and chloride result in a "self-repair" or "self-healing", evidenced by a more compact and uniform product layer, featuring iron oxy/hydroxides and calcium carbonate only (similar to the layers formed in the control, chloride-free cases). In contrast, for steel treated in vesicles-free or "empty" vesicles containing solutions, voluminous iron-oxy/hydroxy chlorides and lower amount of Ca-containing compounds are present.

5 Conclusions

In the framework of a novel approach to corrosion control in reinforced concrete, the application of polymeric nano-particles was investigated. In order to evaluate the difference between "barrier" effects only and self-healing mechanisms, resulting from the "Ca-core" release, the effect of both "empty" and Ca-containing vesicles was studied. In the presence of both vesicle types, corrosion resistance was increased, but the effect was significantly more pronounced when Ca-containing vesicles were involved. The concentration of 0.0024 wt.% vesicles in the model solutions is minimal and therefore self-repair or self-healing of the product layer solely due to the Ca-component is not realistic. The most plausible mechanism is the nature of incorporation of the Ca-containing vesicles in the product layer, enhanced chloride binding effects and adsorption on active sites on the steel surface. The result is a more uniform and stable steel surface layer, initial pitting formation and propagation, but consecutive healing, as evidenced by surface analysis and electrochemical response i.e. largely reduced anodic and corrosion currents and no further pit propagation when Ca-containing vesicles are present.

References

[1] Conway, B.E., *Theory and Principles of Electrode Processes*, The Ronald Press Company: NY, pp. 170–272, 1965.

[2] Broomfield, J., *Corrosion of Steel in Concrete, Understanding, Investigating and Repair*, Spon: London, 1997.

[3] Şahmaran, M., Effect of flexure induced transverse crack and self-healing on chloride diffusivity of reinforced mortar. *Journal of Materials Science*, **42(22)**, pp. 9131–9136, 2007.

[4] Elsener, B., Macrocell corrosion of steel in concrete - implications for corrosion monitoring. *Cement and Concrete Composites*, **24(1)**, pp. 65–72, 2002.

[5] Garces, A., Andrade, M.C., Saez, A. & Alonso, M.C., Corrosion of reinforcing steel in neutral and acid solutions simulating the electrolytic

environments in the micropores of concrete in the propagation period. *Corrosion Science*, **47(2)**, pp. 289–306, 2005.

[6] Broomfield, J.P., Corrosion of steel in concrete. *Uhlig's Corrosion Handbook*, ed. R.W. Revie, Wiley: Hoboken, NJ, pp. 633–648, 2011.

[7] Bertolini, L., Elsener, B., Pedeferri, P. & Polder, R., *Corrosion of Steel in Concrete: Prevention, Diagnosis, Repair*, 1st ed., Wiley: Weinheim, 2004.

[8] Petterson, K., Chloride threshold value and the corrosion rate in reinforced concrete. *Corrosion and Corrosion Protection of Steel in Concrete*, vol. 1, ed. R.N. Swamy, Sheffield Academic Press: Sheffield, p. 461, 1994.

[9] Pedeferri, P., Cathodic protection and cathodic prevention. *Construction and Building Mater*ials, **10(5)**, pp. 391–402, 1996.

[10] Bassi, R. & Davies, H., Testing anti-carbonation coatings for concrete, BRE Information paper IP7/96, 1996.

[11] Buenfeld, N.R., & Zhang, J.-Z., Chloride diffusion through surface-treated mortar specimens. *Cement and Concrete Research*, **28(5)**, pp. 665–674, 1998.

[12] Castle, J.E., Cathodic disbondment. *AIP Conference Proceedings*, **354**, pp. 432–449, 1996.

[13] Bird, C.E, Bond of galvanized steel reinforcement in concrete. *Nature*, **194**, p. 798, 1962.

[14] Herrera, M.J., Martinez, M. & Perez, J.T., Evaluation of electrochemical behavior of two inhibitors for galvanized rebars in alkaline solutions, paper 831, 210th ECS meeting, Cancun, MX, 2006.

[15] International Lead Zinc Research Organisation, Galvanized reinforcement for concrete II. Res., Triangle Park NC, USA, 1981.

[16] Kayali, O. & Yeomans, S.R., Bond of ribbed galvanized reinforcing steel in concrete. *Cement and Concrete Composites*, **22(6)**, pp. 459–467, 2000.

[17] Andrade, C. & Holst, J.D., *Coating Protection for Reinforcement*, Comite Euro-International Du Beton, State of the Art Report, Thomas Telford Publication: London, p. 51, 1995.

[18] Mietz, J., *Electrochemical Rehabilitation Methods for Reinforced Concrete Structures: A State of the Art Report*, Institute of Materials, London, UK, 1998.

[19] Andrade, C., Keddam, M., Nóvoa, X.R., Pérez, M.C., Rangel, C.M. & Takenouti, H., Electrochemical behaviour of steel rebars in concrete: influence of environmental factors and cement chemistry. *Electrochimica Acta*, **46(24–25)**, pp. 3905–3912, 2001.

[20] Gaidis, M., Chemistry of corrosion inhibitors. *Cement and Concrete Composites*, **26(3)**, pp. 181-189, 2004.

[21] Wombacher, F., Maeder, U. & Marazzani, B., Aminoalcohol based mixed corrosion inhibitors. *Cement and Concrete Comp*osites, **26(3)**, pp. 209–216, 2004.

[22] Koleva, D.A., van Breugel, K., Ye, G., Zhou, J., Chamululu, G. & Koenders, E.A.B., Porosity and permeability of mortar specimens incorporating PEO113–b–PS218 micelles. *ACI Materials Journal*, **267**, pp. 101–110, 2009.

[23] Koleva, D.A., Hu, J., van Breugel, K., Milkova, V. & Petrov, P., The influence of tailored nano/micro polymeric aggregates on material properties of cement-based systems. *Materials, Methods and Technologies*, **5(1)**, pp. 63–73, 2011.

[24] Hu, J., Koleva, D.A., Petrov, P. & van Breugel, K., Polymeric vesicles for corrosion control in reinforced mortar: Electrochemical behavior, steel surface analysis and bulk matrix properties. *Corrosion Science*, **65**, pp. 414–430, 2012.

[25] Hu, J., Koleva, D.A., Ma, Y., Schlangen, E., Petrov, P. & van Breugel, K., The influence of admixed micelles on the microstructural properties and global performance of cement-based materials. *Cement and Concrete Research*, **42(8)**, pp. 1122–1133, 2012.

[26] Hu, J., Koleva, D.A. & van Breugel, K., Corrosion performance of reinforced mortar in the presence of polymeric nano-aggregates: electrochemical behavior, surface analysis, and properties of the steel/cement paste interface. *Journal of Materials Science*, **47(12)**, pp. 4981–4995, 2012.

[27] Hu, J., Koleva, D.A., de Wit, J.H.W., Kolev, H. & van Breugel, K., Corrosion performance of carbon steel in simulated pore solution in the presence of micelles. *Journal of the Electrochemical Society*, **158(3)**, pp. C76–C87, 2011.

[28] Hu, J., Koleva, D.A. & van Breugel, K., The influence of PEO_{113}-b-$PS_{780°}$ vesicles on the corrosion performance of carbon steel in simulated pore solution. *ECS Transactions*, **41(16)**, pp. 1–9, 2012.

The corrosion of stirrups and its effect on the seismic fragility of a corroded reinforced concrete column

G. Zhang & B. Li
Nanyang Technological University, Singapore

Abstract

As service rates increase, the corroded rebar stands out as a major problem for existing reinforced concrete (RC) columns in a corrosion-inducing environment. The corrosion of stirrups is more serious than that of longitudinal bars, due to smaller clear concrete cover and smaller diameter. It deteriorates the confinement behaviour and thus decreases the shear resistant capacity, potentially changing the failure modes from ductile flexural failure mode to brittle flexural-shear failure mode, or even shear failure mode. Therefore, the evolution of the seismic damage pattern of existing corroded RC columns and its service-year related fragility analysis are becoming heated research topics in the field of structure hazard prevention research. The objectives of this paper will be to focus on how to use FEM software to numerically simulate the seismic behaviour of corroded columns accounting for materials deterioration and bond degradation. Pushover curves shown that a transverse rebar corrosion level above 10% could reduce the seismic capacity of columns by a great extent. In addition, the stiffness, ductility and energy dissipation capacity declines. It also demonstrated the transition of failure mode from flexure failure to shear failure as the corrosion level of stirrups increased. The fragility analysis based on the FEM results indicated that the RC column with corroded stirrups is much more vulnerable.
Keywords: stirrup corrosion, bond degradation, seismic fragility analysis, FEM, damage pattern, brittle failure.

1 Introduction

The corrosion of reinforcement steel bars (rebar) is a natural electrochemical reaction reinforced concrete (RC) structures have to be faced with. It is exacerbated by exposure to corrosion-inducing environment factors, including de-icing salt, marine salty water, carbon dioxide and sulphur dioxide, among others. The chloride from salt (NaCl) could make hazardously chemical attack on a steel bar by acting as an efficient catalyst in the corrosion process. The corrosion of steel bars in an existing reinforced concrete structure has raised great concerns over its safety and seismic performance among practising engineers, researchers and residents, because a steel bar is the most essential element in RC. Corrosion reduces the effective cross-section area of longitudinal and transverse rebars. As shown in Fig. 1, the volume expansion of the corrosion product will give rise to cover cracking, as a result of which effective concrete area could be reduced.

Figure 1: Corrosion of transverse rebars.

Furthermore, the interaction between concrete and steel bars could be changed fundamentally as the steel bars have an increasing corrosion level. In the primary phase, the expansion of the corrosion products enhances the confinement behaviour of the concrete towards the rebar. Moreover, the particle of corrosion products raises the friction coefficient, which is more obvious for smooth bars. Both described factors contribute to the increase of bond strength in the primary phase. But as more corrosion products accumulate, the chemical cohesion will deteriorate. It may even lead to the reduction of the friction coefficient. For a deformed bar, the ribs could be eradicated, which reduces the mechanical interlocking. Also, the expansion effect of the corrosion products can lead to the cracking of the concrete cover, adversely resulting in the decrease of the confinement. Thus, the bond strength declines as the corrosion level exceeds the boundary value (generally this is 5%).

In experimental studies and engineering investigations on corroded RC, many scholars including Li and Yang [1], Ma *et al.* [2] and Wang and Liang [3] found that the transverse rebar could undergo more severe damage than the longitudinal rebar, and even break at the junction point with the longitudinal rebar. Because of the comparative smaller depth of concrete cover for transverse rebars, the latter are more easily reached by chloride and other corrosion induction chemicals. Besides, the diameter of transverse rebars is smaller than that of the longitudinal reinforcement so transverse reinforcement is more vulnerable to corrosion, especially that at the corner of the column. So, transverse rebar corrosion turns out to start earlier and have higher levels than that of longitudinal rebars.

The corrosion of transverse rebars will seriously degrade the confinement behaviour and thus reduce the shear resistance of the column. Consequently, the damage mode could be shifted from ductile bending failure to brittle bending-shear failure or shear failure. However, up until now, experimental studies and numerical modelling have focused on how corroded longitudinal rebars affect the seismic performance and fragility of an RC beam in general and the corrosion level of transverse rebars has been underestimated. Uncertainties remain on how severely corroded transverse rebars will affect the mechanical strength, seismic capacity and seismic fragility of an RC element column.

Therefore, this contribution will firstly discuss the use of nonlinear finite element analysis (NLFEA) software Abaqus to simulate the behaviour of a corroded RC structure, especially focusing on how to model corrosion-induced cracked concrete, corroded steel rebar, and the bond strength between concrete and rebar, among others. Then, based on these techniques, this paper will present a parametric study on how a highly corroded transverse rebar affects the seismic behaviour of an RC circular column, as well as fragility curves using nonlinear pushover analysis.

2 Methodology for modelling corroded RC

The high nonlinearity of corroded RC structures can be attributed to corrosion-induced cracks in cover concrete, deteriorated steel bar and degraded bond strength between concrete and rebar. Ou *et al.* [4], Sæther and Sand [5] and many other investigators have looked into how corroded RC can be numerically simulated with various finite element analysis software and different approaches. This study utilizes the widely available nonlinear finite element modelling software Abaqus version 6.10.1, which enjoys a great reputation for handling problems with extensive plasticity using a smart solver with automatic increment [6]. This section will explain the challenges in modelling corroded RC and the methods adopted to meet them using Abaqus.

2.1 Cover concrete with corrosion-induced cracking

Bazant [7], Liu and Weyers [8], and Vu *et al.* [9] conducted analytical and numerical studies on how corrosion-induced cracking could be modelled. These approaches are mostly based on the closed-from assumption. According to this assumption, corrosion products replace the concrete-steel interface. The expansion of corrosion products results in the evolution of circumferential stress in the surrounding concrete. The concrete cracking occurs when the circumferential stress exceeds the tensile strength. Different kinds of modification factors were proposed to account for the stiffness degradation of the cracked concrete. Ou *et al.* [4] proposed a softening effect coefficient to account for the softening effect of cracking concrete, as follows:

$$\zeta = \frac{0.9}{\sqrt{1+600\varepsilon_r}} \tag{1}$$

where ε_r is the tensile strain of the cover concrete. The stress strain curve of cracked cover concrete could be expressed as:

$$\sigma = \begin{cases} \zeta f_c' \left[2\left(\dfrac{\varepsilon}{\zeta\varepsilon_0}\right) - \left(\dfrac{\varepsilon}{\zeta\varepsilon_0}\right)^2 \right] & \varepsilon < \zeta\varepsilon_0 \\ \zeta f_c' \left[1 - \left(\dfrac{\dfrac{\varepsilon}{\zeta\varepsilon_0} - 1}{\dfrac{2}{\zeta} - 1} \right)^2 \right] & \varepsilon \geq \zeta\varepsilon_0 \end{cases} \tag{2}$$

where ε_0 is the concrete strain corresponding to the concrete peak stress.

Abaqus has two different material models for concrete: one is damaged plasticity; the other is smeared cracking [6]. Both could describe the stress strain curve by specifying the data points. Damaged plasticity is used in the present modelling of cracked concrete, because it is superior to smeared cracking in terms of convergence rate. As has been described above, softened cover concrete caused by corrosion cracking is simulated by introducing softening factors and modified the concrete stress strain curve.

2.2 Geometry and material strength of corroded rebar

Corrosion initially damages the rebar by eating away steel and reducing the cross section. In experimental studies, the corrosion level is generally measured as the percentage of decreased bar weight relative to the prime bar weight [2, 10]. However, this method fails to describe the true corrosion status by ignoring the position, the depth and area of corrosion pit. This rough way of calculating rebar corrosion level could be idealized as uniform cross section reduction. Even though corrosion is localized and non-uniform by nature, this method has been shown to represent the geometry change of rebar and yielded good modelling results. Lee *et al.* [11] reported that a corroded steel bar has reduced yield and ultimate strength as well as smaller elastic modulus. Regression analysis produced linear equations to calculate the mechanical parameters of a corroded rebar; these are listed in Table 1 where Δc is the corrosion level (%). Fig. 2 shows the reduction factor for each of the parameters of the corroded rebar. Abaqus provides a steel material model with plasticity, explicitly identifying the stress and the corresponding inelastic strain.

Table 1: Mechanical properties of corroded reinforcement.

Mechanical properties parameters	Equation for mechanical properties
Yield strength	$\sigma_{cy} = (1 - 1.98(\Delta c/100))\sigma_y$
Ultimate strength	$\sigma_{cu} = (1 - 1.57(\Delta c/100))\sigma_u$
Elastic modulus	$E_{cs} = (1 - 1.15\Delta c/100))E_s$
Elongation	$\delta_c = (1 - 2.59(\Delta c/100))\delta$

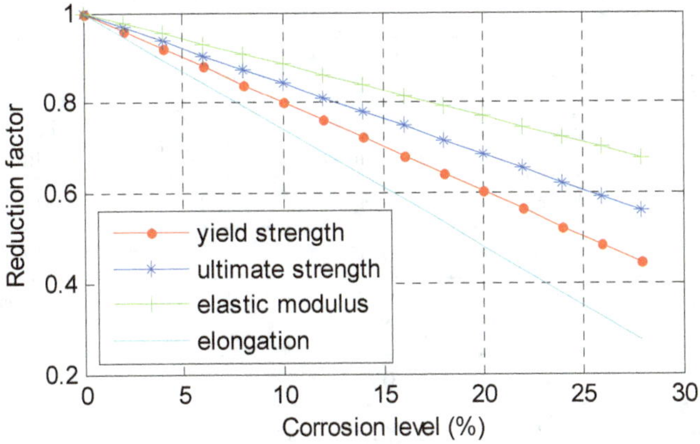

Figure 2: Strength degradation of corroded rebar.

2.3 Simulate bond strength degradation in Abaqus

In modelling concrete, the general practice is to model rebars as truss or beam elements and embed them into concrete which is usually discretized as solid element. This approach has been based on the "perfect bond" assumption that concrete is "perfectly" connected to concrete, which would overestimate the capacity of corroded RC structure.

CEB-FIP 1990 [12] provides guidance on the bond–slip value of confined concrete, as shown in Fig. 3 and Table 2; according to which, the bond strength depends on two major conditions – whether the concrete is confined or not and whether the bond is good. In addition, the ascend branch of the model is defined by the equation:

$$\tau_b = \tau_{\max}\left(\frac{s}{s_1}\right)^{\alpha} \quad \text{when } s \le s_1 \tag{3}$$

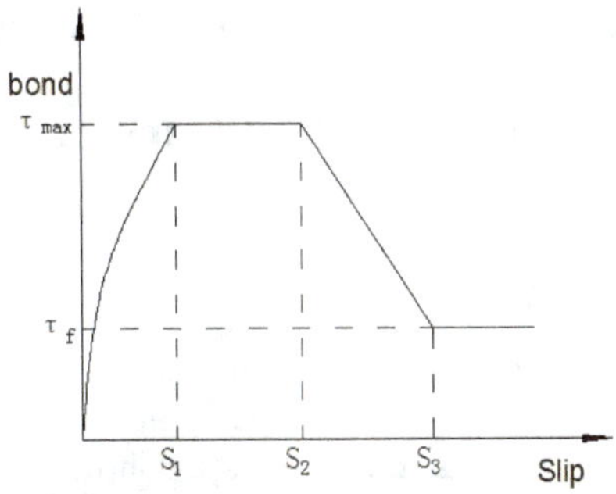

Figure 3: Bond–slip relationship.

Table 2: Parameters for defining bond–slip value.

	Unconfined concrete		Confined concrete	
	Good bond condition	All other bond conditions	Good bond condition	All other bond conditions
S_1 (mm)	0.6	0.6	1	0.6
S_2 (mm)	0.6	0.6	3	0.6
S_3 (mm)	1.0	1.0	Clear rib spacing	Clear rib spacing
α	0.4	0.4	0.4	0.4
τ_{max} (MPa)	$2\sqrt{f_c'}$	$\sqrt{f_c'}$	$2.5\sqrt{f_c'}$	$1.25\sqrt{f_c'}$
τ_f (MPa)	$0.15\tau_{max}$	$0.15\tau_{max}$	$0.4\tau_{max}$	$0.4\tau_{max}$
S_4 (mm)	-	-	$0.5S_3$	$0.5S_3$
τ_y (MPa)	-	-	$0.5\tau_f$	$0.5\tau_f$

Xu [13] tested the bond strength of accelerated carrion specimen on different corrosion levels and conducted regression analysis on the experiment data; they then proposed a relationship between the bond strength reduction factor and different corrosion levels, as follows:

$$\beta = \begin{cases} 1 + 0.5625\eta - 0.3375\eta^2 + 0.055625\eta^3 - 0.003\eta^4 & \eta \le 7\% \\ 2.0786\eta^{-1.0369} & \eta > 7\% \end{cases} \tag{4}$$

where β is the bond strength reduction factor and η the corrosion level.

Abaqus [6] has the functionality to model the bond behaviour between the concrete and reinforcement by defining interaction contact properties with tangential behaviour, normal behaviour and cohesive behaviour. Tangential behaviour could be defined as penalty friction formulation with friction coefficient, shear stress limit, and elastic slip stiffness, among others, which altogether could fully simulate the bond slip behaviour in corroded RC structure.

3 Numerical study on influence of transverse rebars

3.1 Details of modelling specimen

In order to clarify the influence of corroded transverse rebars on the seismic capacity of RC columns, it is essential to look at transverse rebar corrosion level separately. A RC circular column in the experimental study by Ma *et al.* [2] was chosen. The circular column has dimensions as shown in Fig. 4, the concrete cylindrical strength is 20.1 MPa; the clear cover depth is 30 mm; the yield strength and ultimate strength of longitudinal rebar is 373 MPa and 572 MPa, respectively;

the yield strength and ultimate strength of transverse rebar is 327 MPa and 510 MPa, respectively. As has been discussed, researching on how transverse rebar corrosion affects the seismic capacity holds value in understanding corrosion more thoroughly. In order to achieve this goal, a parametric study was designed with transverse rebar corrosion level, longitudinal rebar corrosion level as the control variables, as listed in Table 3. A set of sixteen samples were created, denoted as LxTy, in which x means corrosion level of longitudinal bar and y means corrosion level of transverse bar.

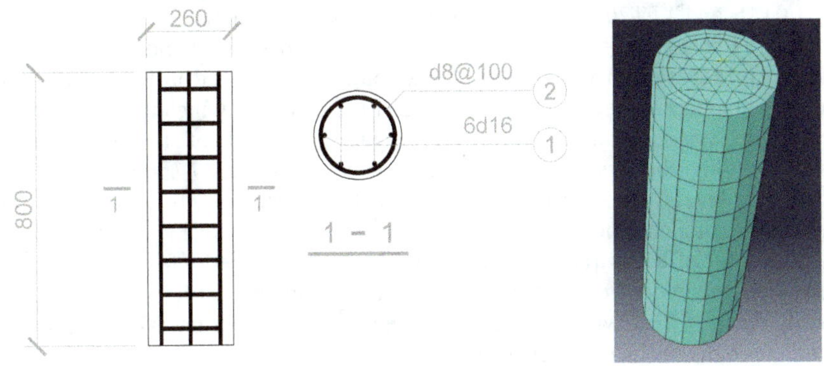

Figure 4: Dimensions of RC circular column, rebar detailing and FEM model (unit: mm).

Table 3: Control variables for parametric study.

Δc_l	Δc_t			
0%	0%	5%	10%	15%
5%	5%	10%	15%	20%
10%	10%	15%	20%	25%
15%	15%	20%	25%	30%

According to Table 3, a total of 16 models were created. As shown in Fig. 4, in the 3D model, concrete is discretized into solid homogenous elements; rebars are modelled using truss elements; the interaction between concrete and a steel bar is simulated by contact properties. The material properties of cracked concrete cover and concrete-rebar contact is introduced in the last section; the truss element section area was reduced according to corrosion level. Also, each element was subject to monotonic pushover analysis. The axial load ratio is 0.15. The maximum displacement at the top of the column is 50 mm, which makes the drift ratio 6.25%.

3.2 Parametric study results

Reaction force F_r and displacement D were retrieved from analysis results, as they could represent the overall seismic capacity. Previous analysis yielded 16 pushover curves, as shown in Figs. 5–7. Specimens with the same longitudinal rebar corrosion level yield at almost the same displacement. In Fig. 5, as the longitudinal corrosion level increases, the columns have smaller maximum strength at smaller displacement. The seismic capacity and elastic behaviour are still controlled by longitudinal bars. Separate investigation reveals that columns with transverse rebar severely damaged tend to have steeper descending branch and the reaction force decreases comparatively faster, which is fully demonstrated by L0T15 and L5T20 in Fig. 6. Columns with high transverse corrosion level apparently sustain more strength loss and ductility loss after yielding because of the sharp reduction of confinement behaviour. As illustrated in Fig. 6, a brittle failure appeared in specimens L0T5 and L5T15, L5T20 in which the longitudinal rebars do not have high corrosion levels. Corrosion of transverse rebar fundamentally changes the failure modes of columns. Fig. 7 shows that as the corrosion level of longitudinal rebars reaches 15%, the differences in the descending branches become minor. As the corrosion level of longitudinal rebars increases, the effect of transverse rebars weakens. As a matter of fact, in the natural corrosion phenomena of RC structures, the corrosion level of longitudinal rebars hardly reaches 15%, while the transverse rebar has a greater probability to suffer from much higher level of corrosion.

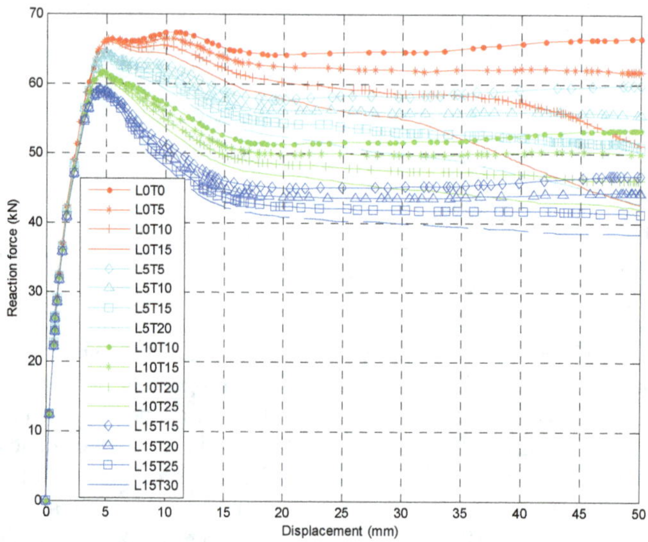

Figure 5: Pushover curves of L0T0–L0T15 and L5T5–L5T20.

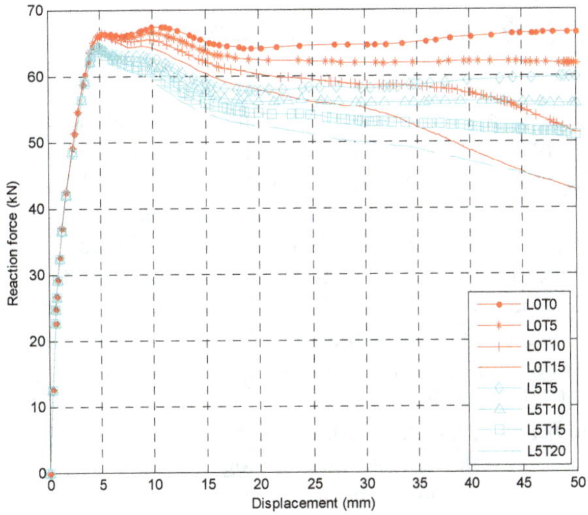

Figure 6: Pushover curves of L0T0–L0T15 and L5T5–L5T20.

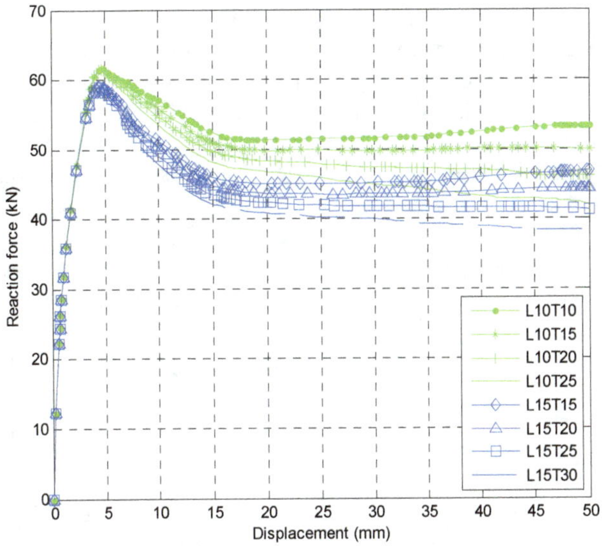

Figure 7: Pushover curves of all specimens.

3.3 Fragility analysis based on nonlinear pushover analysis

The fragility function describes the probability that seismic demand D exceeds seismic capacity C for designated limit states, and it is expressed as:

$$F_c(d) = P[D > C|D = d] = \Phi\left[\frac{\ln(d/m_C)}{\beta_C}\right] = \Phi\left[\frac{\ln(d) - \alpha_C}{\beta_C}\right] \quad (5)$$

where m_C, α_C and β_C are the median, logarithmic mean and logarithmic standard deviation (S.D.) of seismic capacity. Freeman *et al.* [14] put forward the capacity spectrum method (CSM) based on nonlinear static analysis. The capacity spectrum is established by transforming the base-shear displacement pushover curve into acceleration-displacement-response-spectrum (ADRS) format. Then the demand curve is produced based on different structural vibration periods and corresponding influence coefficients. The targeted displacement or displacement of the performance point (PP) is obtained through the comparison between the capacity spectrum and demand spectrum. According to Freeman's research contribution, official standardized fragility derivation procedures have been developed. The Federal Emergency Management Agency (FEMA) [15] has issued HAZUS 99, an applicable standard methodology which uses nonlinear pushover analysis to establish the seismic capacity curve and generate fragility functions. It is a mature and widely employed method in risk estimation of various hazards for diverse infrastructures. HAZUS 99 offers procedures for fragility analysis of a corroded RC column in this research. Limit states are categorized into slight damage, moderate damage, extensive damage and collapse according to the drift angle.

In order to investigate the fragility of circular columns with severely damaged transverse rebar, two cases were considered, L10T10 and L10T20, the fragility curves of which are shown in Fig. 8. This figure shows that L10T20 is more vulnerable to extensive damage and collapse than L10T10, although they have the same probability to have slight and moderate damage. It provides typical examples of RC structures with more severe damaged transverse rebar being at higher risk of seismic attack.

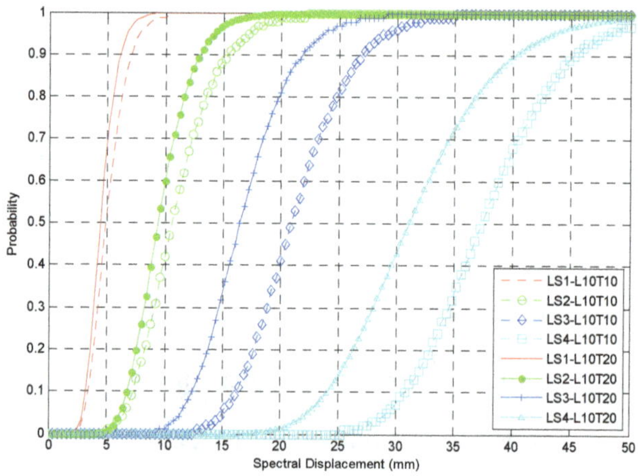

Figure 8: Fragility curves of L10T10 and L10T20.

4 Concluding remarks

This paper suggests a feasible methodology of modelling corroded RC structures using software Abaqus 6.10.1. Corrosion-induced cover cracking, deteriorated rebar and the bond strength between concrete and rebar have been validly simulated with the functionalities in this software. Simulation results have shown that RC columns with severely corrosion damaged transverse rebar have lower seismic capacity, smaller seismic ductility and brittle failure modes. Transverse rebars have greater influence on the behaviour of columns when the longitudinal rebar corrosion level is below 15%. Analysis has demonstrated that high corrosion of transverse rebars should be considered separately because of its changing power in seismic capacity and ductility. Fragility analysis has found that RC structures with higher transverse corrosion level are more vulnerable.

All the columns are tested under monotonic loading so the energy consuming ability is not analysed. This paper only presents the analysis results of columns under monotonic loading when transverse rebar has a higher corrosion level than that of a longitudinal rebar. Research should be carried out on time-variant corrosion level of transverse rebar and longitudinal rebar. A probability model should be developed to describe the corrosion level of transverse rebar as a function of longitudinal rebar corrosion level. Furthermore, further work could focus on how to model corroded columns under cyclic loading and how to generate fragility curves of corroded RC elements.

References

[1] Li, C.Q. & Yang, S.T., Prediction of concrete crack width under combined reinforcement corrosion and applied load. *Journal of Engineering Mechanics*, **137(11)**, pp. 722–731, 2011.

[2] Ma, Y., Che, Y. & Gong, J., Behavior of corrosion damaged circular reinforced concrete columns under cyclic loading. *Construction and Building Materials*, **29**, pp. 548–556, 2012.

[3] Wang, X.H. & Liang, F.Y., Performance of RC columns with partial length corrosion. *Nuclear Engineering and Design*, **238(12)**, pp. 3194–3202, 2008.

[4] Ou, Y.C., Fan, H.D. & Nguyen, N.D., Long-term seismic performance of reinforced concrete bridges under steel reinforcement corrosion due to chloride attack. *Earthquake Engineering & Structural Dynamics*, **42(14)**, pp. 2113–2127, 2013.

[5] Sæther, I. & Sand, B., FEM simulations of reinforced concrete beams attacked by corrosion. *Nordic Concrete Research*, **39(2)**, 2012.

[6] Abaqus 6.10 Documentation, http://www.tu-chemnitz.de/projekt/abq_hilfe /docs/v6.10/

[7] Bazant, Z.P., Physical model for steel corrosion in concrete sea structures – application. *ASCE Proceedings, Journal of the Structural Division*, **105(ST6)**, pp. 1155–1166, 1979.

[8] Liu, T. & Weyers, R.W., Modeling the dynamic corrosion process in chloride contaminated concrete structures. *Cement and Concrete Research*, **28(3)**, pp. 365–379, 1998.

[9] Vu, K., Stewart, M.G. & Mullard, J., Corrosion-induced cracking: experimental data and predictive models. *ACI Structural Journal*, **102(5)**, pp. 719–726, 2005.

[10] Li, J., Gong, J. & Wang, L., Seismic behaviour of corrosion-damaged reinforced concrete columns strengthened using combined carbon fiber-reinforced polymer and steel jacket. *Construction and Building Materials*, **23(7)**, pp. 2653–2663, 2009.

[11] Lee, H.S., Noguchi, T. & Tomosawa, F., Evaluation of the bond properties between concrete and reinforcement as a function of the degree of reinforcement corrosion. *Cement and Concrete Research*, **32(8)**, pp. 1313–1318, 2002.

[12] Comité euro-international du béton, CEB-FIP Model Code 1990: Design Code. No. 213–214. FIB-Féd. Int. du Béton, 1993.

[13] Xu S.H., Degradation model of Reinforced Concrete Structure and Durability Assessment, Doctoral dissertation, Xi'an University of Architecture and Technology, 2003. (In Chinese).

[14] Freeman, S.A., Nicoletti, J.P. & Tyrell, J.V., Evaluations of existing buildings for seismic risk - A case study of Puget Sound Naval Shipyard, Bremerton, Washington. *Proceedings of the 1st US National Conference on Earthquake Engineering*, pp. 113–122, 1975.

[15] Federal Emergency Management Agency (FEMA), National Institute of Building Sciences (NIBS), Earthquake loss estimation methodology: HAZUS 99, Washington, DC, USA, 1999.

Section 2
Cathodic protection

Numerical-analysis-assisted monitoring method for the coating condition on a ballast tank wall

K. Amaya[1], A. Nakayama[1] & N. Yamamoto[2]
[1]*Tokyo Institute of Technology, Japan*
[2]*Research Institute Nippon Kaiji Kyokai, Japan*

Abstract

A numerical-analysis-assisted monitoring method was developed for the coating condition on a ballast tank wall. It was proposed that the coating condition is evaluated by the surface resistance. An identification method was developed to obtain the whole surface resistance from the differential potential induced by the impressed current from an optional anode inside a tank. A differential potential measurement and inverse analysis were introduced to obtain the surface resistance representing the coating condition. The potential measurement and quantitative evaluation were conducted in an actual ship. Verification was performed and there was good agreement between the proposed method and the preliminary visual inspection.

Keywords: corrosion analysis, numerical analysis, surface resistance, inverse problem, coating condition, ballast tank.

1 Introduction

In order to prevent corrosion loss, it is very important to inspect the coating condition inside the ballast tank of ships, such as oil tankers, liquefied natural gas (LNG) and cargo ships. The current inspection standard specifies that the ratio of paint defect surface area to the total surface is visually inspected periodically every two or three years. However, the current visual inspection has some problems, such as the dangerous and dark environment for inspectors, inaccurate evaluation results depending on the inspector's skill as well as the time-consuming process and heavy labour required in order to inspect all hundreds of tank compartments. In order to overcome these issues, a quantitative, safe, economic and efficient monitoring method for the coating condition is extremely desirable.

A ballast tank filled with seawater is easily corroded. Corrosion protection by the paint on the metal surface inside the tank, which improves the insulation for the corrosion current, is applied. The paint has problems with age-related degradation and incipient failure. To protect the metal from corrosion caused by these problems, multiple sacrificial anodes are usually installed in the tank. When seawater is loaded in the tank, the inside surface of the tank becomes cathode and the protective potential works, because of the anode effects. The worse the coating condition, the worse the insulation of the paint becomes and the lower the surface resistance. Therefore, there is the possibility that the coating condition can be evaluated by monitoring the surface resistance.

On the other hand, potential measurements are sometimes conducted for the evaluation of the cathodic protection. Since, however, environmental factors, such as temperature, affect the measured potential and the potential is determined by the number of anodes and the coating condition, the coating condition could not be evaluated directly by the potential measurement.

In this paper, a numerical-analysis-assisted monitoring method was developed for the coating condition on ballast tank walls. It was proposed that the coating condition is evaluated by the surface resistance. An identification method was developed to obtain the whole surface resistance from the differential potential induced by the impressed current with an optional anode. Differential potential measurement and inverse analysis were introduced to obtain the surface resistance representing the coating condition. The potential measurement and quantitative evaluation were conducted in an actual ship. A verification of the proposed methodology was thus performed.

2 Proposed method

2.1 Coating condition and surface resistance

The relationship between potential ϕ [V] and current density i [A/m^2] in an electrolyte near the metal surface is represented by a function called polarization curve expressed as

$$\phi = -f_m(i) \tag{1}$$

This function is normally non-linear and is determined experimentally. The polarization curve depends on the kind of metal, electrolyte and environmental conditions such as temperature, pH, and concentration of oxygen. In this paper, it is noted that the potential ϕ at a certain location in the electrolyte is defined by referring to the metal and has the inverse sign of the potential E employed in electro-chemistry in which the potential is defined to a reference electrode such as SCE or Ag/AgCl.

Considering a small potential change, eqn (1) gives

$$\delta\phi \cong -\frac{\mathrm{d}f_m}{\mathrm{d}i}\delta i = -R\delta i \tag{2}$$

where

$$R \equiv \frac{\mathrm{d}f_m}{\mathrm{d}i} \tag{3}$$

The surface resistance R [$\Omega \cdot m^2$] makes the relationship between the potential change and the current density change. When the resistance is big, this indicates high insulation. Therefore, it is here proposed that the coating condition can be indicated by the surface resistance R. In the case when the non-coating surface resistance is R_0, the ratio of the paint defect area to the total surface area can be evaluated using

$$\alpha = \frac{R_0}{R} \tag{4}$$

2.2 Summary and procedure of proposed monitoring method

The proposed monitoring method is described in this section.

An optional anode and an Ag/AgCl reference electrode for potential measurement are placed in the seawater filled tank to be inspected as shown in Fig. 1. The potential changes at several locations in the tank with respect to the reference electrode are measured in two cases. The first case is that the prescribed current is impressed with an optional zinc anode and the second case is that no current is impressed. In the following, each case is indicated by subscript "ON" and "OFF", respectively. The potential difference

$$\delta\bar{\phi} = \bar{\phi}_{ON} - \bar{\phi}_{OFF} \tag{5}$$

is calculated from the results.

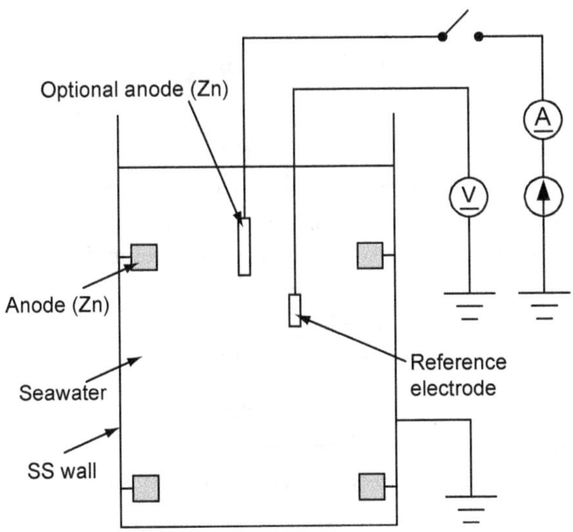

Figure 1: Simple measurement model in ballast tank.

The potential in any point is calculated with various R using a numerical analysis. It is assumed that the surface resistance R is constant in this numerical model. The residual of the measured differential potential $\delta\bar{\phi}$ and the numerically calculated $\delta\phi(R)$ is considered next; it is expressed as

$$F(R) = \sum_{j}^{N} \{\delta\bar{\phi}_j - \delta\phi_j(R)\}^2 \tag{6}$$

where N is the number of the combinations of the potential measurement points and the anode locations.

Obtaining the surface resistance R is to search for R which minimizes eqn (6). A general optimization method, such as Brent's method [1, 2], can be applied for this search.

2.3 Analysis method for potential difference

In this section, the numerical analysis method to calculate the potential $\phi(R)$ or the potential difference $\delta\phi(R)$ for the assumed surface resistance R is described.

First, the mathematical model of the electrical field in the tank is considered. It is assumed that the surface of electrolyte domain Ω is enclosed by $\Gamma(= \Gamma_d + \Gamma_n + \Gamma_m)$, where Γ_m is the metal surface, while the potential and current density are prescribed on Γ_d and Γ_n, respectively. The potential ϕ in a 3-D homogeneous electrolyte domain Ω satisfies the Laplace's equation:

$$\nabla^2\phi = 0 \text{ in } \Omega \tag{7}$$

The general boundary conditions are given by

$$\phi = \phi_0 \text{ on } \Gamma_d \tag{8}$$

$$i\left(\equiv \kappa\frac{\partial\phi}{\partial n}\right) = i_0 \text{ on } \Gamma_n \tag{9}$$

$$-\phi = f_m(i) \text{ on } \Gamma_m \tag{10}$$

where ϕ_0 and i_0 are the prescribed values of the potential ϕ and current density i, respectively, $\partial\phi/\partial n$ is the outward normal derivative, and κ is the conductivity of the electrolyte. The actual boundary conditions for the present numerical analysis are described in Table 1 in section 3.3.

Since the two potentials ϕ_{ON} and ϕ_{OFF} both satisfy the Laplace equation, the potential difference $\delta\phi$ also satisfies the equation

$$\nabla^2\delta\phi = 0 \text{ in } \Omega \tag{11}$$

Eqn (11) implies that in case there are two boundary conditions for one model, the following three conditions are applied instead of eqns (8) to (10):

$$\delta\phi = (\phi_{ON})_0 - (\phi_{OFF})_0 \text{ on } \Gamma_d \tag{12}$$

$$\delta i = (i_{ON})_0 - (i_{OFF})_0 \text{ on } \Gamma_n \tag{13}$$

$$-\delta\phi = f_m(i_{ON}) - f_m(i_{OFF}) \cong \frac{df_m}{di}\delta i \quad \text{on } \Gamma_m \tag{14}$$

In case δi in eqn (14) is small, the approximation becomes better in terms of Taylor expansion. Also, the constant term in the polarization curve, eqn (1), represents the open circuit corrosion potential and can be eliminated; thus, $\delta\phi$ only depends on the surface resistance, defined in eqn (3), according to eqn (2).

The constant term depends on the environmental conditions such as temperature, pH, concentration of oxygen and the reference electrode offset. But the differential method without this term has advantages over the standard method, in case the same reference electrodes are used in the short-time measurement. This formulation easily eliminates the effect of open circuit corrosion potential and reference electrode offset. If the potential or current density is constant in two boundary conditions, the differential boundary conditions are zero according to eqn (12) or eqn (13).

The boundary element method [3] is employed for performing the numerical analysis to solve eqn (11) with eqns (12) to (14). The formulation of the boundary element method is the same as the standard boundary element method applied to eqns (6) to (9) [4–7].

3 Verification of differential method

3.1 Experiment summary

The potential measurement was conducted in a ballast tank of the 23-year-old LNG ship. The LNG ship had ten ballast tanks and each of them was partitioned to approximately ten compartments with bulkheads transversely and vertically. In one of the compartments, the potential was measured by the present method. Five years had passed since the paint-coating of the ballast tank was repaired. The measured compartment was a rectangular parallelepiped (L 3.75 m × W 4 m × H 12.8 m) filled with seawater up to 7.6 m in height. There were eight zinc anode electrodes (L 0.1 m × H 0.1 m × W 0.5 m) in the seawater within the compartment as shown in Fig. 2.

The measurements were conducted with the optional anode (φ15 mm × L150 mm) and both (ON) and (OFF) were measured. The potential along the center line was measured with an Ag/AgCl reference electrode every 0.2 m back and forth between top and the bottom of the seawater. The current from the optional zinc anode was measured. HIOKI 8422-50 logger and HIOKI 3257-50 DMM were used for the potential and current measurement, respectively.

3.2 Measurement results

One set of measurements takes about fifteen minutes. The optional anode was located at a depth of 1.6 m from the seawater surface in the first three sets (Fig. 3). The electrode measuring the potential was attached to the cable on edge. The depth in Fig. 3 is actually the cable length from the ballast tank ceiling and a depth of more than 12.8 m shows that the electrode reached the bottom.

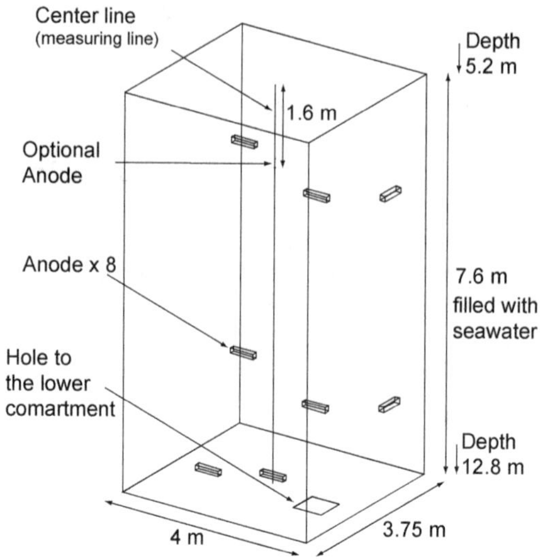

Figure 2: Ballast tank compartment with seawater.

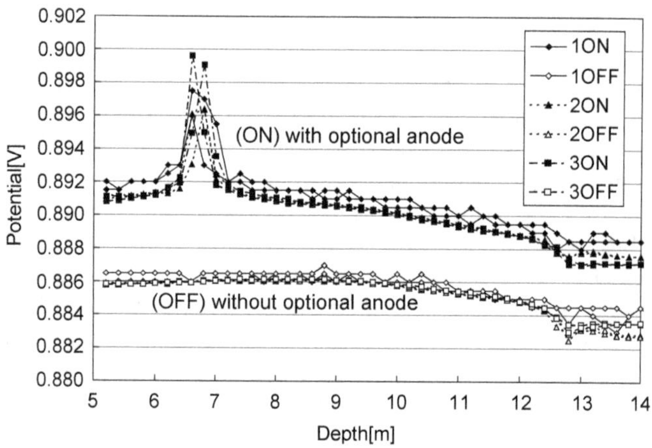

Figure 3: Potential distributions (optional anode depth 1.6 m from seawater surface).

3.3 Numerical analysis and surface resistance identification

In order to verify the proposed differential method, the numerical prediction of the differential potential distribution within the tank is compared with the experimental result.

Firstly, the differential potential distribution within the tank shown in Fig. 2 was measured. The experimental conditions were set to be the same as described in the previous section.

Secondly, the tank is modelled as shown in Fig. 4 for the numerical analysis. The boundary element mesh in Fig. 5 was used and the elements are all constant triangle elements. The number of the elements and nodes are 5670 and 3619, respectively. The boundary conditions are applied as shown in Table 1. The surface resistance of the anodes is set to be 0.16 Ω·m². The total current amount of the optional anode is set to be 30 mA and the access hole boundary condition is set to be $\delta\phi = 0.005$ V from the experiment. The differential potential distributions are calculated for nine cases whose surface resistances R_{SS} are varied.

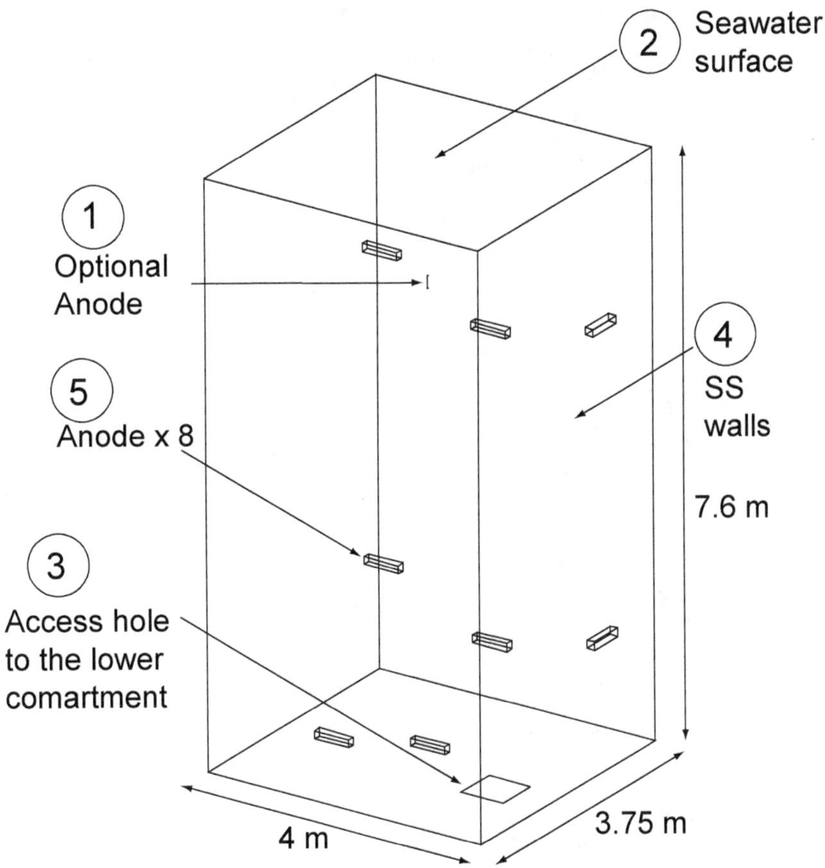

Figure 4: Model of the ballast tank compartment.

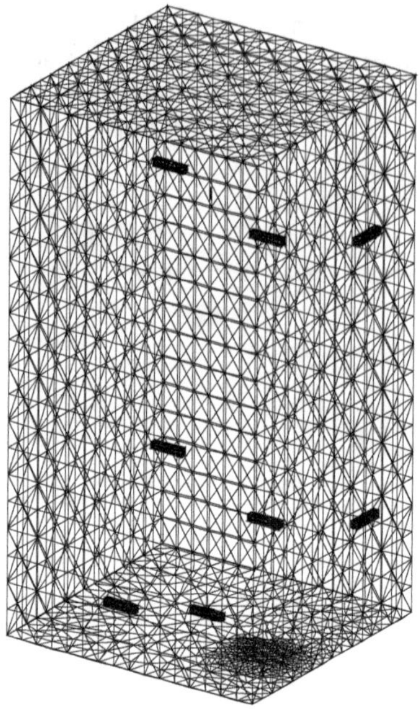

Figure 5: Mesh for the compartment.

Table 1: Boundary conditions of differential model for the compartment.

No.	Boundary condition	Place
(1)	$\delta I = 30$ mA	Optional anode side (ON)
(2)	$\delta i = 0$	Seawater surface
(3)	$\delta\phi = 0.005$ V	Access hole to the lower compartment
(4)	$\delta\phi = -R_{SS}\,\delta i$	SS (walls) $R_{SS} = 1000, 500, 200,$ $100, 50, 20,$ $10, 5$ or 2 [$\Omega\cdot m^2$]
(5)	$\delta\phi = -16\delta i$	Zn (anodes)

The numerical results for nine cases of surface resistance are shown in Fig. 6, which shows that the whole potential is high when the surface resistance is high. The maximum voltage of the sidelobe of the optional anode is 0.005 V.

Thirdly, the surface resistance was estimated by minimizing the residual function

$$F(R_{SS}) = \sum_{j}^{N} \{\delta\overline{\phi}_j - \delta\phi_j(R_{SS})\}^2 \tag{16}$$

The residual function consists of three residuals between measurement and calculated differential potential at the depth of seven meters. The minimization was performed with the brute-force search.

Both numerical and experimental potential distributions are shown in Fig. 7 and they show good agreement. It is noted that a good agreement can be achieved even if the open circuit corrosion potential is unknown, polarization curve is non-linear and unknown and there are offsets of reference electrodes.

The paint defect ratio is lower than 0.2% under the assumption of the surface resistance of bare SS R_0 = 1 Ω·m^2 (see eqn (4)). This result agrees with the outcome of a preliminary visual inspection.

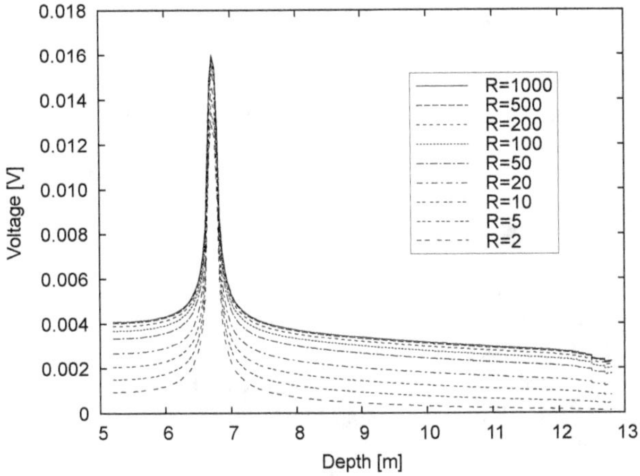

Figure 6: Voltage distributions with various surface resistances.

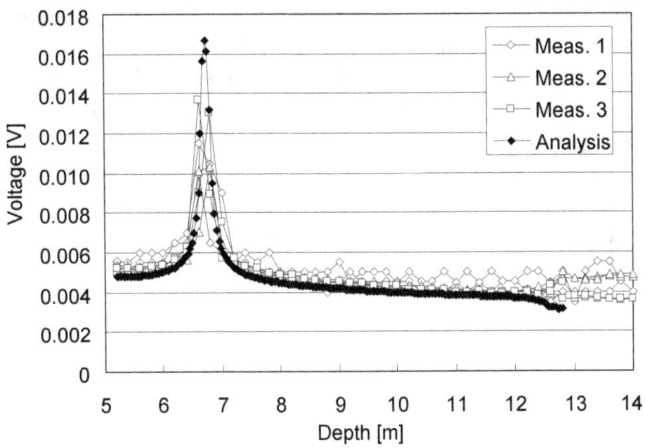

Figure 7: Voltage distributions (depth 1.6 m from seawater surface, SS surface resistance: 500 Ω·m^2.

4 Conclusion

A quantitative monitoring method was developed for the coating condition inside a ballast tank. It was proposed that the coating condition be evaluated by means of the surface resistance. An identification method was developed to obtain the whole surface resistance from the differential potential induced by the impressed current from an optional anode inside a tank. Differential potential measurement and inverse analysis were introduced to obtain the surface resistance representing the coating condition. The potential measurements and quantitative evaluation were conducted in an actual ship.

Both numerical and experimental potential distributions agree well when the differential method is applied. The good agreement can be achieved even if the open circuit corrosion potential is unknown, polarization curve is non-linear and unknown and there are offsets of reference electrodes. The paint defect ratio calculated from the differential method agrees with that obtained from a preliminary visual inspection.

References

[1] Atkinson, K.E., *An Introduction to Numerical Analysis*, 2nd ed., Section 2.8. John Wiley and Sons, 1989.

[2] Press, W.H., Teukolsky, S.A., Vettering, W.T. & Flannery, B.P., *Numerical Recipes in C*. Cambridge University Press: Cambridge, 1992.

[3] Brebbia, C.A., *The Boundary Element Method for Engineers*, Pentech Press: London, 1978.

[4] Amaya, K. & Aoki, S., Effective boundary element methods in corrosion analysis. *Engineering Analysis with Boundary Elements*, **27**, pp. 507–519, 2003.

[5] Aoki, S., Amaya, K., Nakayama, A. & Nishikawa, A., Elimination of error from nonuniform current distribution in polarization measurement by boundary element inverse analysis. *Corrosion*, **54(4)**, pp. 259–264, 1998.

[6] Aoki, S., Amaya, K., Urago, M. & Nakayama, A., Fast multipole boundary element analysis of corrosion problems. *CMES-Computer Modeling in Engineering & Sciences*, **6(2)**, pp. 123–131, 2004.

[7] Purcar, M., Deconinck, J., Van den Bosshe, B., Bortels, L. & Stehouwer, P.J., Numerical 3D BEM simulation of a CP system for a buried tank influenced by a steel reinforced concrete foundation. *WIT Transaction on Engineering Sciences*, **48**, pp. 47–56, 2005.

Boundary element modelling of the impressed current cathodic protection of an offshore platform

X. Wang[1], B. Hou[1] & Z. Lan[2]
[1]*Institute of Oceanology, Chinese Academy of Sciences, China*
[2]*CNOOC Research Institute, China*

Abstract

For most steel structures in a marine environment, corrosion is a big problem as it relates to long-term safety. Cathodic protection is an effective method to resist corrosion in seawater. In most cases, the design of cathodic protection systems depends on the empirical expertise of the corrosion engineers. To achieve the protection potential of the whole structure, excess protective current is usually applied, which leads to the waste of protection material. In this paper, the boundary element method was used to simulate the impressed current cathodic protection of an offshore oil platform to achieve ideal potential and current distribution. The location and the amount of anodes with different current demand were calculated using the commercial software BEASY, which has many applications in the corrosion field. The boundary condition of the model is defined as the parameter of the anode and the cathode of the system, which can be identified from the polarization curve of the material in seawater. A physical model was also built to verify the numerical modelling results, and the experimental results were consistent with the computation predictions. The research proved the efficiency and accuracy of the boundary element method for the impressed current cathodic protection of offshore structures.

Keywords: boundary element method, corrosion, impressed current cathodic protection, offshore platform.

1 Introduction

As the demand of energy grows rapidly, the offshore oil and gas exploitation has become an important assurance of energy supply. Cathodic protection is an effective technology to prevent a steel structure from suffering severe corrosion in the marine environment. Generally, the operation lifetime of an offshore oil platform is about 20 to 30 years, and the cathodic protection system is also designed for the same lifetime as that of the structure. However, in many marine oil exploiting fields, the offshore platform in operation usually exceeds its design life because of the rich petroleum reservoir and the inaccurate prediction of the mining rate. According to statistical data, 35% of the 20-year old platforms need to continue to serve the production; this demands complete structural health evaluation and the necessary reinforcement to guarantee the structure's safety. The cathodic protection system is one of the important parts of such reinforcement, and repairing or renovating is needed for the original system to accommodate the need of future operation life. On the other hand, for most of the offshore platforms built 10 to 30 years ago, the sacrificial cathodic protection system was the most common method to resist the marine corrosion. The consumption rate of the anodes of the system sometimes exceeds the original design rate because of the effect of marine environment factors, poor anode quality and inadequate anode current.

Though the impressed current cathodic protection (ICCP) and sacrificial anode cathodic protection are both used in offshore structures, they have different features. For those offshore structures which require extending or retrofitting their cathodic protection, the impressed current system is an economic and simple choice which has great advantages over the sacrificial anode system. Firstly, a large amount of galvanic anodes are usually installed on land before launching to the designated area; if all these anodes are retrofitted underwater, it would incur a great expense, while the ICCP can reduce most of the installation work. Secondly, as the structural surface has changed considerably compared with the original steel surface because of calcareous deposits and coating failure or breakage, it is difficult to evaluate the accurate current demand for protection, and once the amount and the location of the anodes has been determined, the current cannot be adjusted, while the ICCP can conveniently adjust the output current according to the monitored potential and current. So for marine structures which need retrofitting, the ICCP is the optimal choice.

However, the selection of cathodic protection parameters is a big problem that causes concern among corrosion engineers. For a long time over the last decades, the design depended on the experience of engineers, especially with regard to the location and the amount of the anodes. With the dramatic advances in numerical simulation and computing technology, computer modelling methodology was applied to the cathodic protection field. Among such modelling, the boundary element method (BEM) has shown great advantages with respect to accuracy for large marine structures. Compared to other numerical analysis methods, such as the finite element method (FEM) [1, 2] and the finite difference method (FDM),

BEM analysis can effectively solve three dimensional problems for structures of large size with less meshing requirements [3–6].

In this contribution, the BEASY software [7], developed on the basis of the boundary element method, was used to aid the design of an ICCP system and a physical jacket platform model was built in a seawater tank. The computed results and experimental data were compared to check the validity of the numerical analysis.

2 Experimental work

2.1 The ICCP principle

Metal corrosion is an electrochemical reaction with oxidation and reduction reaction on different electrodes. For steel in seawater, the following reactions occur:

$$Fe_{(s)} \rightarrow Fe^{2+}_{(aq)} + 2e^- \quad (1)$$

$$O_{2(g)} + 2H_2O_{(l)} + 4e^- \rightarrow 4OH^-_{(aq)} \quad (2)$$

$$2Cl^- \rightarrow Cl_2 + 2e^- \quad (3)$$

In these reactions, Fe with solid state converts to Ferrous II ions with aqueous state, which means the loss of the steel. To inhibit the oxidation reaction, some measures have been taken to prevent the corrosion. For the cathodic protection, a type of metal with more negative potential or a power source connect with steel and provide additional electrons. In an ICCP system for a steel structure, a DC power (or rectifier converting AC power to DC power) and a type of non-consumable anode are required. The principle of ICCP for a steel structure in marine environment is depicted in Fig. 1.

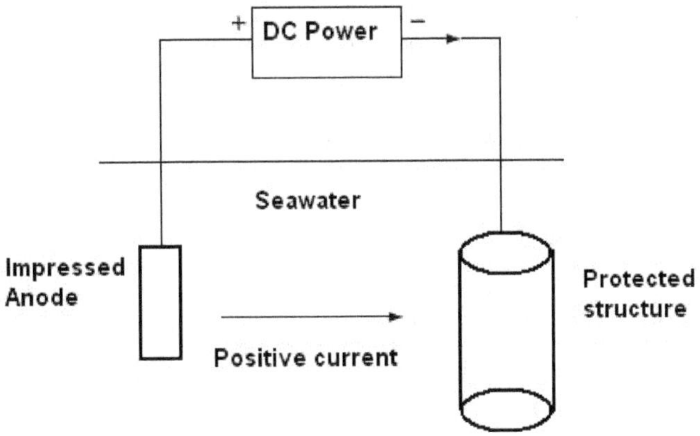

Figure 1: Diagram of impressed current cathodic protection.

In practice, the design of ICCP usually refers to some standards or criteria, such as NACE RP0176 and DNV RPB401. The most important point of the design is that overprotection and under protection must be avoided. The commonly accepted protection potential is from −850 mV to −1050 mV vs. silver/silver chloride electrode.

The selection of proper protection potential and current density is very important since this concerns the final distribution of ICCP. The current density for ICCP is dependent upon environmental factors such as pH value, salinity, oxygen content, coating integrity, marine fouling, sea currents, pressure and calcareous deposit among others.

The numerical simulation provides an effective method to help advance understanding of the distribution of the potential and current density for marine structures. As the offshore installation cost is very high, the simulation can model the real-time corrosion protection condition in advance to reduce unreasonable problems caused by the design.

2.2 ICCP experiment

To simulate the cathodic protection condition of a marine structure, a scaled-down physical model of a jacket platform was constructed referring to a real platform of China National Offshore Oil Corporation (CNOOC). The jacket platform had a symmetrical design with four legs and three levels in the vertical direction. The legs and the support tubes of the platform had different diameter. The form and dimensional details of the platform are shown in Fig. 2. The platform was made of steel without any coating on the surface. All the terminals of tubes were sealed to avoid the ingress of seawater.

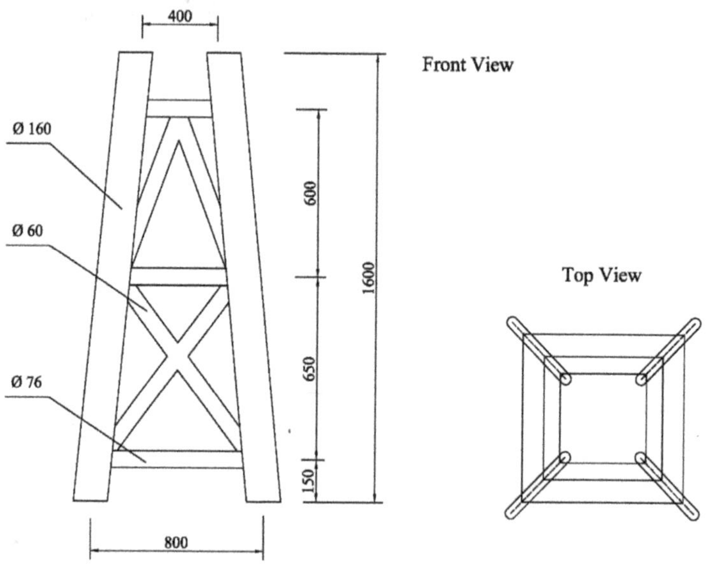

Figure 2: Detailed dimensions of the platform (mm).

The platform was placed in a marine environment modelling tank, which was as described in an earlier publication [8]. The modelling tank was large enough to place the jacket platform within it. An accurate control system was designed to model the flow of the tide through a paddle wave maker and a siphon tube. The solution in the tank was fresh seawater pumped from the sea.

For the design of an ICCP system for a marine structure, some factors are needed including seawater resistivity, total protective current required, the anode type and the design life.

The total current requirement (I_{req}) for an offshore structure can be calculated as the product of the total exposed area (S) of the structure and the current density (i_j) for protection; this is expressed as:

$$I_{req} = i_j \cdot S \tag{4}$$

After the total current has been decided, the major problem is how to distribute the current to the whole structure, so the arrangement of the anodes is very important.

To investigate the effect of the distance between anode and structure, the anode was placed at a distance of 20 cm and 60 cm from the platform.

The electrochemical polarization experiment was conducted using a traditional three-electrode cell setup. The working electrode was steel with dimensions 1 cm × 1 cm × 1 cm and the exposed area was 1 cm^2. The reference electrode was saturated calomel electrode (SCE) with Luggin capillary, and the counter electrode was rectangular platinum sheet. The working electrode was abraded gradually using 600, 800 and 1200 grit Sic paper, degreased by acetone, washed with distilled water and dried in air.

Electrochemical experiments were carried out using a Parstat 2273. The polarization curve measurement performed a scan starting from −200 mV to +200 mV vs. open circuit potential at a scan rate of 0.5 mV/s.

3 Results and discussion

3.1 Boundary element method

The boundary element method applied to the corrosion analysis solves the Laplace equation [8]:

$$\nabla^2 \phi = 0 \tag{5}$$

subject to the boundary conditions illustrated in Fig. 3.

Some assumptions are made leading to simplifications of the corrosion conditions. Γ is the entire surface of the electrolyte domain Ω, comprising Γ_1, Γ_2, Γ_{3a} and Γ_{3b}; ϕ and I represent, respectively, the potential and current density at a point; ϕ_0 and I_0 are the initial constant values of potential and current density, respectively; $f_a(\phi_a)$ and $f_c(\phi_c)$ are relationships between the potential and current density of anode and cathode, respectively. For the ICCP system, the boundary condition parameter of anode is specified as a constant potential or constant current value, and the cathode parameter can be measured using linear polarization method.

The BEASY software, which was designed for solving the Laplace equation and the corresponding computation algorithm, has been used in the corrosion field for several years. The key points in obtaining an ideal computation result are the construction of the model and the application of the third boundary condition as described in Fig. 3.

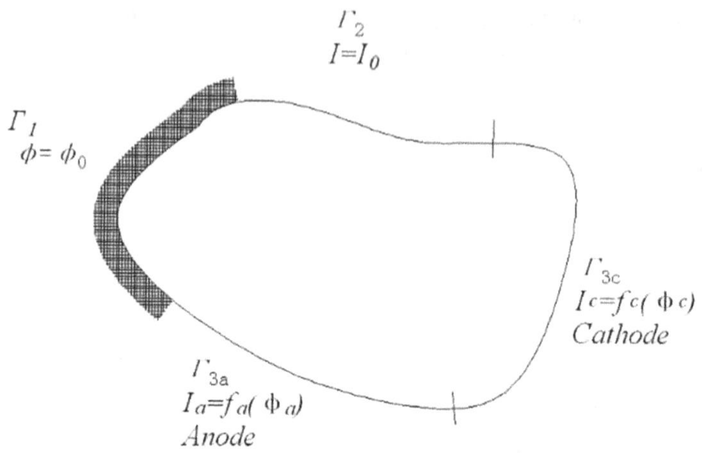

Figure 3: Boundary conditions for BEM.

3.2 Numerical modelling

The first step in the cathodic protection modelling is the development of the numerical model in a graphical user interface. The model was built using the coordinates of every point in the design drawing sheet. In the numerical model, all the dimension parameters were introduced into the BEASY software, including the cross-support tubes in the vertical and horizontal direction, including the diameters of every tube. On this basis, the anode model was added, with the accurate dimension, shape and location on the model. The computer model of the platform is shown in Fig. 4.

After completing the building of the computer model, the polarization data obtained from an electrochemical experiment were digitized into the software polarization database; this constituted the third boundary condition in the computer modelling. Besides the parameters mentioned above, the environmental conditions of the seawater solution were also very important, since the electrical conductivity and the temperature would affect the potential and current distribution of the cathodic protection system.

In the experimental investigation, three impressed current anodes were placed on each side of the jacket, that is, a total of 12 anodes. To evaluate the influence of the distance between anode and jacket platform on the cathodic protection, the anodes were arranged at different distances from the platform.

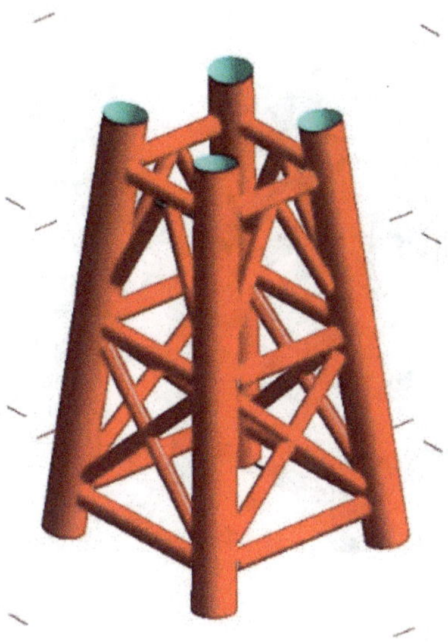

Figure 4: Computer model of the platform.

The calculated results for the potential distribution over the jacket are depicted in Fig. 5. With the anodes at a distance of 20 cm from the jacket, the value of the potential was in the range from −979.3 mV to −1068.9 mV, while with the anodes at a distance of 60 cm from the jacket, the value of the potential ranged between −980.9 mV and −1075.4 mV. The potential contours can visually reflect the potential distribution as well as the effect of anode location and number on the jacket. The actual potential distribution on the platform was measured to verify the numerical results. For 20 typical points on the platform, the results showed that the difference between numerical modelling and measured results was less than 12 mV, which was sufficiently small to confirm the effectiveness of the boundary element method for ICCP modelling.

4 Conclusions

In this research, the potential distribution over a jacket platform with impressed current cathodic protection in seawater was modelled using the boundary element method. The effect of different arrangements of the amount and location of the anodes on the cathodic protection was also assessed. BEM modelling was shown to be suitable for analysing the cathodic protection system of an offshore jacket.

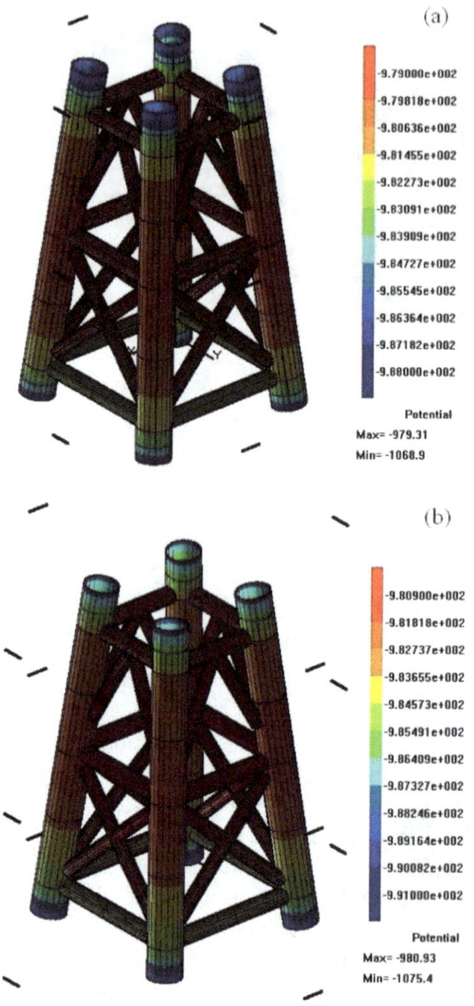

Figure 5: Potential distribution of the platform with ICCP: anodes at a distance of (a) 20 cm, (b) 60 cm.

Acknowledgements

The authors acknowledge the financial support of the International S&T Cooperation Program of China (2010DFR50860) and National Key Technology R&D Program (Grant No. 2012BAB15B01).

References

[1] Sharland, S.M., Jackson, C.P. & Diver, A.J., A finite-element model of the propagation of corrosion crevices and pits. *Corrosion Science*, **29(9)**, pp. 1149–1166, 1989.

[2] Zhou, C., Wang, C. & Song, Y., Evaluation of cyclic oxidation of thermal barrier coatings exposed to NaCl vapor by finite element method. *Materials Science and Engineering: A*, **490(1–2)**, pp. 351–358, 2008.

[3] DeGiorgi, V.G. & Wimmer, S.A., Geometric details and modeling accuracy requirements for shipboard impressed current cathodic protection system modeling. *Engineering Analysis with Boundary Elements*, **29(1)**, pp. 15–28, 2005.

[4] Aoki, S. & Amaya, K., Optimization of cathodic protection system by BEM. *Engineering Analysis with Boundary Elements*, **19(2)**, pp. 147–156, 1997.

[5] Amaya, K. & Aoki, S., Effective boundary element methods in corrosion analysis. *Engineering Analysis with Boundary Elements*, **27(5)**, pp. 507–519, 2003.

[6] Aoki, S. & Kishimoto, K., Prediction of galvanic corrosion rates by the boundary element method. *Mathematical and Computer Modelling*, **15(3–5)**, pp. 11–22, 1991.

[7] BEASY cathodic protection software user guide manual, version 10, CM Beasy limited.

[8] Lan, Z., Wang, X., Hou, B., Wang, Z., Song, J. & Chen, S., Simulation of sacrificial anode protection for steel platform using boundary element method, *Engineering Analysis with Boundary Elements*, **36(6)**, pp. 903–906, 2012.

Calculation of the protective current density distribution of a cathodic protection system with galvanic anodes in terms of double-layer electrolyte

A. Mujezinović[1], Ai. Muharemović[2], I. Turković[1]
& Ad. Muharemović[3]
[1]*University of Sarajevo, Bosnia & Herzegovina*
[2]*BH Telecom d.d. Sarajevo, Bosnia & Herzegovina*
[3]*Independent System Operator, Sarajevo, Bosnia & Herzegovina*

Abstract

This paper considers the effect of the discontinuity of electrolyte electrical conductivity on the distribution of potential and protective current density in cathodic protection systems with galvanic anodes. Its aim is to present a simultaneous application of both analytical and numerical models for the calculation of the distribution of protective potential and current density in the cases of homogeneous and double-layer electrolyte. For non-linear boundary conditions at the electrode surface (secondary distribution of protective current density), the indirect boundary element method is used, because of the complexity of calculation for which collocation at the point method was used. The calculation is further complicated due to the nonlinearity of boundary conditions at the electrode surfaces. In order to show the importance of this analysis, calculations for the observed system as well as of errors caused by neglecting the boundary discontinuity of the soil conductivity are provided. Based on these calculations and error analyses, the impact of the double-layer electrolyte on the correct calculation of the cathodic protection system with galvanic anodes is evaluated. This paper also provides the analysis when the double-layered nature of the electrolyte can be practically ignored, which is of great importance to designers of these systems.

Keywords: cathodic protection, protection current density, single-layer electrolyte, double-layer electrolyte.

1 Introduction

Cathodic protection (CP) is a technique that prevents corrosion of underground metallic structures. The factors that may affect the parameters of the modern CP systems are indispensable data for project documentation. They describe a multi-disciplinary phenomenon that complicates the definition of CP system parameters. The potential distribution of anode strings is determined for each specific case and it depends on the real polarization curves. The polarization curves, in the mathematical sense, define the boundary conditions for determining the value of the electrode potential and the current density. Calculation of the electrode potential and the current density can be done by numerical techniques. The boundary element method (BEM) is mostly used for the calculation of CP system parameters. The application of BEM for the calculation of the protective potential and the current density of long pipelines, with either larger or smaller insulation damage, is well covered by Riemer and Orazem [1]. Riemer developed a calculation method that takes into account the potential drop in the pipeline, i.e. the attenuation that is not balanced over the entire pipeline [2]. Adey *et al.* incorporated a full 3D model for calculating the potential of the CP system, also including neighboring objects that have connections to the anodes [3]. Peratta *et al.* applied the multi-layer (ML) BEM method for the calculation of the parameters of a CP system in multi-layer soil [4]. Liu and Wang proposed a line model-based fast BEM for the calculation of parameters of the cathodically protected pipelines placed in multi-layered soil [5].

The calculations of the distribution of both the potential and protective current density form the basis of a CP system project. The factors that complicate the calculation are the non-linear boundary conditions on electrode surfaces. In addition, the calculation needs to take into account the occurrence of any boundary discontinuity of electrical soil conductivity. The electrolyte, in which the metallic pipeline is buried, is mostly inhomogeneous (layered) and has two or more layers of different values of electrolyte conductivity [6]. Based on what is mentioned above, it becomes clear that the exact analytical solution of this problem is only partial and superficial.

2 Theory and implementation

This paper indicates that there is a substantial difference between methods used for determining the parameters of CP system, as well as in the treatment of boundary conditions, i.e. between:

- The exact analytical method that treats the primary distribution of protective current density, taking into account the characteristics of electrolyte and geometry of the system [7, 8], and
- The numerical method that treats the secondary distribution of protective current density, i.e. taking into account the kinetic effects that are specific to the electrode surface and result in nonlinear boundary conditions [9, 10].

2.1 Primary distribution of the protective current density

2.1.1 Homogeneous electrolytes

As previously mentioned, when modeling the primary distribution of the protective current density, only the electrical and geometrical characteristics of corrosive elements are relevant. In this section of the paper, the authors analyze an example of galvanic anode placed in a homogeneous electrolyte. If the DC source is installed as galvanic anode at some location in the electrolyte with coordinates (x', y', z'), then the potential φ at any point P in the electrolyte with coordinates (x, y, z), at distance r, can be defined as:

$$\phi(x,y,z) = \frac{1}{4\pi\gamma_u} \int_S \frac{J(x',y',z')}{r} \mathrm{d}S \,, \tag{1}$$

where: γ_u is the electric conductivity of the electrolyte, $J(x, y, z)$ the surface density of the protective current, r the distance between the source and observation points and dS the integration path.

In order to symplify the mathematical model, the observed anode string, which is shown in Fig. 1, is equivalented with one anode, having total length equal to the sum of the lengths of individual anodes. When solving the primary distribution of the protective current density of CP systems with galvanic anodes, it is necessary to add a set of equations that take into account the boundary conditions at the interface between anode strings and the electrolyte. The simplest case is the one that considers the potential at the complete outer surface of the anode string as a constant value (this condition can only be used in the primary distribution because

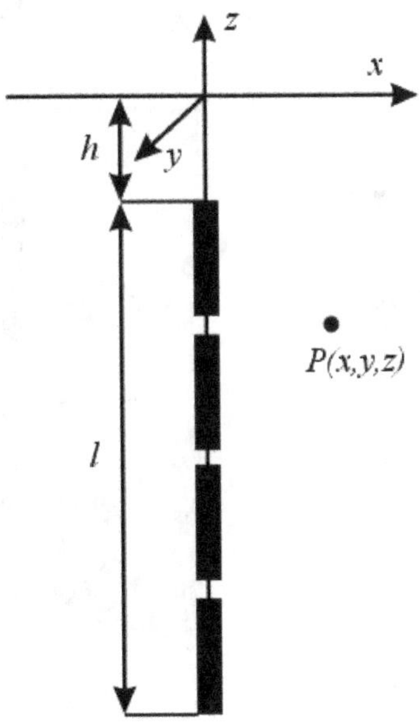

Figure 1: Anode string.

of neglecting the polarization on the surfaces of the galvanic anodes). Similarly, the potential change in the electrolyte is constant as well. The boundary conditions can be written as:

$$\varphi = const \tag{2.a}$$

$$\frac{\partial \varphi}{\partial \vec{n}} = const \tag{2.b}$$

$$\vec{n} \times \left(\frac{\vec{J}_u}{\gamma_u} - \frac{\vec{J}_a}{\gamma_a} \right) = 0 \tag{2.c}$$

$$\vec{n} \times \left(\vec{J}_u - \vec{J}_a \right) = 0 \tag{2.d}$$

where $\vec{n}$ is the normal unit vector on the boundary surfaces, γ_a the electrical conductivity of air, $\vec{J}_a$ the current density in air at discontinuity boundary of electrical conductivity, $\vec{J}_u$ the current density in electrolyte at the boundary of electrical conductivity discontinuity.

Because of its low electrical conductivity (10^{-18} S/m), the air is treated as an insulator. Therefore, the mathematical analysis of the problem considers the mirror source at a distance h away from the electrolyte/air boundary, with the same protective current density as the original, placed in the electrolyte (Fig. 2(a)). The

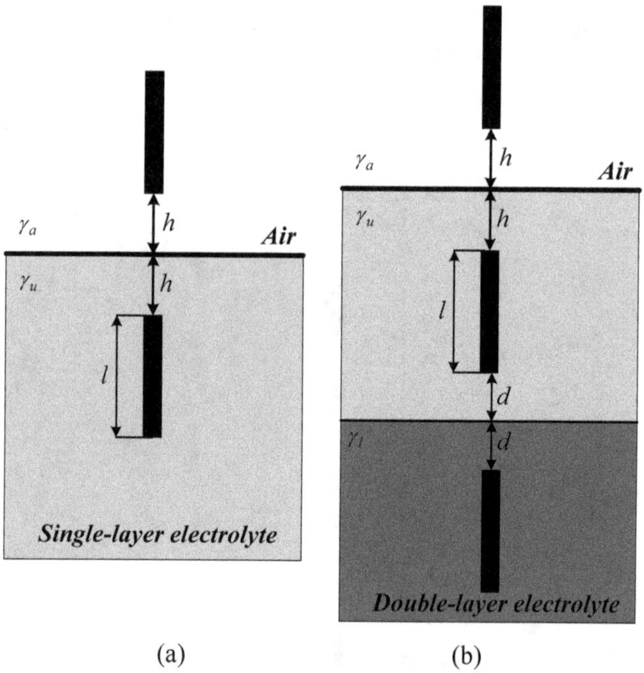

Figure 2: Equivalent anode in (a) single-layer electrolyte and (b) double-layer electrolyte.

expression for calculating the potential in the electrolyte, in the vicinity of the anodic string, is:

$$\phi = \frac{I}{4\pi l \gamma_u}\left(\ln\frac{z+l+h+\sqrt{(z+l+h)^2+r^2}}{z-l-h+\sqrt{(z-l-h)^2+r^2}} + \ln\frac{z-h+\sqrt{(z-h)^2+r^2}}{z+h+\sqrt{(z+h)^2+r^2}} \right) \quad (3)$$

where: l is the total length of the equivalent anode string, h the depth of burying anode string, I the total current intensity and r can be expressed as follows:

$$r = x^2 + y^2. \quad (4)$$

2.1.2 Double-layer electrolytes

Soil as the electrolyte is mainly composed of more layers, which differ in chemical and geological composition. When expecting a large difference in conductivity of the electrolyte layers, it is advisable to use a double-layer model of the electrolyte. In a case of simple geometry of electrodes, such as those considered in the paper, the easiest way to find a solution is by using a method of images.

In order to estimate the exact scalar value of the electric potential of the equivalent anodic strings, it is necessary to determine the position of equivalent galvanic strings, as well as their current densities. Furthermore, it is necessary to set appropriate boundary conditions. The boundary conditions applicable in this situation are almost identical to the boundary conditions for a homogeneous electrolyte. The equivalent anode string, as well as its images, has constant potential, the potential change within each medium is constant and on all the boundaries between the media boundary conditions given by eqns (2.c) and (2.d) apply. The double-layer model considers high-order mirror images that have a direct impact on the potential of the equivalent anode (Fig. 2(b)). The impact of the high-order image attenuates as the distance between the galvanic string and image increases, i.e. as the order of the image increases. Therefore, when calculating the potential of the anode surface, it is good enough to consider second order images. The effects of mirror images can be simply calculated as a sum of effects of individual images. The easiest way of calculating the potential of the equivalent galvanic string, buried in the double-layer electrolyte, is by using the following relations:

$$\phi(x,y,z) = \frac{I}{4\pi l \gamma_u}\left[f_1 + \alpha f_2 + \beta f_3 + \alpha\beta\left(f_4 + f_5\right)\right], \quad (5)$$

where:

$$\alpha \approx 1 \quad (5.a)$$

$$\beta = \frac{\gamma_u - \gamma_l}{\gamma_u + \gamma_L} \quad (5.b)$$

$$f_1 = \ln\frac{z+l+h+\sqrt{(z+l+h)^2+r^2}}{z-l-h+\sqrt{(z-l-h)^2+r^2}} \quad (5.c)$$

$$f_2 = \ln \frac{z - h + \sqrt{(z-h)^2 + r^2}}{z + h + \sqrt{(z+h)^2 + r^2}} \tag{5.d}$$

$$f_3 = \ln \frac{z + 2l + h + 2d + \sqrt{(z+2l+h+2d)^2 + r^2}}{z - 2l - h - 2d + \sqrt{(z-2l-h-2d)^2 + r^2}} \tag{5.e}$$

$$f_4 = \ln \frac{z - 2d - h - l + \sqrt{(z-2d-h-l)^2 + r^2}}{z + 2d + h + l + \sqrt{(z+2d+h+l)^2 + r^2}} \tag{5.f}$$

$$f_5 = \ln \frac{z + 3h + 3l + 2d + \sqrt{(z+3h+3l+2d)^2 + r^2}}{z - 3h - 3l - 2d + \sqrt{(z-3h-3l-2d)^2 + r^2}}, \tag{5.g}$$

where γ_l is the soil conductivity of the lower layer of double layer soil and d the distance between anode and lower layer of electrolyte.

The calculation results of the potential distribution in the vicinity of the anode string, according to eqns (3) and (5), are presented below. The total length of the equivalent anode string is 6 m and it is buried in a single-layer of soil (Fig. 3(a)) with soil resistivity ρ_u = 50 Ωm. In the second example (Fig. 3(b)), the anode string, with the same characteristics, is buried in the upper layer of double-layer soil. The soil resistivity of the upper layer is ρ_u = 50 Ωm and that of the lower layer ρ_l = 1000 Ωm. The total thickness of the upper layer is 10 m and the depth of anode string burial is 2.8 m. When modeling this example, second order images are taken into account.

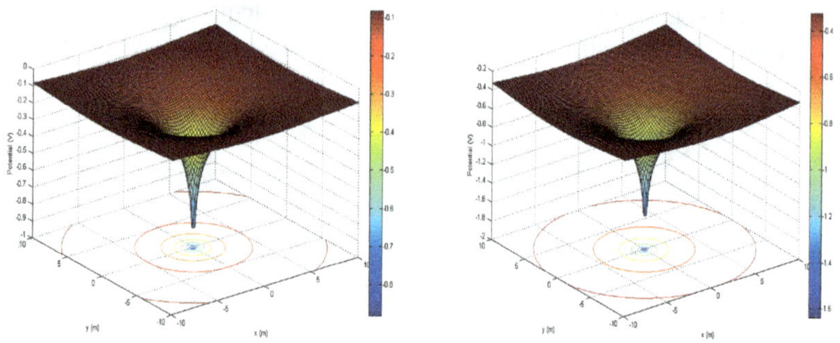

(a) (b)

Figure 3: Diagram of the potential distribution of anode string in (a) single-layer soil and (b) double-layer soil.

2.2 Secondary distribution of protective current density

When calculating the secondary distribution of the protective current density, the kinetics of the electrode processes must be taken into account. Processes of hydrogen separation, dissolving of metal and oxygen reduction, simultaneously occur on the cathode surface. As a result of electrochemical reactions, the expression for the polarization current density on the cathode surface can be written as:

$$j_C = j_{0,Fe} \times 10^{\frac{\phi-\phi_{Fe}}{\beta_{Fe}}} - j_{\lim,O_2}\left(1+10^{\frac{\phi-\phi_{O_2}}{\beta_{O_2}}}\right)^{-1} - j_{0,H_2} \times 10^{\frac{\phi-\phi_{H_2}}{\beta_{H_2}}}, \quad (7)$$

where j_C is the sum of all partial current densities for cathode surface (bare steel), $j_{0,Fe}$ the current density corresponding to the metal dissolution reaction, $j_{\lim,O_2}$ the threshold current density of oxygen reduction, j_{0,H_2} the current density corresponding to the reaction of hydrogen separation, ϕ_{Fe}, ϕ_{O_2}, ϕ_{H_2} are the corrosion potentials for the corresponding reactions, β_{Fe}, β_{O_2}, β_{H_2} are Tafel's coefficients (slopes) and ϕ is the potential difference of interface metal/electrolyte.

Galvanic anodes are made from materials that are more electrically active than the cathode materials. Therefore, the dominant reaction on the anode surface is oxygen reduction. Mathematically, this can be defined as follows:

$$j_A = j_{\lim,O_2}\left(10^{\frac{\phi-\phi_{Zn}}{\beta_{Zn}}} - 1\right), \quad (8)$$

where j_A is the current density for zinc anode, φ_{Zn} the corrosion potential of zinc and β_{Zn} the Tafel's coefficient (slope) for zinc.

The diagram in Fig. 4 presents the boundary conditions on the electrode surfaces of the CP system with galvanic anodes. It is noticeable that these are highly nonlinear boundary conditions, especially on a cathode surface. The values of the protective potential of the cathode surface depend on the operating time of the CP system. At the beginning, i.e. at the time of CP commissioning, the cathode and anode surfaces get on the potential defined by points A and B, respectively. In this situation, the voltage drop in the electrolyte is defined by $j_1 r_{p1\Sigma}$. After some time, i.e. after polarization of steel and increase of the resistance of the anodic string, the potentials of the cathode and anode are defined by points A_1 and B_1, respectively.

2.2.1 Electrode potential and current density calculation

When analyzing CP systems, finding the values of the electrode potential and protection current density on the boundary surface is of particular interest. Therefore, BEM is an ideal method for solving this problem. Electrode potential and current density of the electrode surface of the CP system placed in single or double-layer soil can be calculated by using following integral equation:

$$c(q)\phi(q) + \int_S \phi(p)\frac{\partial G(p,q)}{\partial \vec{n}}\mathrm{d}S = -\rho_u \int_S j(p)G(p,q)\mathrm{d}S + \phi_\infty, \quad (9)$$

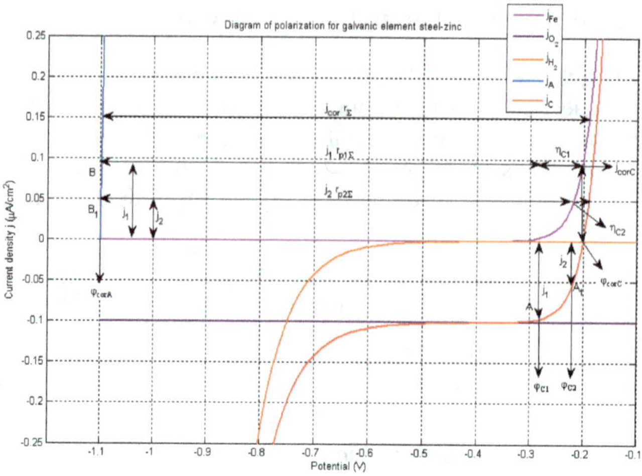

Figure 4: Polarization diagram of galvanic element steel/zinc (vs. Cu/CuSO$_4$).

where p is the field source point, q the observation point, $c(q)$ a constant, $j(p)$ the surface current density, $G(p,q)$ the Green's function, $\varphi(q)$ the electric potential at observation point, $\varphi(p)$ the electric potential at source point, φ_∞ is a constant potential on the boundary at infinity, ρ_u the soil resistivity of upper layer and S the boundary surface.

The corresponding Green's functions are:

- For a homogeneous soil:

$$G(p,q) = \frac{1}{4\pi}\left(\frac{1}{r_{p1,q}} + \frac{1}{r_{p2,q}}\right), \tag{10}$$

- For double-layer soil:

$$G(p,q) = \frac{1}{4\pi}\left[\frac{1}{r_{p1,q}} + \frac{1}{r_{p2,q}} + \sum_{n=1}^{\infty} \beta^n \left(\frac{1}{r_{p3,q}^n} + \frac{1}{r_{p4,q}^n} + \frac{1}{r_{p5,q}^n} + \frac{1}{r_{p6,q}^n}\right)\right], \tag{11}$$

where $r_{pi,q}$ is the distance between the i-th field source and observation points. The first field source point represents the original one while the others represent the images of the field source point.

Eqn (11) is the Green's function for double-layer soil when both source and observation points are in the upper layer. For the other situations, such as when the source and observation points are in different layers or when they are both in the lower layer of double-layer soil, Green's function has a different form that can be found in Berberovic *et al.* [11].

Integral eqn (9) is solved using the direct boundary element method. The electrode surfaces are discretized using 2D biquadratic boundary elements. The collocation method is applied at a point, so that the weight function is a Dirac delta function. Applying the collocation method at the point, an integral expression for a boundary element can be written as:

$$c(q_i)\phi(q_i)+\sum_{e=1}^{n_e}\int_{-1}^{1}\int_{-1}^{1}\sum_{i=1}^{9}[\phi_i(p)N_i(\xi_p,\eta_p)]\det[J(\xi_p,\eta_p)]\frac{\partial G(\xi_p,\eta_p;q)}{\partial \vec{n}}\mathrm{d}\xi\mathrm{d}\eta$$
$$=-\rho_u\sum_{e=1}^{n_e}\int_{-1}^{1}\int_{-1}^{1}\sum_{i=1}^{9}\left[j_i(p)N_i(\xi_p,\eta_p)\right]\det[J(\xi_p,\eta_p)]G(\xi_p,\eta_p;q)\mathrm{d}\xi\mathrm{d}\eta\,, \tag{12}$$

where $N(\xi,\eta)$ is the shape function, ξ, η are Gauss points, $\det[J(\xi,\ \eta)]$ is the determinant of the Jacobean matrix and n_e the number of boundary elements.

In order to solve this system of equations, it is necessary to add another integral equation [12], as follows:

$$\int_S j(p)\mathrm{d}S=0\,. \tag{13}$$

After the application of the collocation method at a point to eqn (13), the latter can be written:

$$\sum_{e=1}^{n_e}\int_{-1}^{1}\int_{-1}^{1}\sum_{i=1}^{9}\left[j_i(p)N_i(\xi_p,\eta_p)\right]\det[J(\xi_p,\eta_p)]\mathrm{d}\xi\mathrm{d}\eta=0\,. \tag{14}$$

The previously obtained systems of eqns (12) and (14) can be expressed in matrix form [2]:

$$\begin{bmatrix} H_{aa} & H_{ac} & -1 \\ H_{ca} & H_{cc} & -1 \\ 0 & 0 & 0 \end{bmatrix}\begin{Bmatrix} \phi_a \\ \phi_c \\ \phi_\infty \end{Bmatrix}=\begin{bmatrix} G_{aa} & G_{ac} \\ G_{ca} & G_{cc} \\ A_a & A_c \end{bmatrix}\begin{Bmatrix} j_a \\ j_c \end{Bmatrix}, \tag{15}$$

where $[H]$ and $[G]$ are matrices of coefficients, $\{\varphi\}$ is the column vector of potential and $\{j\}$ the column vector of current density. In eqn (15), indices c and a represent cathode and anode surface, respectively.

To solve eqn (15) it is necessary to apply an iterative solution technique due to the nonlinear boundary conditions. By solving this system of eqn (15), the values of the protection current density and potential on the electrode surfaces can be obtained.

2.2.2 Calculation of the potential on the soil surface

The mathematical model presented in the previous section is suitable for the calculation of electric potential and current density distribution on the electrode surfaces of the CP system. For the calculation of the electric potential distribution on the soil surface, Fredholm's integral equation of the first kind can be applied:

$$\varphi(q)=\rho_u\int_S j(p)G(p,q)\mathrm{d}S \tag{16}$$

Fredholm's integral equations of the first kind are solved using the indirect boundary element method. Integral equation (16) is solved in almost identical way as the integral equations (9) and (13). After the collocation method is applied at a point, eqn (16) can be rewritten in the following form:

$$\varphi(q_i)=\rho_u\sum_{e=1}^{n_e}\int_{-1}^{1}\int_{-1}^{1}\sum_{i=1}^{9}[j_i(p)N_i(\xi_p,\eta_p)]\det[J(\xi_p,\eta_p)]G(\xi_p,\eta_p;q)\mathrm{d}\xi\mathrm{d}\eta \tag{17}$$

The eqn (17) can be expressed in matrix form:

$$\begin{bmatrix} G_{as1} & G_{cs1} \\ G_{as2} & G_{cs2} \\ G_{as3} & G_{cs3} \end{bmatrix} \begin{Bmatrix} j_a \\ j_c \end{Bmatrix} = \begin{Bmatrix} \varphi_{s1} \\ \varphi_{s2} \\ \varphi_{s3} \end{Bmatrix}, \tag{18}$$

where index si represents the i-th colocation point on the soil surface.

As can be noted in eqn (18), for the calculation of the electric potential distribution on the soil surface, the current density distribution on the electrode surface must be known.

Fig. 5 shows the distribution of the protective potential of the CP system with galvanic anodes of the non-isolated steel pipeline, for homogeneous soil with soil resistivity $\rho_u = 50\ \Omega\text{m}$. Fig. 6 shows the same CP system in double-layer soil with a soil resistivity of the upper layer $\rho_u = 50\ \Omega\text{m}$, thickness of the upper layer $h = 10$ m and soil resistivity of the lower layer $\rho_l = 1000\ \Omega\text{m}$. The pipeline has a diameter of 4 m, and is buried at a depth of 3.75 m. Anode strings are placed on both sides of the pipeline at a distance of 5.5 m away from the pipeline axis. The burial depth of the anode strings is 2.8 m, and its length is 6 m.

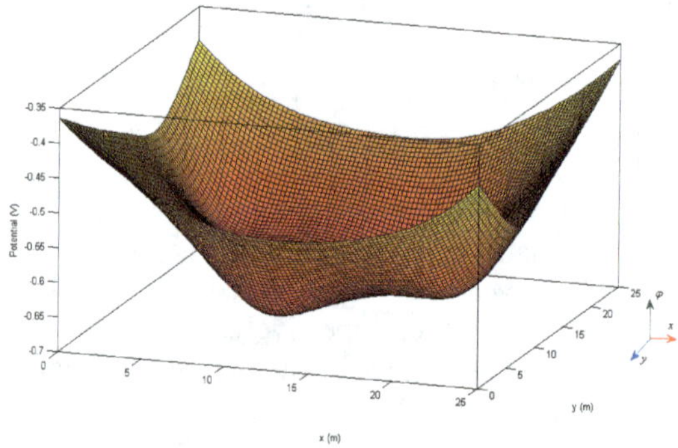

Figure 5: Potential distribution of CP system in single-layer soil.

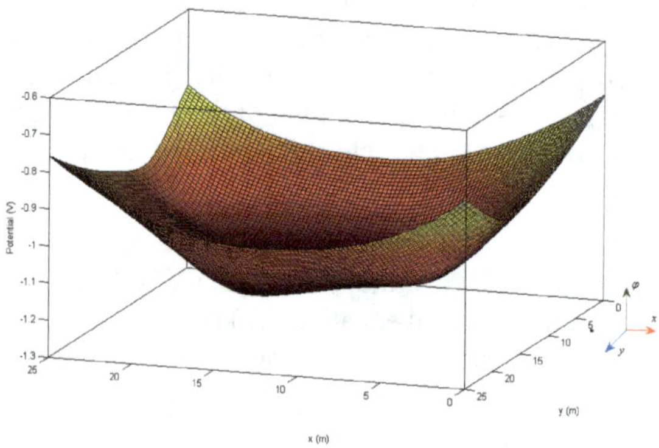

Figure 6: Potential distribution of CP system in double-layer soil.

Fig. 7 provides a comparison of protective potential values for single-layer and double-layer soil. It can be noticed that the absolute value of the protective potential increases as the electrical resistivity of the lower layer increases. In such situations, neglecting the multiple layers of soil leads to overprotection of a protected object and high financial cost.

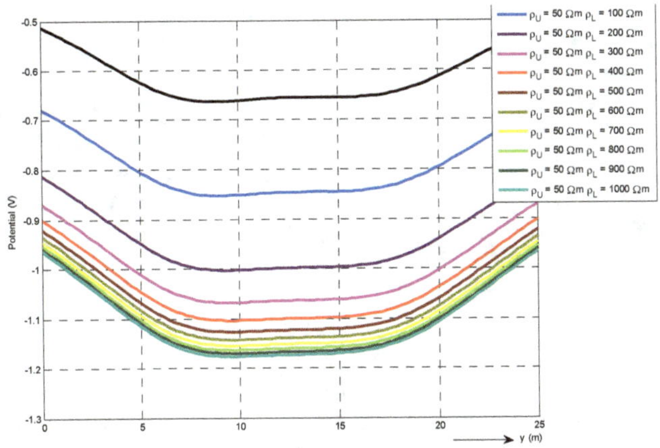

Figure 7: Comparison of the equipotential lines of the CP system for cases of single-layer and double-layer soil for x = 12.5 m; 0 < y < 25 m.

Fig. 8(a) gives the value of the percentage error of the potential, in the case of neglecting multiple layers of soil. It is noticeable that with the increase of soil resistivity, the percentage systematic error also increases, by over 40%. In these situations, from a technical point of view, neglecting multiple layers has a positive effect, because an object becomes overprotected. On the other hand, from an economic point of view, the cost of equipment increases. Therefore, for the large differences in the values of soil resistivity, between neighboring layers, it is necessary to take into account the multiple layers of electrolyte. Fig. 8(b) shows how the percentage systematic error changes with the change of thickness of the upper layer (the layer in which protected object and anode strings are placed). The considered double-layer soil had an upper layer with variable thickness and soil resistivity $\rho_u = 50\ \Omega m$, while the soil resistivity of the lower layer is $\rho_l = 1000\ \Omega m$ in the first case, and $\rho_l = 500\ \Omega m$ in the second case. In both situations it is noticeable that the error decreases with increasing thickness of the upper layer of soil.

3 Conclusion

The main task in determining the optimal CP system with galvanic anodes is to define technically the correct distribution of the protective current density and potential. Usually, the static current field of the CP system is calculated by solving

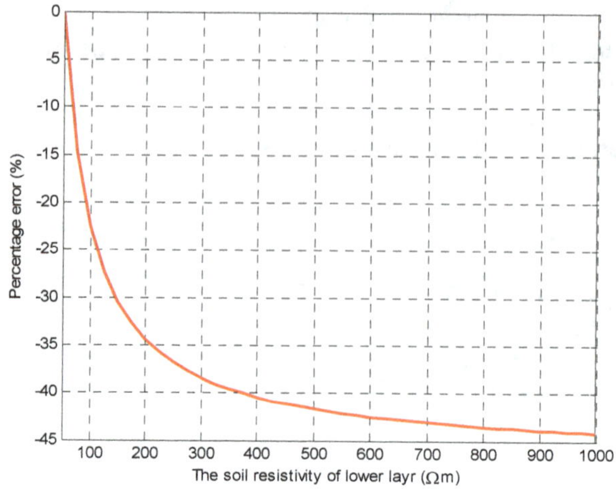

(a)

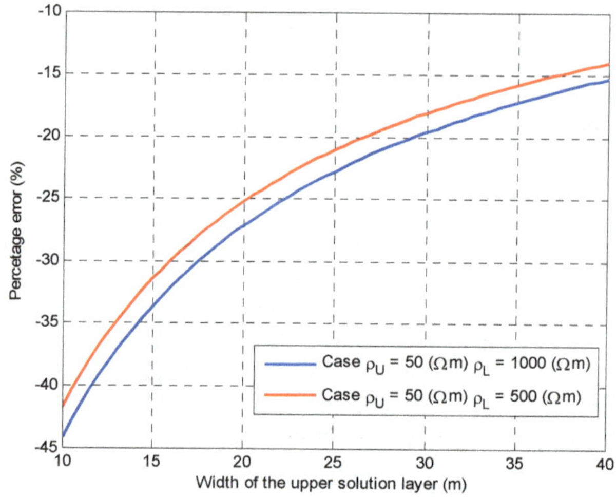

(b)

Figure 8: Percentage errors in the calculation of potential of the CP system.

the integral field equation. When applying the CP system to underground steel structures, geometries as well as nonlinear mathematical interpretation of electrochemical reactions are very complex. Therefore, the integral field equation can only be solved using numerical techniques. Also, considering the fact that the electrolyte is inhomogeneous, the paper analyzes the effect of discontinuity of electrolyte conductivity on the distribution of the protective potential of the CP system with galvanic anodes. This problem has been solved using the indirect boundary element method. The paper also presents an analysis of the systematic errors that appear when neglecting the fact that the electrolyte is composed of more layers. It turned out that errors in the protective potential calculation can occur and can exceed 40%. Also, in a case of double-layer soil with an upper layer of large thickness, the soil can be modeled as single-layered. In this case, the soil resistivity of a single-layer is equal to the soil resistivity of the upper layer in a double-layer model. This conclusion follows from the fact that the percentage systematic error decreases with increasing thickness of the upper layer.

References

[1] Riemer, D.P. & Orazem, M.E., Modelling coating flaws with nonlinear polarization curves for long pipelines. *Modelling of Cathodic Protection Systems*, ed. R.A. Adey, WIT Press: Southampton, pp. 225–257, 2006.

[2] Riemer, D.P., *Modelling Cathodic Protection for Pipeline Networks*, Ph. D. Dissertation, University of Florida, Gainesville, FL, 2000.

[3] Adey, R.A., Brebbia, C.A. & Niku, S.M., Application of boundary elements in corrosion engineering. *Topics in Boundary Element Research, vol. 7: Electrical Engineering Applications*, ed. C.A. Brebbia, Springer: Berlin, pp. 34–64, 1990.

[4] Peratta, A.B., Baynham, J.M.W. & Adey, R.A., Computational modelling of cathodic protection systems for pipelines in multi-layer soil. *WIT Transactions on Engineering Sciences*, **65**, pp. 35–46, 2009.

[5] Liu, L.Q. & Wang, H.T., A line model – based fast boundary element method for the cathodic protection analysis of pipelines in layered soils. *Computer Modeling in Engineering & Sciences*, **90(6)**, pp. 439–462, 2013.

[6] Barić, T., Boras, V. & Nikolovski, S., Analysing the effect of discontinuity of electrical conductivity of soil on ground resistance. *Energija Journal of Energy*, **55(4)**, pp 454–473, 2006.

[7] Haznadar, Z. & Stih, Z., *Electromagnetic Fields, Waves and Numerical Methods,* IOS Press: Amsterdam, 2000.

[8] Baeckmann, W., Schwenk, W. & Prinz, W., *Handbook of Cathodic Corrosion Protection*, Gulf Publishing Company: Houston, Texas, 1997.

[9] Muharemovic, A., Turkovic, I., Muharemovic, A., Tasakovic, S. & Mujezinovic, A., Calculation methods of cathodic protection system parameters with vertical anode zinc string. *20th International Expert Meeting*, Power engineering, Maribor, Slovenia, 2011.

[10] Muharemović, A., Zildžo, H. & Letić, E., Modelling of protective potentials distribution in cathodic protection systems using coupled BEM/FEM method. *WIT Transactions on Modelling and Simulations*, **47**, pp. 105–113, 2008.

[11] Berberovic, S., Haznadar, Z. & Stih, Z., Method of moments in analysis of grounding systems. *Engineering Analysis with Boundary Elements*, **27(3)**, pp. 351–360, 2002.

[12] Santiago, J.A.F. & Telles, J.C.F., A solution technique for cathodic protection system with dynamic boundary conditions by the boundary element method. *Advances in Engineering Software*, **30(9-11)**, pp. 663–671, 1999.

Computational modelling of cathodic protection systems for pipelines in multi-layer soil

A. B. Peratta, J. M. W. Baynham & R. A. Adey
CM BEASY Ltd, UK

Abstract

Computational modelling of cathodic protection (CP) systems involving thin multi-layer media represents a real challenge in terms of accuracy and efficiency required in the numerical calculation. In the case of CP for transmission pipelines, these long metallic structures are usually buried a distance H (approximately a metre or so) below ground level and extend horizontally typically more than ten thousand times H. A number of impressed current anode beds are distributed along the pipeline, providing protection against the corrosion of the structure. In addition, the vertically stratified nature of the soil needs to be considered in the model, in order to obtain more accurate representation of the environment. This is particularly relevant when considering the effect of different types of rocks, soil porosity, or water saturation, at different depths. This type of scenario requires three-dimensional modelling involving a thin multi layered electrolyte, with a typical aspect ratio (lateral extension to thickness) of the order 10^4 to 10^6. The paper presents an efficient and accurate computational approach based on the boundary element method for simulating the level of protection against corrosion of the pipeline as well as current densities and electric potential in different points of the soil. The resulting modelling approach is then applied to assessing real case scenarios. The simulation approach considers the non-linear electrode kinetics on the metal surfaces in the form of polarisation data and also the internal resistance of the pipeline and other electrical connections involved in the CP system. Example applications are presented showing how the model can be used to predict the "signatures" associated with different defect types in the pipe coating.

Keywords: cathodic protection, multi-layer, boundary element method, transmission pipelines.

1 Introduction

One of the most effective ways to control corrosion in metallic structures embedded in an electrolyte, such as the case of underground or undersea transmission pipelines, is by means of a combination of coating and impressed current cathodic protection (ICCP) systems. Field surveys such as the direct current voltage gradient (DCVG), and monitoring systems are aimed at different aspects of the assessment of the corrosion control system. Field measurements require skilled technicians, are often expensive and difficult to obtain. In addition, different sections of the pipeline are not always accessible to the surveyors and, in most cases, the noise in the potential measurements can mask developing defects in the pipeline coating. Moreover, the interpretation of field data and its correlation to the performance level of the corrosion control system is not always straightforward.

The direct computational modelling of CP systems offers a variety of tools for processing the information collected from the field and complementing it, allowing better interpretation of field measurements, as well as enabling the correlation against the CP design parameters and their impact on the observable magnitudes.

By direct CP simulation, it is meant the process of predicting field results such as:

- ON and OFF potentials,
- potential gradients and currents fields in the soil, as well as
- over potential and current density along the pipeline;

provided that the ICCP design parameters, such as, for example:

- geometrical arrangement and material properties of pipelines and anodes,
- electrical resistivity distribution of the soil,
- type and location of the electrical connections between rectifiers, anodes and structure,
- coating breakdown factor along the pipeline

are given as input data.

The direct simulation of ICCP systems by computational modelling has achieved in the past few years quite a mature state producing reliable results for industrial environments. This type of simulation is particularly useful at the design stage for the analysis of "what-if" case scenarios including the problems of interference with foreign CP systems or metallic structures.

The purpose of this work is to present an efficient 3D direct simulation tool based on the boundary element method, applied to two case scenarios involving transmission pipelines. The paper is focused on the effects in the potential measurements introduced by variable breakdown factors associated with the coating.

This document is organised in the following way. Section 2 describes modelling approach and provides pointers to the relevant literature involved in the method. Section 3 introduces a simple case for study where the effects of localised damages in the coating along the pipeline can be detected. Section 4 shows a more realistic case scenario involving three parallel pipelines running along the same corridor. Finally, section 5 establishes the conclusions.

2 Direct modelling

The direct modelling consists of obtaining results of electric potential and current density at any point in the electrolyte and on the surfaces of electrodes. Under most common situations, this requires solving the steady state charge conservation equation in the electrolyte in 3D space given by:

$$\nabla \cdot \mathbf{j} = 0 \ , \quad \mathbf{x} \in \Omega \tag{1}$$

where

$$\mathbf{j} = -\sigma(\mathbf{x}) \ \nabla u(\mathbf{x}) \tag{2}$$

represents current density, σ is the electrolyte conductivity, $u(\mathbf{x})$ is the potential field, ∇ is the 3D Laplace operator and Ω represents the integration domain (electrolyte).

Eqns (1) and (2) can be solved together with the corresponding boundary conditions which are usually prescribed by imposing polarisation curves at the electrode surfaces, isolating conditions at ground level, and/or fixed potentials at any known equipotential surfaces in the electrolyte (if any).

The boundary element method (BEM) [1] has been widely used to solve Laplacian equations and in particular simulate cathodic protection systems for underground and offshore structures [2–4]. The most significant advantages of the method are first that the formulation is based on the fundamental solution of the leading partial differential operator in the governing equation, and second that it requires only mesh discretisation on the boundaries of the problem. The former aspect confers high accuracy, while the latter substantially simplifies the pre-processing stage of the model, since volume discretisation is not needed.

The forward modelling of long transmission pipelines involves considering the soil as a thin film electrolyte (see Fig. 1), since the pipeline span (L) is much bigger than the soil depth (h) relevant to the modelling. In addition, the soil is generally stratified in one or many layers along the vertical (z) direction. This thin film stratified integration domain is very difficult to solve with standard modelling techniques such as FEM or BEM. Therefore, in order to be able to solve this type of integration domains without the need for extraordinarily high computational resources, a "multi-layer" BEM has been developed (ML-BEM).

The idea behind the ML-BEM is that the stratified nature of the medium is packaged into the corresponding Green's function. In other words, the BEM is applied in the same way as in the case of the homogeneous electrolyte, except that the Green's function for the homogeneous Laplace equation given by

$$G(\mathbf{x}_i, \mathbf{x}_j) = 1/(4\pi\, r)$$

is replaced by the multi-layer Green's function given by:

$$G(\mathbf{x}_i, \mathbf{x}_j, m, n) = \frac{1}{4\pi\sigma_m} \sum_{k=1}^{N_{appr}} \frac{\alpha_{ijmn}}{\left\| \mathbf{x}_i - \mathbf{x}_j + \mathbf{g}_{ijmn} \right\|} \tag{3}$$

where $\mathbf{x}$ denotes the 3D position vector and $r = \|\mathbf{x}_j - \mathbf{x}_i\|$; the sub-indices i and j stand for the source and field point, respectively while m and n indicate the layer of the source and field points, respectively; α_{ijmn} is a weight coefficient and $\mathbf{g}_{ijmn}$ denotes a displacement vector. The Green's function written in this way can be regarded as one produced by a weighted method of images. The summation is performed over the number of approximating terms N_{appr}.

The calculation of the weight and displacement vectors goes beyond the scope of this paper and can be found in Peratta *et al.* [5] and citations therein. Finally, the Green's function (3) replaces the $1/r$ kernel used for homogeneous regions and the same BEM strategy can be employed.

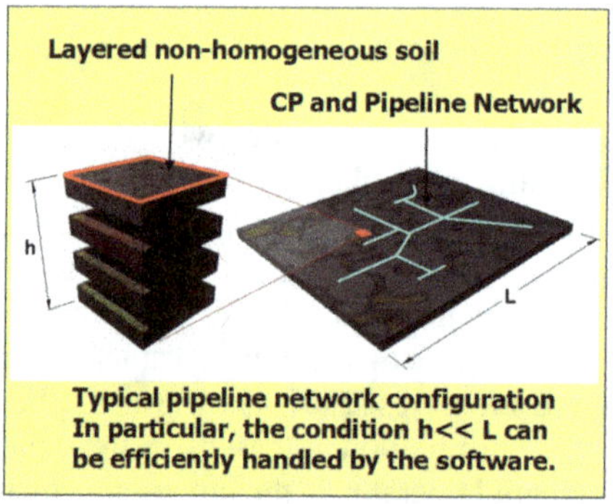

Figure 1: The soil (electrolyte) considered as thin film stratified medium.

3 Case study 1

Fig. 2(a) shows a 16 km length of pipeline, composed of 5 electrically connected subsections. At the connection joints between subsections there are four rectifiers delivering constant current to 4 anodes A1 to A4. The table in Fig. 2(b) shows the *x-y* coordinates of the anode in the ground. In the figure, *Dz* represents the vertical dimension of the anode, which extends down from the ground surface. Details of the pipeline are shown in Table 1. The total current delivered by the anodes is 40 A (10 A each). The polarisation curve used for steel is shown in Fig. 3.

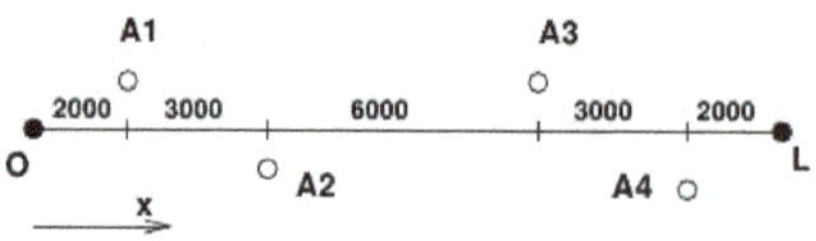

(a)

Anodes	x	y	Dz	diam
A1	2000	-200	0-50	0.2
A2	5000	+200	0-50	0.2
A3	11000	-200	0-50	0.2
A4	14000	+200	0-50	0.2

(b)

Figure 2: (a) Pipeline layout, (b) relevant anode parameters; dimensions in m.

Table 1: Pipeline specifications.

PIPELINE		
L	16 km	Length
z	4 m	Depth
D_e	1 m	External diameter
t	0.01 m	Wall thickness
σ_s	5.74713×10^6 S/m	Steel conductivity
ρ	$5.59453 \times 10^{-6}\ \Omega/m$	Resistance per unit lenght

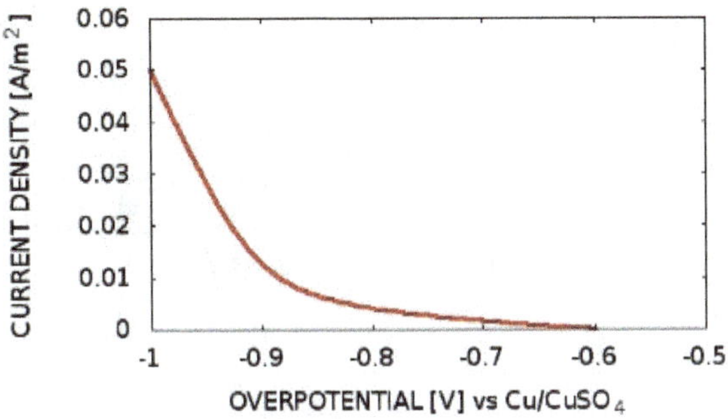

Figure 3: Polarisation curve for the metallic structure.

The soil model consists of two layers, with depths and conductivity as shown in Table 2. The top of the first layer and the bottom of the second layer are insulating. The system of coordinates is such that the pipeline is oriented in x direction from $x = 0$ to $x = 16{,}000$ m; the z direction indicated depth ($z = 0$ represents ground level, and $z = 4$ m is the centreline of the pipeline). Note that models with this type of boundary conditions are not easily solved with standard techniques such as half space or method of images [6].

Table 2: Soil properties.

Layer	z_{min} to z_{max} (m)	Conductivity (S/m)
1	0 to –30	0.02
2	–30 to –100	0.007

Fig. 4 shows the over potential along the pipeline (which is almost the same as the off-potential calculated above the pipeline at ground level) for different conditions of the pipeline. The red curve corresponds to a uniform breakdown factor BF0 = 1 (100%, bare steel) for the whole pipeline. The green curve corresponds to the case of a uniformly partially damaged coating with BF0 = 0.02 (2%). The black dotted curve corresponds to undamaged uniform coating characterised by uniform BF0 = 0.01 (1%). The breakdown factor (BF) is a convenient method for representing general degradation of the pipeline coating. A 10% BF implies that the coating has degraded to the point where an area equivalent to 10% of the surface area is bare metal.

The solid black curve shows the pattern of results produced by 3 localised coating defects at 6,000, 8,000 and 10,000 m (with local breakdown factor BF equal to 0.02 (2%), 1 (100%) and 1 (100%), respectively) on the pipeline which has an otherwise uniform breakdown factor of 1%. The length of each defect along the pipeline is 2 m. It can be observed that the effect of the local damage on the over potential is much localised, and that there is no observable general shift of over-potential. This is because in this case the change of current density is small at regions away from the damage.

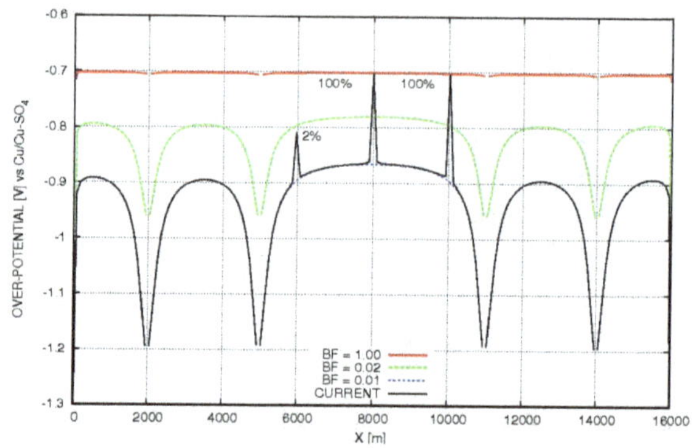

Figure 4: Comparison between different coating damages in the over potential distribution along the pipeline.

The pattern observed for the case of 19 localised defects is shown in Fig. 5. The distribution of defects is shown in Fig. 6 in terms of the breakdown factor as a function of length along the pipeline.

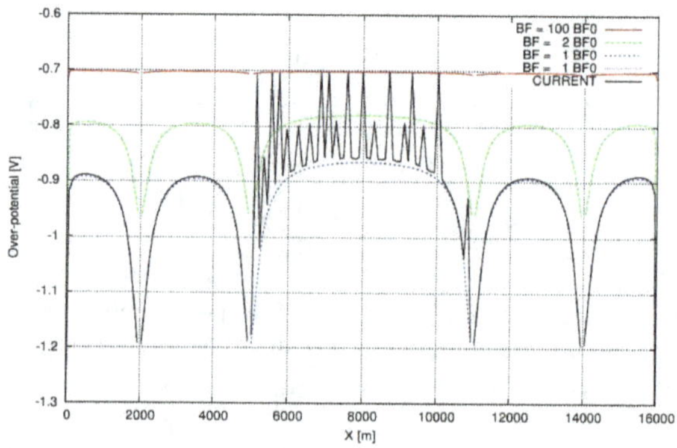

Figure 5: "Signature" over potential along the pipeline produced by 19 defects distributed as shown in Fig. 6.

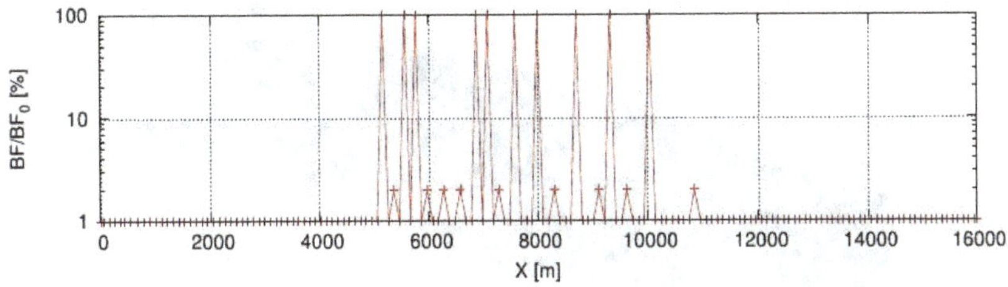

Figure 6: BF along pipeline corresponding to the pattern shown in Fig. 5.

The method of using breakdown factors in the model to represent local breakdown in the quality of the coating is adequate when the area of damage is relatively large (i.e. larger than a few metres) and the model is capable of predicting what the potential and gradient signatures generated near the pipeline are. This provides a methodology of characterising the size and extent of the defect from its signature. However, for more distinct local holidays in the coating, a different type of model is required as it is necessary to model the variation of the field around the pipe as discussed in the next section.

The simulation tool is also capable of modelling in more detail the effect of localised damage and the pattern/signature of potential and potential gradient at ground level produced by small defects in the coating. The rest of this section shows the simulation results applied to the case of a 0.2 × 0.2 m^2 zone of damaged coating at x = 10,000 m. The situation is illustrated in Fig. 7 where a three-dimensional model of the pipeline is shown. This model is different from the previous one in that the elements consider the variation of the coating quality around the pipe.

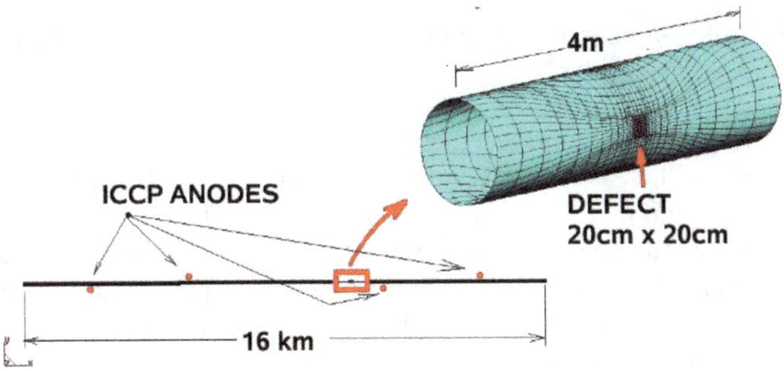

Figure 7: Detailed representation of a local defect.

Fig. 8 shows the over potential near the holiday. The rest of the pipeline (total length 16 km) is also included in the model and it is considered to have a uniform coating breakdown factor of 0.1%. In this case, each ICCP anode injects 2 A to the system and an increased number of elements are used in the region that contains the holiday.

The on-potential at ground level with respect to the pipeline can be observed in Fig. 9.

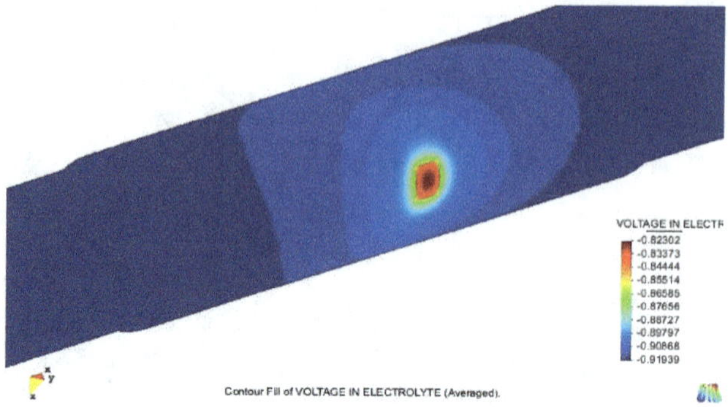

Figure 8: Overpotential near a lateral holiday.

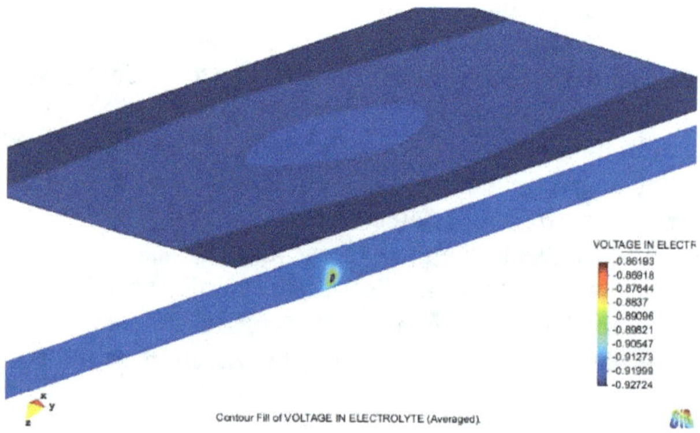

Figure 9: Soil-to-pipe potential at ground level and electrolyte potential on the pipeline.

Fig. 10 represents the surface distribution of transversal current density at ground level when the CP system is switched on. The observation plane (at ground level) covers a region of 30×15 m^2 centred at x =10,000 m.

It is interesting to compare the different types of pattern produced by the different types and location of defects in the pipeline. In particular, the contrast of voltage and voltage gradient profiles with respect to the case of a healthy pipeline (no holiday at all) is shown in Figs. 11 and 12.

Fig. 11 on the left shows a comparison of longitudinal voltage gradients at ground (z = 0) level at y = 0.5 m (the pipeline is in the plane y = 0, at depth z = 4 m). The figure compares the case of a uniformly coated pipeline (no defect) and the same pipeline with the holiday (defect). Fig. 10 on the right shows the same comparison but with the transversal voltage gradient at different y positions.

Fig. 12 compares the potential profiles for the pipeline with and without holiday. The picture on the left shows the soil-to-pipe on-potential at ground level above the pipeline (i.e. y = 0, z = 0). The corresponding over potential profile of each pipeline is shown on the right (z = 4 m, y = 0.5 m).

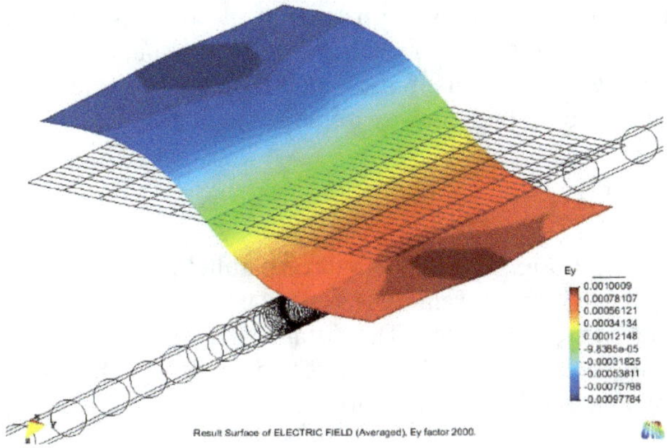

Figure 10: Surface distribution of y component of current density at ground level.

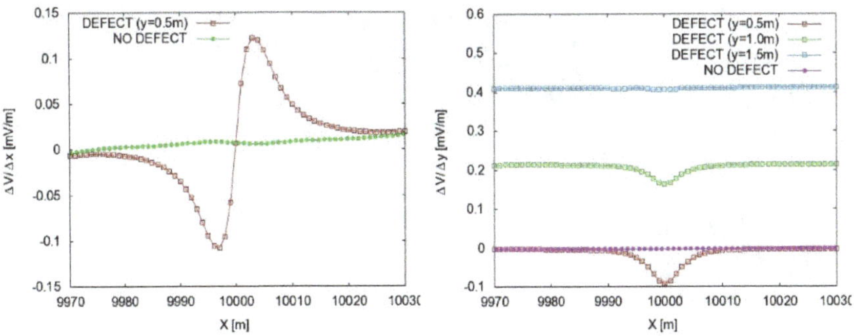

Figure 11: Voltage gradient in x (left) and y (right) direction at ground level.

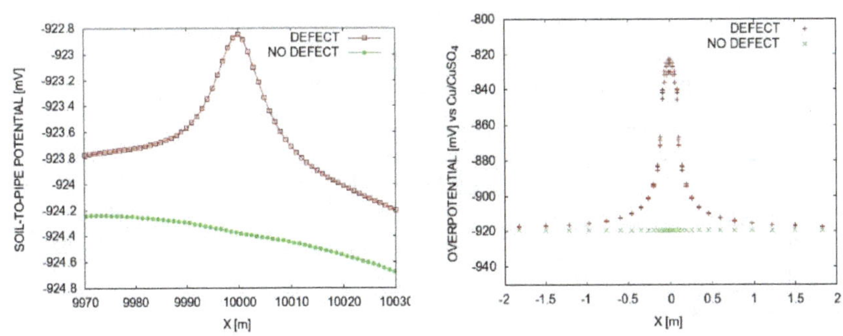

Figure 12: (Left) Soil-to-pipe potential. (Right) Over potential on pipeline near the defect.

The results demonstrate how the pattern of electrical signatures observed at ground level due to the presence of small defects on the coatings can be determined in addition to the attenuation of the potential and potential gradients as the observation point moves farther away from the coating.

4 Case study 2

The model shown in Fig. 13 considers three parallel pipelines of approximately 70 km in length sharing the same CP system which consists of 11 ICCP deep anodic beds delivering 10 A each. The distance between pipelines is 10 m in the local normal direction. The model considers two soil layers (0.02 S/m from ground level to –35 m, and 0.005 from –35 m to –100 m).

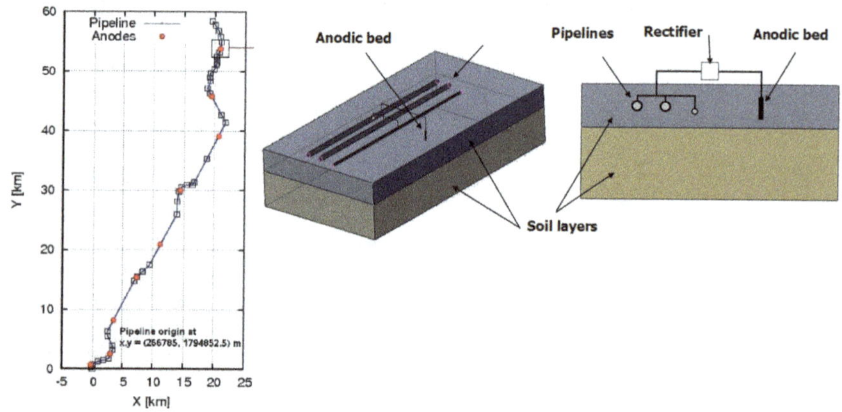

Figure 13: Schematic representation of case study 2. (Left) Plan view: the dots indicate the location of the ICCP anodes. (Middle and right) Detail of electrical connections between anode and the pipelines at each connection point.

Table 3: Relevant pipeline properties for case study 2.

Pipeline name		A	B	C
Type		NPS6 SCH80s	NPS16 SCH	NPS10 SCH80
Pipeline length (m)	*L*	68 442 m	68 442 m	68442 m
Ext. diameter (inch)	*De*	6.625″	16″	10.750″
Wall thickness (inch)	*t*	0.432″	0.656″	0.594″
Metal resistivity (Ohm·m)	*rho*	1.74×10^{-7}	1.74×10^{-7}	1.74×10^{-7}
Breakdown factor	BF	Uniform = 50%	Uniform 3%	Uniform 100%

Table 4: Anodes location co-ordinates (m).

Name	x	y	Name	x	y
A1	266,543.4	1,795,525.1	A7	285,301.0	1,829,520.0
A2	269,839.0	1,797,394.0	A8	287,517.0	1,833,957.0
A3	270,400.0	1,803,073.0	A9	286,274.0	1,840,641.0
A4	274,181.0	1,810,219.0	A10	287,682.0	1,848,656.0
A5	278,114.0	1,815,776.0	A11	286,435.0	1,853,240.0
A6	281,253.0	1,824,888.0			

Fig. 14 (left) shows the over potential of the three pipelines along their developed length. In this case, the difference between over potential and OFF potential is negligible.

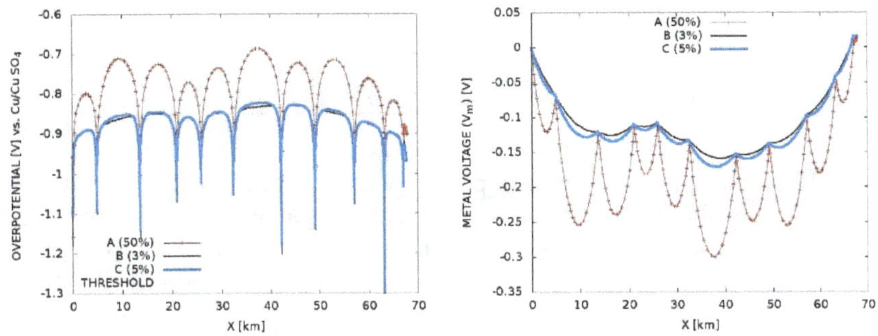

Figure 14: (Left) Over potential along the three pipelines. (Right) Metal voltage potential along the pipelines.

The ohmic drop in the return path due to finite conductivity of the metal is not negligible in large distances and usually affects the CP parameters. In this example, the voltage difference at different points of the structure is approximately 300 mV, as observed in Fig. 14 (right). The maximum variations are registered in pipeline A due to its smaller cross sectional area which introduces a higher resistance per unit length.

The simulation has been repeated considering different qualities of the coating for pipeline C. The effects of the quality of the coating of pipeline C on the over potential profile along the neighbouring pipelines can be observed in Fig. 15. The top figure shows the case of variable breakdown factor estimated from a CIS survey, in this case, the BF ranges from 0 to approximately 35%. The bottom figure represents the case of pipeline C with no coating at all.

The results obtained demonstrate a number of features. First, the feasibility of the modelling approaches for estimating the quality of the coating using data from survey data which is generally spiky and noisy. Second, they reveal how the quality of the coating of one of the pipelines can affect the level of protection in the neighbouring structures.

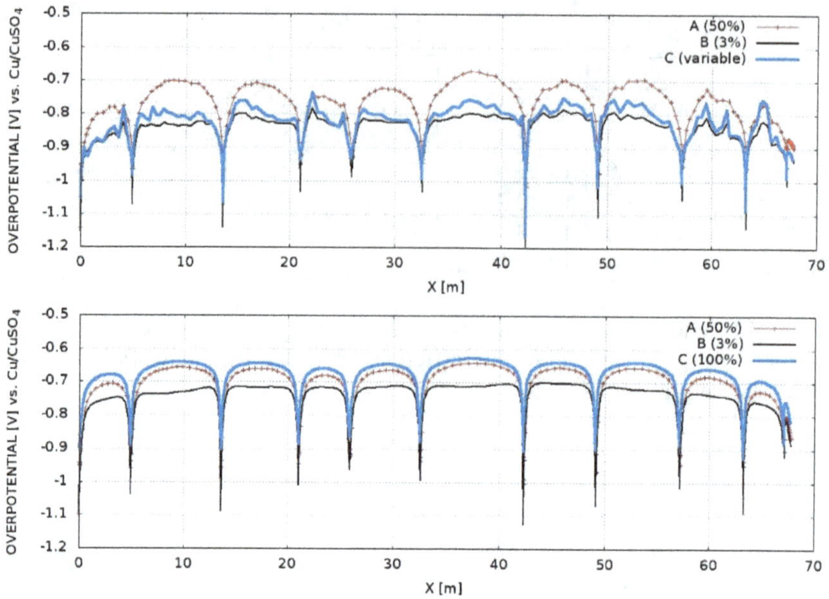

Figure 15: Over potential along pipelines with different coating qualities of pipeline C. (Top) Variable breakdown factor estimated from close interval survey (CIS) measurements. (Bottom) Pipeline C with 100% BF.

5 Conclusions

A BEM based modelling tool has been developed for the direct simulation of CP systems involving pipelines in multi-layer media.

The formulation can represent several layers of geological regions in the soil and does not require the mesh discretisation of the interface between layers. As a consequence, the computational cost of the simulation does not increase with the number of layers.

Two case scenarios were presented in order to illustrate some of the applications of the tool. The first one provides comprehensive results on the impact of the distribution of localised damage to the coating on the over potential from a global point of view. In addition, the effects of a localised coating damage on the local patterns of voltage gradient, soil-to-pipe potential at ground level have been obtained, including both localised degradation and holidays in the coating.

In the second case, the interference between three 70 km pipelines running parallel along the same corridor has been modelled. The results obtained demonstrate the model's ability to represent the case where the damage in the coating of one of the pipelines affects the level of protection of the others.

The modelling tool has been used to investigate cases of interference due to varying quality of coating breakdown factor.

Once calibrated, the simulation becomes a powerful tool for analysing different types of scenarios, to conduct optimisations on existing CP systems, to provide better interpretation of survey data, and to improve the survey techniques themselves. In particular, the modelling approach can be applied to interpret survey data in regions where access is difficult by using adjacent measurements in conjunction with the model to simulate the data required.

References

[1] Brebbia, C.A., Telles, J.C.F. & Wrobel, L.C., *Boundary Element Techniques – Theory and Application in Engineering*, Springer Verlag: Berlin, Heidelberg, NY, Tokyo, 1984.

[2] Riemer, D.P. & Orazem, M.E., Modelling coating flaws with non-linear polarization curves for long pipelines. *Advances in Boundary Elements, vol. 12: Corrosion and Cathodic Protection Modelling and Simulation*, ed. R.A. Adey, WIT Press: Southampton, pp. 225–259, 2005.

[3] Riemer, D.P. & Orazem, M.E., Application of boundary element models to predict the effectiveness of coupons for accessing cathodic protection of buried structures. *Corrosion*, **56**, pp. 794–800, 2000.

[4] Adey, R.A. & Baynham, J., Design and optimization of cathodic protection systems using computer simulation. *Corrosion 2000*, Paper 00723, Houston, Texas, NACE International, 2000.

[5] Peratta, A.B., Baynham, J.M.W. & Adey, R.A., A computational approach for assessing coating performance in cathodically protected transmission pipelines. *Corrosion 2009*, Paper 6595, Atlanta, Georgia, NACE International, 2009.

[6] Jackson, J.D., *Classical Electrodynamics*, 2nd ed., John Wiley & Sons: NY, Chichester, Brisbane, Toronto, Singapore, 1975.

Numerical modelling of cathodic protection systems for deep-well casings

A. B. Peratta, J. M. W. Baynham & R. A. Adey
CM BEASY Ltd, UK

Abstract

This work is focused on the 3D simulation of cathodic protection (CP) systems for metallic deep-well casings immersed in stratified soil. The soil is considered as a multi-layered electrolyte with electric conductivity, which varies from layer to layer. Given a CP system, the aim of the simulation is to predict the level of protection (including the normal current entering the structure) against corrosion along the well casing. It is common practice to fill the annular space between soil and metallic structure with cement along the whole depth. The article presents a modelling approach which incorporates this feature. In addition, a unique CP system may be used to protect more than one well. In this case, the simulation is used to analyse problems of interference and load unbalance to the electrical resistance in the power lines and flow lines. The modelling approach is aimed at simplifying the model design, especially for dealing with a multi-layered electrolyte, and improving the accuracy over existing traditional techniques, which entail a detailed and time consuming representation in the model of the interfaces between different layers of the electrolyte. The effects of different scenarios of electrolytes on the CP system are considered.
Keywords: cathodic protection, boundary element method, deep-well casings.

1 Introduction

The predictive simulation of cathodic protection (CP) systems applied to well casings provides a number of advantages at the design, monitoring and optimization stages of the project. Accurate modelling capabilities help to gain better understanding of the relationship between different CP variables, the infrastructure to be protected and its surrounding environment.

An effective method to monitor corrosion in buried structures is to measure the pipe/soil potential along its span. With horizontal pipelines, this is technically feasible; however, the access to data on the local CP performance in well casings is very limited and can only be obtained using expensive logging tools.

Deep-well casing depths vary usually from a few hundred metres to a few kilometres. Along its depth, the casing may penetrate different geological regions with varying physical properties including water content, porosity, permeability and salinity among others. Therefore, from the modelling point of view, the casing can be considered as exposed to a stratified (multi-layered) electrolyte of varying electrical conductivity. Moreover, from layer to layer the polarization properties of the metallic structure may vary in addition to the characteristic conductivity, due to the corrosivity of the medium.

The electrical conductivity of the soil and its distribution are key parameters, which determine how current flows between the anode and the metallic structure, and therefore strongly influence the levels of corrosion protection along the casing. The modelling of CP systems involves predicting the current and potential fields at any point in the electrolyte and at surfaces of electrodes. Under most common situations, this requires solving the steady state charge conservation equation in the electrolyte in 3D space given by:

$$\nabla \cdot \mathbf{j} = 0 \ , \ \mathbf{x} \in \Omega \tag{1}$$

where

$$\mathbf{j} = -\sigma(\mathbf{x}) \ \nabla V_e(\mathbf{x}) \tag{2}$$

represents current density, σ is the electrolyte conductivity, V_e is the potential in the electrolyte measured against remote earth, and Ω represents the integration domain (electrolyte).

The boundary element method (BEM) [1] has been widely used to solve Laplacian equations and in particular simulate cathodic protection systems for underground and offshore structures [2–6]. The most significant advantages of the method are first that the formulation is based on the fundamental solution of the leading partial differential operator in the governing equation and second that it requires only mesh discretisation on the boundaries of the problem. The former aspect confers high accuracy, while the latter substantially simplifies the pre-processing stage of the model, since volume discretisation is not needed. The standard BEM is traditionally aimed at solving homogeneous electrolytes. In the case of non-homogeneous conditions, it is common practice to combine BEM with domain decomposition or multi-region (MR) technique. In this way the electrolyte, considered as piecewise homogeneous, is represented as a collection of sub-regions, each one of them with homogeneous conductivity. Then, neighbouring regions are connected to each other by prescribing continuity of potential and normal current density through their common interface [3].

In this contribution, a new technique for computing accurately and efficiently 3D problems involving deep-well casings immersed in multi-layered soil is used and compared against the traditional MR. The goal of the multi-layer (ML) approach is to avoid including in the model the interfaces between different layers, by employing a fundamental solution specifically designed for multi-layer materials. The major advantages over the more traditional MR approach are mainly the reduction in the engineering time to prepare a model, since interfaces do not need to be included in the model. Also the solving time is minimised, since only the degrees of freedom representing the well casing and the anodes need to be considered. An important consequence of this feature is that the number of layers will not significantly affect the computational cost of the calculation. Moreover, a common limitation in BEM is that the distance between two elements of the mesh must not be too small in comparison to the characteristic length of the largest element, otherwise accuracy and stability of the solution is compromised. This translates into a practical limitation in modelling thin layers of electrolyte, or models in which the thickness of the layer is small in comparison with its lateral extension. However, the ML approach does not suffer from this limitation, and therefore allows the end user to include thin layers without major problems of accuracy or high computational cost.

The article is organised as follows: section 2 provides briefly some background information on the ML. Section 3 presents a comparison between MR and ML results in a particular case scenario. Section 4 demonstrates the modelling capabilities of subsurface thin layers. This type of modelling is difficult and time consuming with the standard MR approach but rather straightforward to conduct with the ML. Section 5 presents a method for modelling the cement in the annular space between the outer tube and soil. Section 6 presents the case of a single anodic bed protecting two distant casings. Overall conclusions are given in section 7.

2 Multi-layer modelling approach

The modelling of CP systems protecting well casings is intrinsically three dimensional (3D). Hence 3D BEM kernels are required. The ML approach incorporates the stratified nature of the medium into the corresponding Green's function. Then BEM is applied in the same way as in the case of the homogeneous electrolyte, except that the Green's function for the homogeneous Laplace equation given by

$$G(\mathbf{x}_i, \mathbf{x}_j) = \frac{1}{4\pi r}$$

is replaced by the multi-layer Green's function given by:

$$G(\mathbf{x}_i, \mathbf{x}_j, m, n) = \frac{1}{4\pi\sigma_m} \sum_{k=1}^{N_{appr}} \frac{\alpha_{ijmn}}{\left\| \mathbf{x}_i - \mathbf{x}_j + \mathbf{g}_{ijmn} \right\|} \quad (3)$$

where $\mathbf{x}$ denotes the 3D position vector, the sub-indices i and j stand for the source and field point, respectively; $r = \|\mathbf{x}_j - \mathbf{x}_i\|$; m and n indicate the layer of the source and field points, respectively; a_{ijmn} is a weight coefficient and $\mathbf{g}_{ijmn}$ denotes a displacement vector. The summation is performed over the number of approximating terms N_{appr} in the multi-layer Green's function. The calculation of the weight and displacement vectors goes beyond the scope of this contribution and can be found in refs. [2, 3, 7–9] and their citations. Finally, any standard BEM strategy can be employed. In this case, collocation direct BEM has been adopted.

3 Case study: comparison between MR and ML

A number of comparison tests and validations of the ML approach have been performed using as reference the MR technique. This section introduces a conceptual model used to compare the results and calculation performance of both methods. The conceptual model consists of a single well casing and one ICCP anodic bed in stratified soil consisting of 7 layers. The well is 1750 m deep and consists of 4 sections of different diameter. The specification of the main metallic tube relevant to the simulation is shown in Table 1 where z is the depth co-ordinate, D_{ext} the tube outside diameter, t its wall thickness and R its electrical resistance.

Table 1: Main tube specifications.

Section	Soil layer	z_{min}	z_{max}	D_{ext}	t	R
		m	m	m	m	Ohm/m
1	1	0	300	0.35	0.012	1.37×10^{-5}
2	2	300	500	0.25	0.012	1.94×10^{-5}
3	2,3,4	500	1200	0.175	0.012	2.83×10^{-5}
4	4,5,6,7	1200	1750	0.15	0.012	3.34×10^{-5}

Fig. 1 (right) illustrates the conceptual model. The numbers on the left-hand side indicate the conductivity of each layer, while the scale on the right-hand side indicates the z coordinate of each interface. The dot circle labelled as "0" in the figure represents the origin of the xyz system of coordinates assumed in this work. The anode bed, represented by a 30-cm diameter by 8 m long cylinder, is located 75 m apart from the well in x direction.

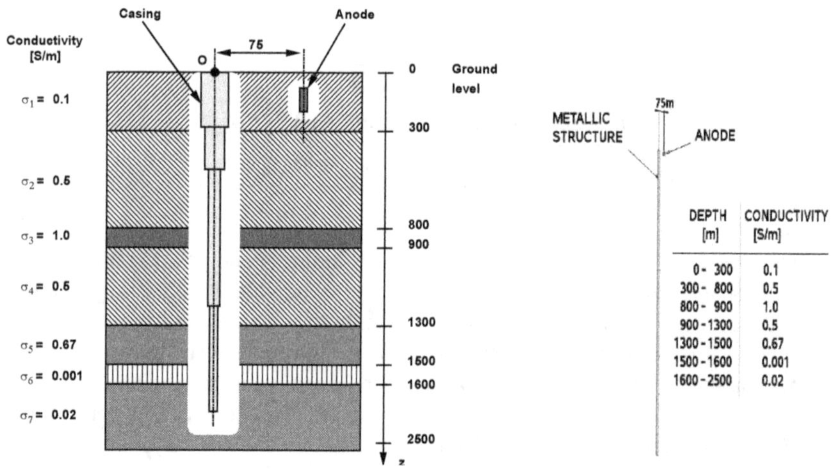

Figure 1: Conceptual model of MR (left). ML model (right).

The size of the linear system of equations that needs to be solved in this case would be $N_{DOF} \times N_{DOF} = 3339 \times 3339$, where N_{DOF} is the number of degrees of freedom. The top of the anode is at 30 m deep and the total current injected into the CP system is $I_A = 10$ A. The scenario posed for this case study is similar to the one presented in earlier works apart from minor variations. The discretisation mesh of the MR model is shown in Fig 2. The interfaces between layers need to be discretised and incorporated in the model. On the other hand, the ML requires the discretisation of the casing and anode only, thus avoiding elements on the interfaces and bounding box.

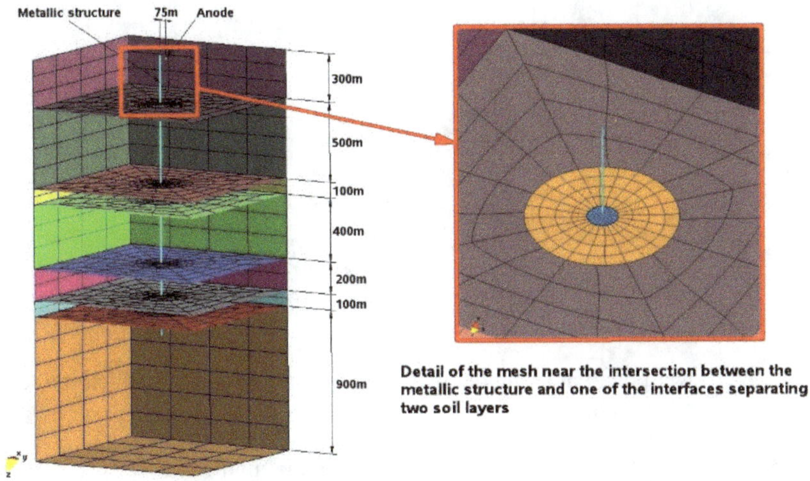

Figure 2: Mesh discretisation for the MR model.

In this example the casing was discretised with 177 tube elements and the anode with 10 elements. Hence, the total number of degrees of freedom is 187 and the corresponding linear system of equations is of size 187×187. The ML model is shown in Fig. 1 (right). A comparison between MR and ML for the over potential along the structures is shown in Fig 3. More detailed results of the comparison can be found in Peratta *et al.* [3].

4 Effect of thin layers

One of the major advantages of the ML is that it is not limited with regard to the number of layers or horizontal extent or thickness of each layer. In order to illustrate this feature and to provide better understanding of the effects of the conductivity of thin layers on the over potential distribution on the whole structure, an uppermost subsurface layer 30 m thick with variable conductivity has been added to the model. This type of study is more difficult to conduct with the MR approach in view of the fine discretisation needed for the interface of the thin layer.

The over potential as function of depth for different sub-surface conductivities in layer 0 is shown on the left of Fig. 4. It can be seen that the conductivity of the soil in the first layer does not substantially affect the over potential anywhere else apart from the first 30 m in contact with that layer. Each curve corresponds to a different soil conductivity in layer 0 which extends from ground level to 30 m. The depth is represented in logarithmic scale, since the main changes of over potential occur in the first few soil layers starting from ground level.

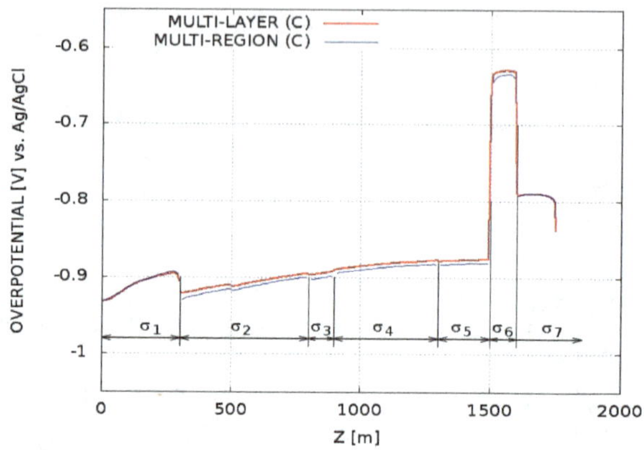

Figure 3: Over potential along depth. Comparison between ML (thick red) and MR (thin blue).

Fig. 4 on the right shows the average over potential in the region $0 < z < 30$ m of the casing for different electrolyte conductivity of the first layer. Lower conductivity in the subsurface layer translates into more negative over potentials.

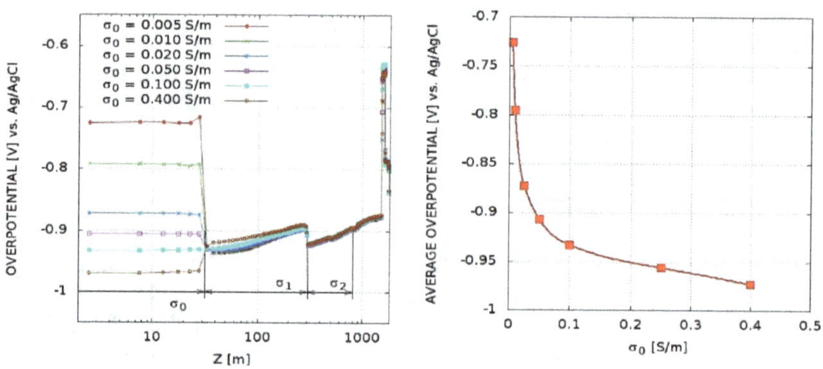

Figure 4: Over potential along the well casing in the 8-layer soil model.

5 Cement layer

This section considers the influence of a cement layer surrounding the casing from the electrical conductivity point of view, i.e. any other effect apart from the ohmic drop has not been considered.

Fig. 5 (left) shows an arbitrary cross section of the well under consideration. The current density in the concrete is considered to be oriented in the radial direction. Under these circumstances, the concrete is introduced into the model as a passive medium which introduces an additional ohmic-type voltage drop in the electrolyte. The idea is to derive the equivalent resistance offered by the concrete and use it to adjust the original polarization curve of the steel, as illustrated in Fig. 5 (right). The original polarization curve of the steel shown in red is corrected with a linear ohmic drop (dashed line). The resulting polarization curve used in the model is represented by a thick black continuous line. Assuming that the current density vector j has only radial component, its magnitude at the internal radius r_2 varies linearly with the voltage drop across the cement according to:

$$j(r_2) = k_c r_2 \ \log(r_1/r_2)\Delta\phi \, ,$$

where $\Delta\phi = \phi(r_1) - \phi(r_2)$ is the potential difference across the cement and k_c the cement electrical conductivity. For practical convenience, the surface distribution of ohmic resistance ρ_s is defined as

$$\rho_s = (r_2/k_c)\log(r_1/r_2)$$

Then the over potential V of the equivalent polarization contemplating the IR drop across the concrete is corrected according to:

$$V = V - \rho_s j \, .$$

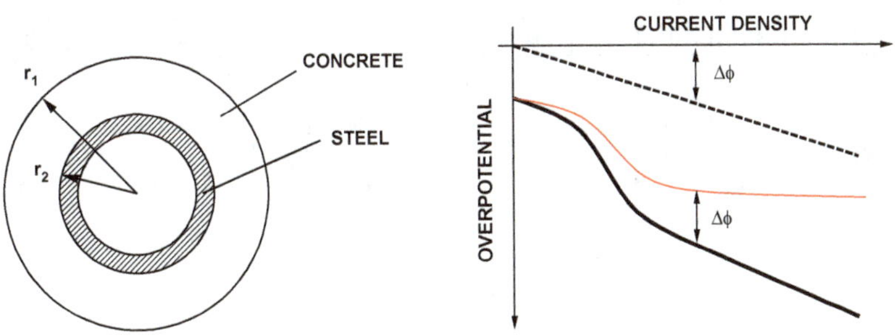

Figure 5: Cross-section of the well casing (left). Adjustment of the original polarization curve due to the ohmic drop in the cement (right).

In order to investigate the effect of the concrete surrounding the metallic case, the same model of well casing described above was considered in the following scenarios named “BARE”, “WET”, “DRY” and “MIX”.

- “BARE”: This scenario considers the case of bare steel (no concrete at all).
- “WET”: The metallic case is inside a cement cylinder of external diameter 0.9 m. The concrete is considered highly porous and water has infiltrated.
- “DRY”: Same as wet, but with lower conductivity.
- “MIX1”: This scenario assumes that the pipe sections 1, 2 and 3 are inside a cylindrical column of dry cement, while section 4 ($1200 < z < 1750$ m) is considered to be damaged and therefore with the conditions of the wet case.

The two types of concrete considered are shown in Table 2.

Table 2: Cement conductivity.

Type of concrete	Electric conductivity
Wet	0.025 S/m
Dry	0.004 S/m

The details of each case scenario are listed in Table 3 and the relevant polarisation curves used are shown in Fig. 6.

Table 3: Case scenarios.

Section	Span		r_1	r_2	$h = r_1 - r_2$	ρ_s		
	z_{min}	z_{max}				WET	DRY	MIX1
	m	m	m	m	m	Ohm·m^2	Ohm·m^2	Ohm·m^2
1	0	300	0.45	0.175	0.05	2.87	17.95	17.95
2	300	500	0.45	0.125	0.1	2.78	17.38	17.38
3	500	800	0.45	0.0875	0.1375	2.49	15.56	15.56
4	1200	1750	0.45	0.075	0.15	2.33	14.59	2.33

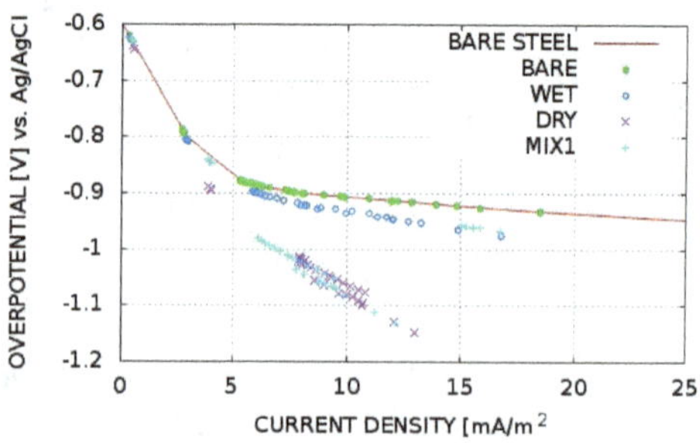

Figure 6: Polarisation curves.

The external diameter of the cement lining is kept constant and equal to 0.45 m, hence the thickness of the cement wall *h* varies along depth.

As a consequence, the resulting polarization curve is modified with different *IR* drop resistances depending on the sector of the well considered.

Fig. 7 shows a comparison of normal current density, over potential, metal voltage and axial current density for the different scenarios.

These results illustrate how the performance of the CP system is affected by the electric resistivity of the cement layer, which depends on a variety of other physical and chemical properties. In particular, the change of resistivity due to porosity and water saturation has significant effect on the distribution of overpotential along the well casing.

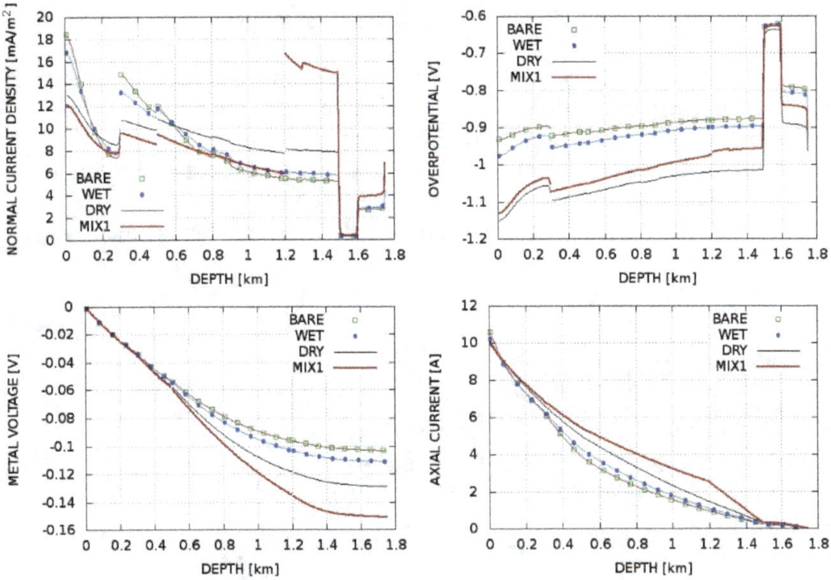

Figure 7: Normal current density (top left), over potential (top right), metal voltage (bottom left) and axial current density (bottom right) in function of depth for "bare", "wet", "dry" and "mix" scenarios.

6 Multiple wells

The physical situation of this case scenario is shown in Fig 8. The model corresponds to the bare steel case and includes the *IR* drop in the connections to the power supply but not those in the associated flow lines. The two wells are protected with a single ICCP anode.

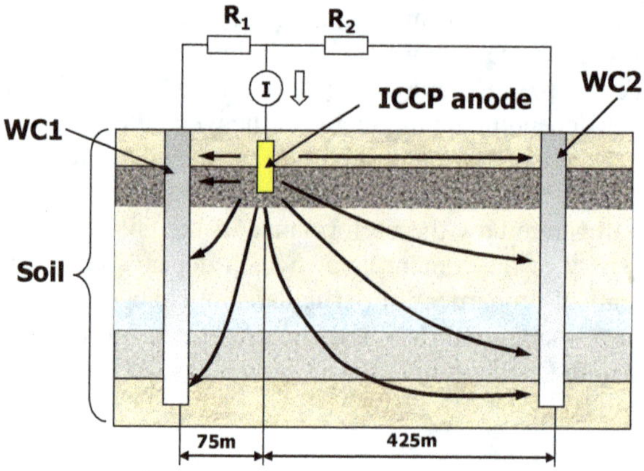

Figure 8: Two well casings connected to a unique ICCP anode.

The anode is 75 m apart from WC1 and 425 m apart from WC2. The electrical resistance of the power supply cable is represented with the resistors R1 and R2 of 0.4 and 1.44 Ohm, respectively. Therefore, the unbalance in the currents flowing to the two wells is not only due to the different resistance in the return path circuit but also due to the ohmic *IR* drop in the soil.

Fig. 9 shows the results of normal current density, over potential, metal voltage and axial current flowing along each well casing.

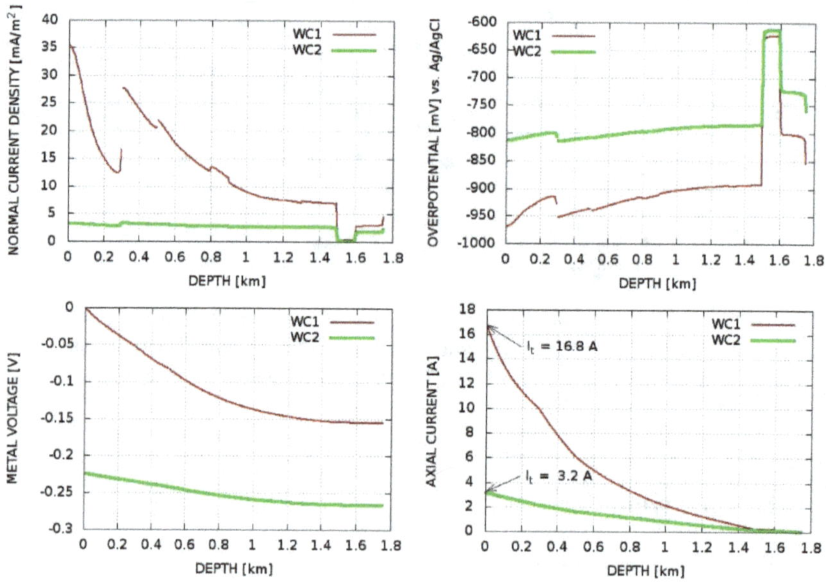

Figure 9: Comparison of normal current density, overpotential, metal voltage and total axial current for the two well casings connected to the ICCP anode.

As expected, the normal current density on the well casing more distant from the anode (WC2) is significantly lower in comparison to the one closer (WC1). In addition, the over potential on the former is on average approximately 100 mV lower than on the latter, except in the less conductive soil layer at depth between 1500 and 1600 m where the difference in over potential is negligible.

In this case, none of the well casings show anodic behaviour but with different conditions this may occur and would be predicted by the model.

7 Conclusions

A new approach for the numerical modelling of CP systems applied to well casing structures in vertically stratified soil has been developed. A significant advantage over the traditional multi-region technique is the reduction of the number of degrees of freedom. In addition, the modelling preparation is simplified, as interfaces between soil layers do not need to be modelled at all.

The multilayer (ML) approach can easily represent thin layers of electrolyte without additional mesh discretisation thus substantially reducing the work of the user.

A feature of the ML is that the computational cost of the calculation does not increase with the number of layers. The results obtained with the ML are in good agreement with the traditional multi-domain approach.

The most important benefit of the ML approach is the simplicity in its application to well casing applications as the data can be prepared very easily and the calculations made within a few minutes. Consequently, sensitivity studies on proposed new designs can be quickly completed.

The model can be used to optimise the use of measurement techniques such as surface *E*-Log*I* measurements and Corrosion and Protection Evaluation Tool (CPET) logs to improve the cost effectiveness and reliability of CP designs [10].

An approach for incorporating the cement that fills the annular space between the metal and the soil has been proposed and the influence of different qualities of the cement on the CP performance has been quantified. In cases where there is doubt about the quality of the cement in a particular layer, the model can be used to assess its impact and identify effective corrosion control options.

The model has been applied to multiple well situations where there is the possibility of interference between the wells. Results have been presented where not only the ohmic drop in the electrolyte is considered but also the *IR* drop in the power supply cables, flow lines and other associated equipment and structures is modelled.

References

[1] Brebbia, C.A., Telles, J.C.F. & Wrobel, L.C., *Boundary Element Techniques – Theory and Application in Engineering*, Springer Verlag: Berlin, Heidelberg, NY, Tokyo, 1984.

[2] Peratta, A.B., Baynham, J.M.W. & Adey, R.A., A computational approach for assessing coating performance in cathodically protected transmission

pipelines. *NACE Corrosion 2009 Conference*, Atlanta, Georgia, Paper 6595, 2009.

[3] Peratta, A.B., Baynham, J.M.W. & Adey, R.A., Advances in cathodic protection modelling of deep well casings in multi-layered media. *NACE Corrosion 2009 Conference*, Atlanta, Georgia, Paper 6555, 2009.

[4] Riemer, D.P. & Orazem, M.E., Modelling coating flaws with non-linear polarization curves for long pipelines. *Advances in Boundary Elements, vol. 12: Corrosion and Cathodic Protection Modelling and Simulation*, ed. R.A. Adey, WIT press: Southampton, pp. 225–259, 2005.

[5] Riemer, D.P. & Orazem, M.E., Application of boundary element models to predict the effectiveness of coupons for accessing cathodic protection of buried structures. *Corrosion*, **56**, pp. 794–800, 2000.

[6] Adey, R.A. & Baynham, J., Design and optimization of cathodic protection systems using computer simulation. *Corrosion 2000*, NACE-00723, NACE International, Orlando, Florida, 2000.

[7] Seremet, V.D., *Handbook of Green's Functions and Matrices*, WIT Press: Southampton, Boston, 2002.

[8] Ymeri, H., Nauwelaers, B. & Maex, K., Computation of conductance and capacitance for IC interconnects on a general lossy multilayer substrate. *Active and Passive Electronic Components*, **24**, pp. 87–114, 2001.

[9] Smedes, T., van der Mejis, N.P. & van Gendered, A.J., Boundary element methods for 3D capacitance and substrate resistance calculations in inhomogeneous media in a VLSI layout verification package. *Advances in Engineering Software*, **20**, pp. 19–27, 1994.

[10] Roche, M., Vittonato, J. & Jebara, M., Cathodic protection modelling of deep wells casing by 3D software simulation: Comparison with E-Logl and CPET data. *NACE Corrosion 2008*, Paper 08273, 2008.

Optimization of ship's cathodic protection system to minimize electric and magnetic signature by mathematical simulation

S. Xing[1,2], J. Wu[1,2] & Y. Yan[1,2]
[1]*State Key Laboratory for Marine Corrosion and Protection, PR China*
[2]*Luoyang Ship Material Research Institute, PR China*

Abstract

In this contribution, the protective effect and distribution of magnetic field of cathodic protection (CP) systems designed by experience were calculated by boundary element method (BEM) and the calculation results agreed with the measured results when the ship berthed at port, which suggested that the calculation results are accurate. The performance of a CP system was optimized by moving the position of the anodes as well as changing the output current of anodes. The objective was to design the system so that the electric and magnetic field was minimized and adequate protection was provided. The calculation results show that when the anodes were symmetrically installed at the 130th and 232nd frame, the corrosion related magnetic field and total output current were reduced by about 40% and 8.6% respectively.
Keywords: cathodic protection, mathematical simulation, boundary element method, electric and magnetic field.

1 Introduction

Coatings combined with impressed current cathodic protection (ICCP) is the most common means for the protection of shipboard. They interact with each other to protect shipboard. Coatings provide primary corrosion protection by isolating the hull metal from the seawater, while ICCP provides secondary corrosion protection in those areas where the paint is damaged or degrade. The protective effect is directly related to the ICCP configurations; an incorrectly designed ICCP system would not only adversely influence the protective effect but also the distribution of underwater electrical potential (UEP) and corrosion related magnetic (CRM) field, which are known as the signatures of the ship.

Electric fields and coincident magnetic fields, arise around a ship due to the current flow from the ICCP system. UEP and CRM signatures exist even in the absence of a cathodic protection system. They are caused by the galvanic potential differences between the metallic structures in contact with the seawater. For example, hull and propeller provides a sufficient driving potential to create an electric field. These electromagnetic fields take the form of steady electric and steady magnetic fields arising from the steady flow of current around the hull of the vessel. Modern mines can detect these fields and use them to detect and classify passing ships.

Research [1–4] suggests that design and optimization of the ICCP system is an effective means of preserving the integrity of a vessel and of controlling the signatures. The design of an optimized ICCP system requires accurate placement of the anodes and reference electrodes on the ship hull. Many shipboard ICCP systems currently in use were designed using relationships and engineering judgment. However, the protection effect of the CP system designed by experience cannot be known and sometimes it might lead to under or over protection and high electric signature. Today computer simulation techniques based on BEM like BEASY have enabled the electric and magnetic fields generated by the galvanic interaction of the ship metallic structure and ICCP system to be predicted. Thus, BEASY provides a tool to predict changes in the protection level of the ship and the electric field in the seawater caused by the ICCP system

In this paper, BEASY was used to predict and optimize the distribution of potential and electromagnetic field signature.

2 Experimental work

2.1 Model

A vessel with a length of 125 meters was investigated. The ship had 250 frames at intervals of 0.5 m and the frames were assigned the numbers 0 to 250 from bow to stern. The geometry of interest in the boundary element model was the wetted surface of the hull and major appendages. The ship has two propellers and two rudders; the propellers were made of nickel-aluminium-bronze alloy (NAB) and modelled as solid disks with equivalent surface area as the real propellers. The shaft is made of carbon steel and the propellers and shafts were assumed to be uncoated because of turbulence engendered by propeller movement. The ship hull and rudders are also made of carbon steel, which were coated to prevent corrosion. The ICCP system evaluated included four anodes and a centre controlled power supply. The BEM model of half of the ship is shown in Fig. 1.

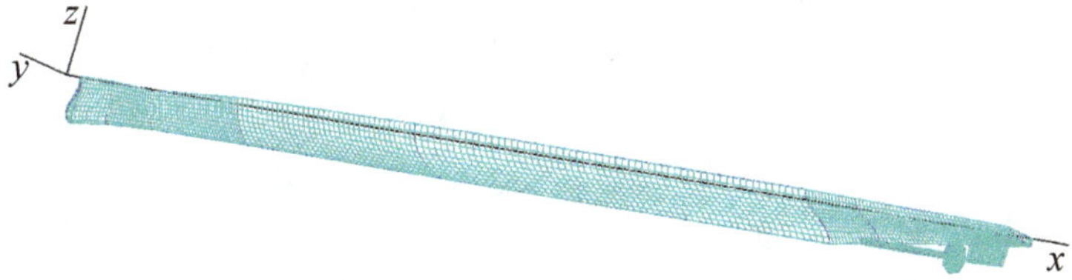

Figure 1: BEM model of the ship.

2.2 Boundary conditions

The cathodic polarization curves of uncoated and coated carbon steel are shown in Fig. 2(a) and the polarization curve of NAB is shown in Fig. 2(b); these curves were set as the boundary conditions of the ship model. During calculation, the polarization curve of coated steel was set as the boundary condition of ship hull and rudder, the polarization curve of uncoated steel was set as the boundary condition of shafts and the polarization curve of NAB was set as the boundary condition of propellers. A box of 2500 meters was created as the infinite boundary where the current density was zero.

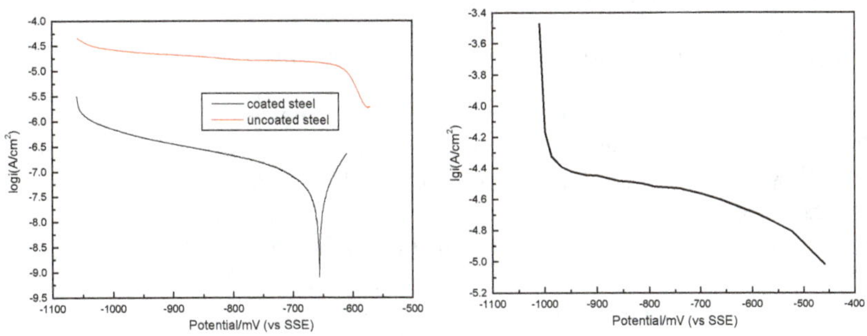

(a) Carbon steel (b) NAB

Figure 2: Polarisation curves used as boundary conditions.

2.3 ICCP optimization

The optimization process required the problem to be posed in the form of an objective function, design variables and constraints. The sum of the squares of the differences between the potential on the element of the surfaces and the potential target value of the surfaces was used as the objective function to be minimized [5, 6]:

$$Obj = \sum_{i=1}^{n_s} \sum_{j=1}^{n_e} (\phi_{ij} - \phi_{\text{target},i})^2 \tag{1}$$

subject to the following constraints on the surface of the cathode Γ_C [3]:

$$g_i = \frac{\phi_i - \phi_{\max,i}}{\left|\phi_{\min,i}\right|} \leq 0 \quad \text{on } \Gamma_C, \quad i = 1, 2, \ldots, n \tag{2}$$

$$g_j = \frac{\phi_{\min,j} - \phi_j}{\left|\phi_{\min,j}\right|} \leq 0 \quad \text{on } \Gamma_C, \quad j = 1, 2, \ldots, n \tag{3}$$

where n_s is the number of surfaces, n_e is the number of elements per surface, ϕ_{target} is the target potential per surface, $\phi_{max,j}$ and $\phi_{min,j}$ are the maximum and minimum protection potentials, respectively.

2.4 Magnetic fields

The CRM signature generated by the currents flowing in the sea water could be calculated by solving the equation:

$$\nabla^2 \mathbf{A} = -\mu \mathbf{J} \tag{4}$$

for the vector potential **A** [1] where **J** is the current density vector and μ the permeability of the medium. The magnetic field is then simply **B** = curl **A**.

3 Results and discussion

3.1 Performance of initial ICCP system on shipboard

On the vessel, a single zone, four anodes ICCP system was used to protect the hull and appendages. The anodes were symmetrically installed at the 142nd frame and 242nd frame and the initial input current was 8 A and 10 A, respectively. A saturated sulphate reference electrode (SSE) was installed at the 186th frame. With the reference electrode reading of −850 mV (vs SSE), the potential distribution on the hull is shown in Fig. 3. The calculation results indicate that the hull, propellers and rudders were under protected, the potential range in the hull was from −846 mV to −1000 mV. The current output of anode at the 142nd frame and the 242nd frame was 5.74 A and 7.40 A, respectively.

At the same time, the practical ship protective potential of the 37th, 72nd, 128th, 168th and 228th frame was measured when the ship berthed at port by Cu/$CuSO_4$ reference electrode. Compared with the measured results, the calculation results showed the same trend as the practical measured data and were near the same values as the measured results, as shown in Fig. 4. Thus, the calculation results appear to be accurate.

The CRM field distribution at sea level is shown in Fig. 5, where the x axis is parallel with the keel of the ship and the y axis is vertical to the keel. The keel of the ship was at the position y = 0 m. The calculation results indicated that the magnetic field position with most strength was the region around the 142nd frame where anodes were installed and then was the position of stern around the 240th frame, while the normal of the magnetic field at the two positions was inverted. In Figs. 5, 7 and 9, the same colour denotes the same strength of the magnetic field and constant colour represents the same gradient of magnetic field.

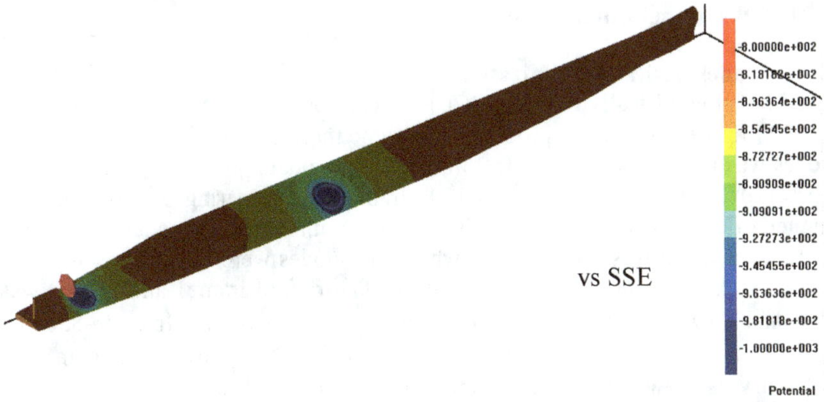

Figure 3: Potential plot of vessel protected by initial ICCP system.

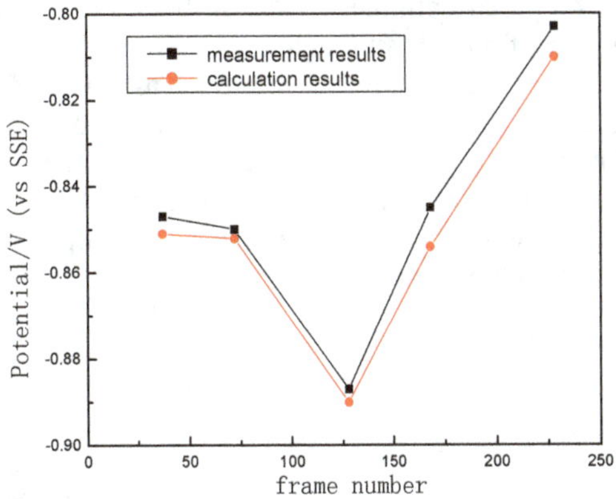

Figure 4: Potential vs frame number.

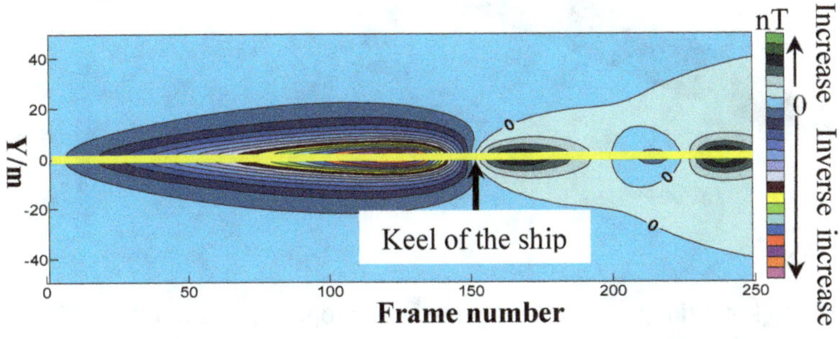

Figure 5: CRM field generated by initial ICCP system.

3.2 ICCP system optimization

3.2.1 Anodes position optimization

Based on an empirically designed ICCP system, one group of anodes was set at the 232th frame because the propellers consume more current. The other group of anodes was initially set at the 115th frame, the candidate positions were set at the 115th, 120th, 130th, 135th, 142nd and 150th frame. By comparing the potential and magnetic field distribution of the ICCP system with different anodes setting, the best ICCP system was identified, in which the anodes were installed at the 130th frame and 232nd frame. When the initial input current of frame1 and frame2 was 5.7 A and 7.4 A, respectively, and with a reference electrode reading of −850 mV vs SSE, the potential distribution of ship hull ranged from −821 mV to −1000 mV as shown in Fig. 6. The current output of anode at the 130nd and 232nd frame was 5.32 A and 7.51 A, respectively.

After optimization, the CRM field distribution at sea level is shown in Fig. 7. In that figure, it can be seen that the position of maximum magnetic field strength changed to the 130th frame, the position of relative strong magnetic field changed to the 160th frame. Compared with the CRM distribution of initial ICCP system, by moving the position of anode, although the CRM field strength increased at stern, the magnetic field along the ship was improved as indicated in Fig. 8, which shows the magnetic field distribution at $y = 0$ from bow to stern. The magnetic signature increase between 150th frame and 200th frame may be related to the distance increase between the two groups of anodes. The distance increase resulted in the current density increase between them, which led to the magnetic signature increasing. Through anode position optimization, the magnetic field strength was reduced by about 30%.

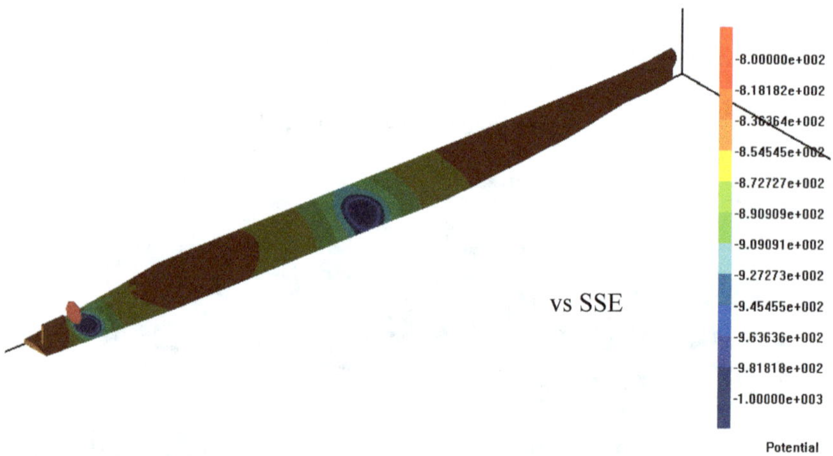

Figure 6: Potential plot of vessel protected by optimized anodes position ICCP system.

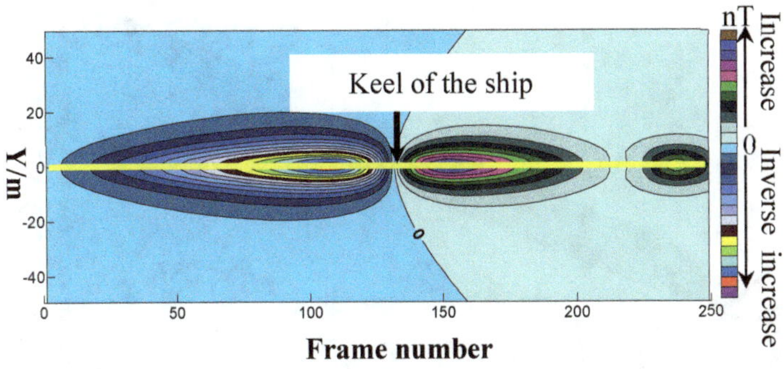

Figure 7: CRM field distribution after anodes position optimization.

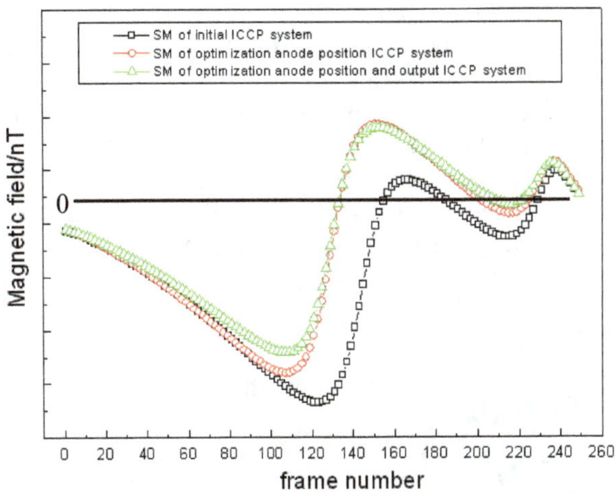

Figure 8: CRM field distributions at $y = 0$ m.

3.2.2 Anodes output current optimization

Once the position of anodes was fixed, the optimization of anode output current was carried out. The purpose of optimizing the current was to minimize the corrosion related electromagnetic signature while the hull was provided with a sufficient cathodic protection effect. The initial input current of frame1 and frame2 was 5.3 A and 7.5 A, respectively. After current optimization, the obtained potential distribution on the hull is shown in Fig. 9. Compared with Figs. 3 and 5, the protective potential obviously changed in a more positive direction after current optimization, but the protective potential remained lower than −800 mV, while the current output of anodes at the 130th and 232nd frame was minimized to 5.08 A and 6.93 A, respectively. Thus, through anode position and output current optimization, the output current was reduced by 8.6%.

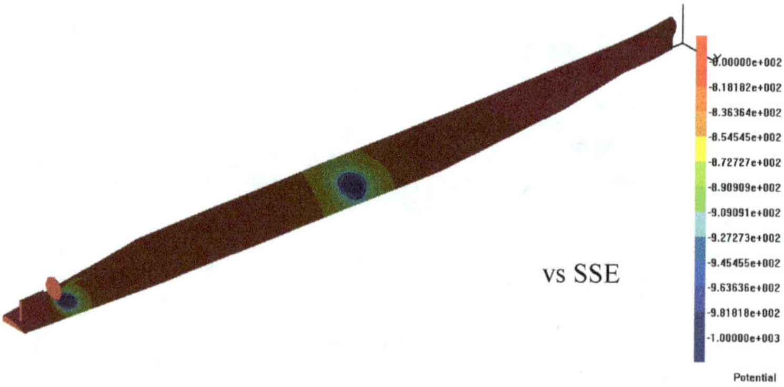

Figure 9: Potential distribution after anodes position and output current optimization.

The magnetic field was further reduced after current optimization as shown in Fig. 10. Compared with the magnetic field generated by the initial ICCP system, the strength of magnetic field was reduced by about 40%. Like the magnetic field of the ICCP system optimized with respect to the anode position, the magnetic field around the position where the anode was installed was the strongest. In order to exhibit more intuitively the magnetic field distribution of initial and optimized ICCP systems, the magnetic signature at $y = 0$ is shown in Fig. 8. In this figure, it can distinctly be seen that the magnetic field along the ship was step by step reduced by position and current of anodes optimization.

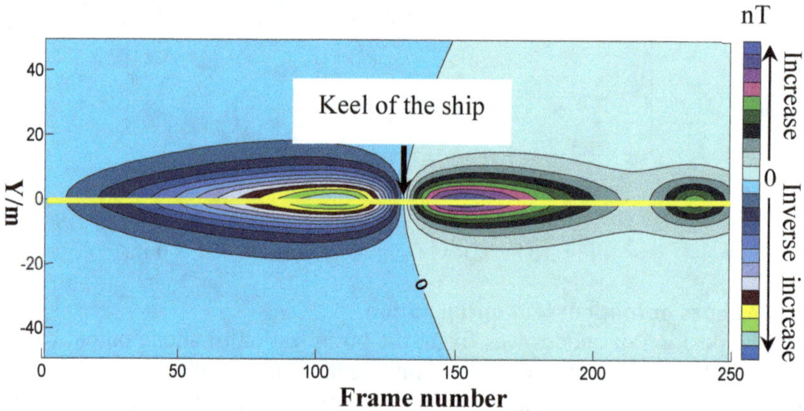

Figure 10: CRM field distribution after anodes position and current optimization.

4 Conclusions

An analysis based on BEM can reasonably predict and optimize the potential distribution and minimize the magnetic signature of a vessel.

Anode position optimization is an effective approach to control CRM signature; for the ship considered, the signature was least when the anodes were symmetrically installed on the 130th frame and the 232nd frame.

The magnetic signature was reduced by about twenty-five percent through the applied anode position and output current optimization.

References

[1] Santana Diaz, E., Adey, R., Baynham, J. & Pei, Y.H, Optimisation of ICCP systems to minimise electric signatures. *MARELEC Conference*, Stockholm, 2001.

[2] Adey, R. & Baynham, J., Predicting corrosion related electrical and magnetic fields using BEM, *UDT Europe 2000: Undersea Defence Technology Conference and Exhibition*, London, 2000.

[3] Santana Diaz, E. & Adey, R., Optimisation of the performance of an ICCP system by changing current supplied and position of the anode. *WIT Transactions on Modelling and Simulation*, **32**, pp. 475–485, 2002.

[4] Santana Diaz, E. & Tims, R., A complete underwater electric and magnetic signature scenario using computational modelling. *Proceedings of the Marine Electromagnetics Conference (MARELEC)*, Amsterdam, 2006.

[5] Santana Diaz, E. & Adey, R., Optimising the location of anodes in cathodic protection systems to smooth potential distribution. *Advances in Engineering Software*, **36(9)**, pp. 591–598, 2005.

[6] Santana Diaz, E. & Adey, R., Predicting the coating condition on ships using ICCP system data, *Tri-Service Corrosion Conference*, Las Vegas, 2003.

The influence of coating damage on the impressed current cathodic protection effect

J. Wu[1,2], S. Xing[1,2] & F. Yun[1,2]
[1]*State Key Laboratory for Marine Corrosion and Protection, PR China*
[2]*Luoyang Ship Material Research Institute, PR China*

Abstract

A one-hundredth scale hull model in linear scaled resistivity was used to simulate the cathodic protection potential distribution of a hull with different coating damage extent and position. The potential distribution of the hull was simultaneously measured by a PXI modular instrument and a program written in Labview 7.1. At the same time the potential and current density of the vessel with a 2%, 4% and 6% damage were calculated by the boundary element method. The results obtained from the physical scale model (PSM) method agreed with calculation results. The impressed current cathodic protection (ICCP) effect and its induced electric signature were simultaneously influenced by the coating damage extent and position. The damage extent mainly influenced the ICCP output current and x axis's electric signature and the damage position mainly influenced the electric signature of the y axis. The calculation results showed that the electric signature increased by about 35.7% due to 6% coating damage.

Keywords: cathodic protection, physical scale model, computer simulation, electric signature.

1 Introduction

Corrosion damage is a major factor in ship maintenance and availability. Coating combined with an impressed current cathodic protection (ICCP) system is the most common means of shipboard corrosion control. The two methods interact with each other to protect the shipboard. Coatings provide primary corrosion protection by isolating the hull metal from the seawater, while ICCP systems protect the hull by applying an external source of current to the ship where the paint is damaged or degraded.

The cathodic protective effect is related to the ICCP design and coating state. Before 1980s, the effect of an ICCP system, designed by experience, could not be exactly known. In the past decade, considerable effort has been spent on the development of the rapid estimate and design methodologies which have a scientific rather than designer expertise basis. Two such methodologies are the physical scale model (PSM) method and computational modeling using boundary element techniques [1].

The PSM technique has been developed to assess the performance of ICCP systems installed on shipboards since 1980s. It has been reported that directly proportional dilution of the seawater results in a similar potential with reference to miniature electrodes placed along the side of the model as the actual potential on a full-size ship [2]. Apart from evaluation, this technique could help in examining the relationships between the coating state and ICCP protection effect and optimize the position of anodes and reference electrodes initially; thus, it has been used to design ICCP systems installed on US Navy ships. But the PSM involves only physical scale not chemical scale, so it is impossible to overcome the change in polarization behavior in a scaled resistivity electrolyte [3]. In recent years, computer simulation technology has been developed for calculating potential and current density distribution in electrochemical systems. A computer simulation technique can rapidly yield the potential and current density distribution on the hull with different coating state to estimate the impact of coating damage extent and position on an ICCP system. A primary limitation of computer simulation is the need for accurate and appropriate polarization response data [4], while the PSM does not require polarization response data. The concept of a unified approach between computational simulation and PSM is very attractive since it uses the strengths of each method to compensate for the weaknesses of the other [4]. The unified approach uses validated computational tools to examine hull geometry, boundary conditions and other parameters to define a relatively small set of experiments that can be completed in a time and cost efficient manner.

According to experience, coating state plays an important role in the distribution of cathodic protective potential and electric signature, but there were few reports about how the coating damage influences the potential and signature distribution. In order to control the signatures and to preserve the integrity of a vessel it is essential to predict the impact of coating damage extent and position on the electric fields and potential distribution. In this paper, PSM and boundary element method (BEM) technologies (BEASY10.0) were used to study the coating damage influence on protective potential and electromagnetic signature.

2 Experimental work

2.1 Physical scale model

A vessel with a length of 125 m was investigated. The ship had 250 frames at intervals of 0.5 m and the frames were designated 0 to 250 from bow to stern. The ship scale model was a one/hundredth scale hull model with the same materials as the actual ship and the model was immersed in seawater diluted to the same scale ratio. The scale model is shown in Fig. 1. The definition of coating damaged position and area shown in Table 1 was based on observations of ships in dry-dock after a period of service time. Throughout, the tests were performed at room temperature, potential measurements were made with eight Ag/AgCl reference electrode fitted with Luggin capillaries.

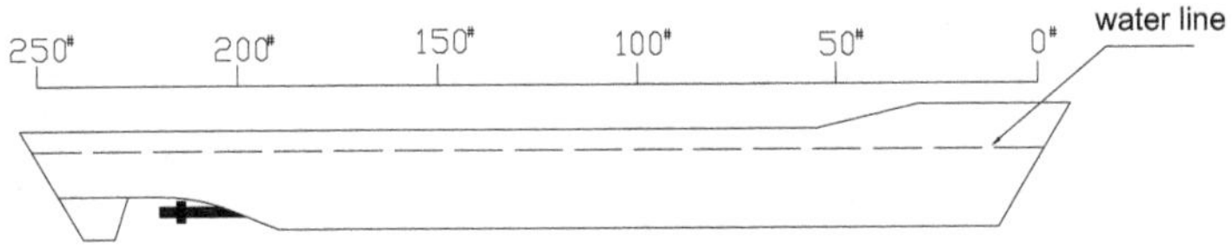

Figure 1: Physical scale model of vessel.

Table 1: Shipboard coating damage extent and position.

Coating state	2% damage	4% damage	6% damage
Damage position	50th frame (symmetrical)	120th and 200th frame (symmetrical)	50th, 120th and 200th frame (symmetrical)

2.2 Measurement system

The protective potential of different positions on vessel was simultaneously gathered by National Instruments and Labview 7.1 program.

2.3 Potential measurements

The scale model was immersed in sea water hold in a tank with dimensions 1.8 m × 0.6 m × 0.9 m and polarized to −850 mV for two days for the formation of calcium and magnesium deposition, then the model was moved into diluted sea water with the conductance of 379 μS/cm. The potential of 10th, 30th, 50th, 70th, 90th, 110th, 130th, 150th, 170th, 190th, 210th, 230th and 250th frame was measured with the reference electrode potential set at −850 mV.

2.4 Computer simulation

The ship has two propellers and two rudders. The propellers were made of a nickel-aluminium-bronze alloy (NAB) and modeled as solid disks with equivalent surface area as the real propellers. The shaft was made of carbon steel and the propellers and shafts were assumed to be uncoated because of turbulence engendered by propeller movement. The ship hull and rudders were also made of carbon steel and they were coated to prevent corrosion. The ICCP system evaluated included four anodes and a centre controlled power supply. The BEM model of half of the ship is shown in Fig. 2.

The cathodic polarization curves of uncoated and coated carbon steel are shown in Fig. 3(a) and the polarization curve of NAB is shown in Fig. 3(b); these were set as the boundary conditions of the ship model.

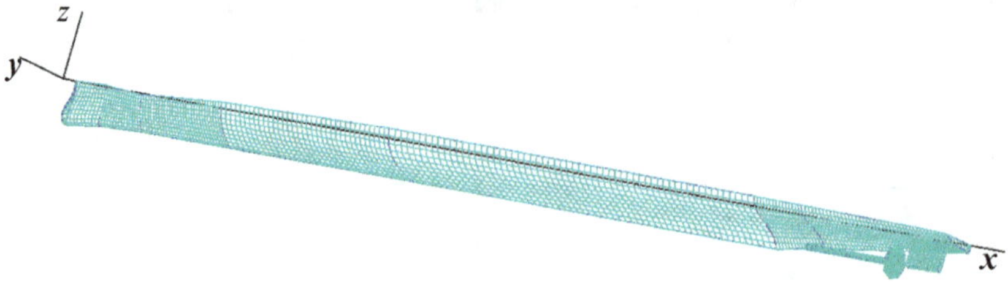

Figure 2: BEM model of ship.

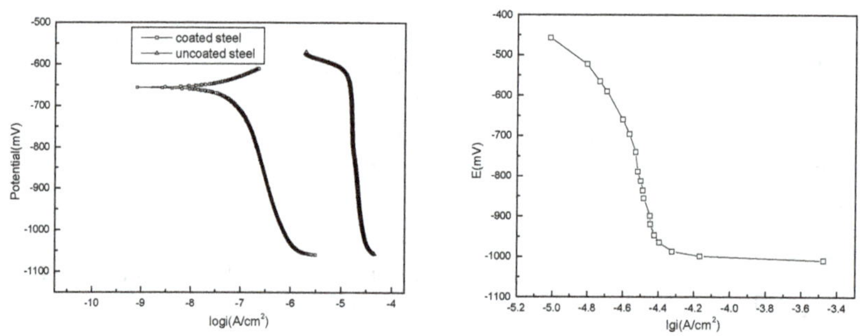

(a) Carbon steel (b) NAB

Figure 3: Boundary conditions.

3 Results and discussion

3.1 Potential simulation

Former studies indicated that the ICCP system would perform best when the anodes are symmetrically installed at the 130th and the 232nd frames and the reference electrode is installed at the 186th frame. The hull of ship was under sufficient protection when the coating was intact, as shown in Fig. 4. The PSM measured data and computational results agreed well with each other except in the anode zones. The difference between PSM and BEM in the anode zones was most likely due to the change in polarization of the materials in the diluted seawater, because the PSM supposes that dilution of the electrolyte does not significantly change the polarization behavior of the materials [5].

When the coating was symmetrically damaged at the 50th frame and the damage area was about 2% of the whole underwater area, the potential distribution measured by experimental PSM and calculated by computer simulation is shown in Fig. 5. According to PSM experimental results and calculation data, the ICCP system can still provide enough cathodic protection to the hull with the potential ranging from −800 mV to −1100 mV with the reference electrode reading of −850 mV. Because of the coating damage and damage positions being far away from the anodes, the potential at coating damage positions changed more positive in comparison with the area elsewhere.

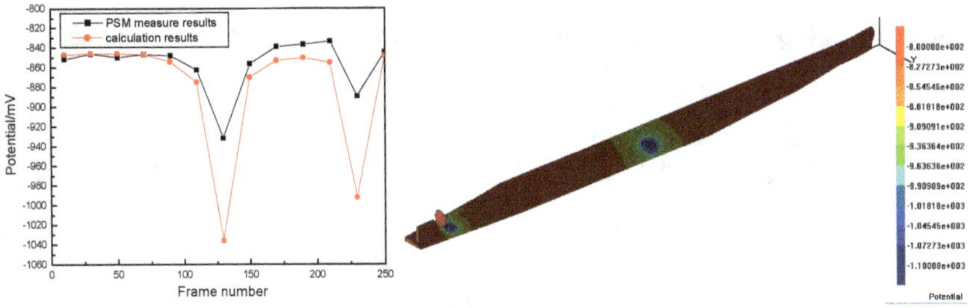

Figure 4: The potential (mV) distribution of ship hull when coating perfect.

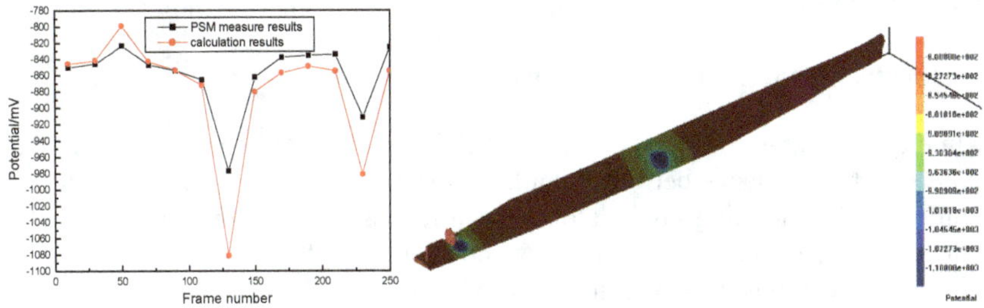

Figure 5: The potential (mV) distribution of ship hull when coating 2% damage.

Even though the coating damage area increased, it had a small influence on the potential distribution, as shown in Fig. 6. The hull was under protected and there was no potential wave in the coating damage regions. According to ohm law, the resistance is directly proportional to length, so if the coating damage position was neighboring the anode, the potential of the uncoated steel could be easily polarized to the designed level even though the uncoated steel consumed more current than the coated steel.

Similarly, the coating was symmetrically damaged at the 50th, 120th and 200th frames and the damage area was the same at each location; the potential variation at the 50th frame was the biggest among the coating damage regions as shown in Fig. 7. So for the ICCP system, when the coating damage extent was less than 6%, the protective potential was mainly influenced by the coating damage position, not the damage area.

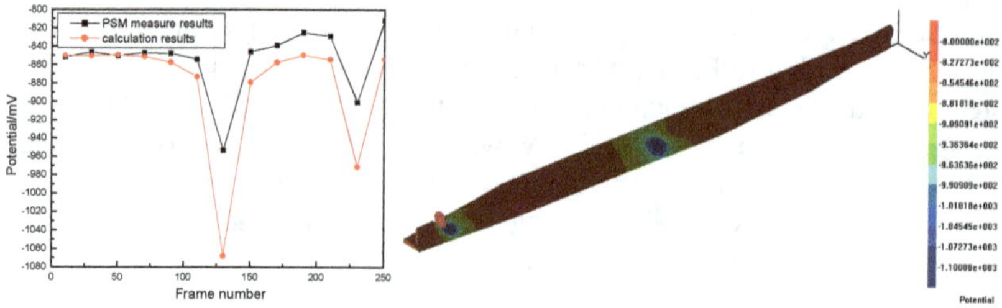

Figure 6: The potential (mV) distribution of ship hull when coating 4% damage.

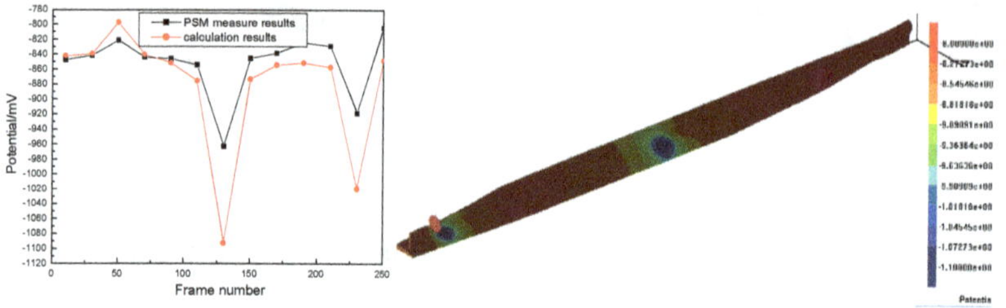

Figure 7: The potential (mV) distribution of ship hull when coating 6% damage.

3.2 Electromagnetic signature simulation

A line of sampling points was defined at a depth of 10 m beneath the sea level, and the vertical distance between simulation points and keel was about 10 m. The solution at the sampling points along this line was used to compute the electric signature. The underwater potential is shown in Fig. 8 and the current density along *x*, *y* and *z* axis is shown in Fig. 9. According to the calculation results, the potential gradient nearby the anodes increased with the coatings damage increase and the electric signature increased by about 35.7% when the coating was 6% damaged compared with the perfect coating. Due to the coating damage positions far away from anodes, the coating damage of 50th frame resulted in potential gradient increase in the bow region.

It is well known that the polarization current of uncoated steel is obviously larger than that of coated steel. So, if some regions' coating of hull is damaged, the ICCP system must increase the output current to maintain the hull potential under the critical potential, and then the total output current increases with damage extent increase. But the current density distribution along *x*, *y* and *z* axis at the same time was influenced by the damage position. From Fig. 9, it can be seen that, the *x* axis's current density was mainly influenced by the damage extent, the current density along *y* axis was mainly influenced by the damage position and the *z* axiṡ's current density was influenced by both damage extent and position. That is to say, the coating damage extent may result in the electric signature increase along the vessel and the bow's coating damage may induce the electric signature to increase in the vertical orientation.

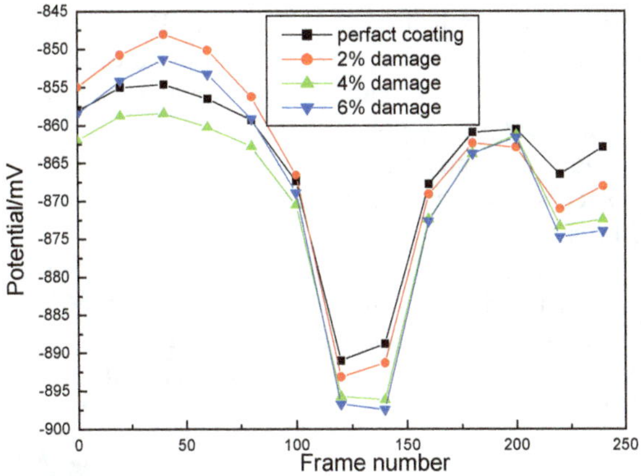

Figure 8: Underwater potential distribution along the keel.

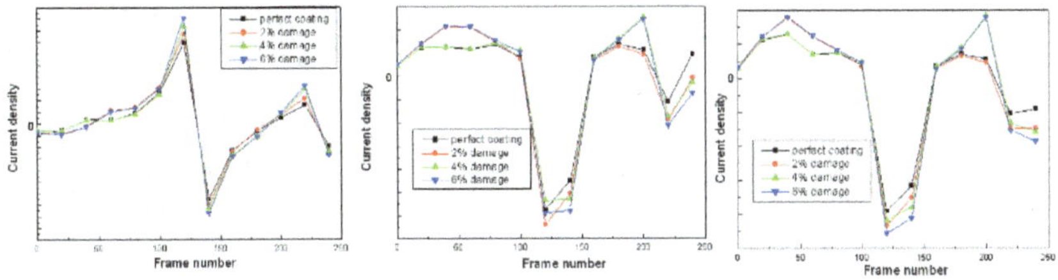

(a) x axis (b) y axis (c) z axis

Figure 9: Current density distribution along the keel.

4 Conclusions

According to the PSM and computer simulation studies, the ICCP protection effect and its induced electric signature were simultaneously influenced by the coating damage extent and position. The potential gradient and the electric signature increased with the increase of coating damage area and the electric signature increased by about 35.7% when the coating damage area extended by 6%. The damage extent mainly influenced the ICCP output current and x axis's electric signature and the damage position mainly influenced the electric signature of the y axis.

References

[1] DeGiorgi, V.G, Thomas, E.D, Lucas, K.E. & Kee, A., A combined design methodology for impressed current cathodic protection systems. *WIT Transactions on Modelling and Simulation*, **15**, pp. 335–345, 1996.

[2] Ditchfield, R.W., McGrath, J.N. & Tighe-Ford, D.J., Theoretical validation of the physical scale modelling of the electrical potential characteristics of marine impressed current cathodic protection. *Journal of Applied Electrochemistry*, **25**, pp. 54–60, 1995.
[3] Gurrappa, I., Physical and computer modeling for ship's impressed-current cathodic-protection systems. *Corrosion Prevention and Control*, **41(2)**, pp. 40–45, 1994.
[4] DeGiorgi, V.G., Wimmer, S.A., Hogan, E. & Lucas, K.E., Modeling the experimental environment for shipboard ICCP systems. *WIT Transactions on Modelling and Simulation*, **32**, pp. 439–447, 2002.
[5] Gage, N. & Mart, P., Further cathodic protection studies of naval ships [J], *CAP*, pp. 1–13, 2002.

Recovering current density from electric potential data

J. Irša, A. N. Galybin & A. Peratta
Wessex Institute of Technology, UK

Abstract

In this paper, an inverse problem for galvanic corrosion modelled by the two-dimensional Laplace's equation was studied. The considered problem deals with experimental measurements of electric potential, where due to lack of data, numerical integration is impossible. The problem is reduced to the determination of unknown complex coefficients of approximating functions, which are related to the known potential and unknown current density. Employing continuity of those functions along subdomain interfaces and using condition equations for known data leads to an over-determined system of linear algebraic equations which are subjected to experimental errors. Reconstruction of current density is unique. The reconstruction contains one free additive parameter which does not affect current density. The method is useful in situations where limited data on electric potential are provided.

Keywords: current density, potential measurements, reconstruction, complex variables.

1 Introduction

The evaluation of field of current density is essential in problems of galvanic corrosion. In many cases the direct measurement of current density is not feasible, while the electric potential can be obtained from experimental measurements. This is particularly true in the case of cathodic protection systems in general, where many surveying techniques (for example, DCVG and CIS for underground structures) rely in potential measurements at different points at the electrolyte in order to identify the current distribution along the metallic structures.

In these cases, it is common that the region of interest for collecting data is not accessible to the operator and in order to exploit the available measurements some indirect method is required.

There are many studies that employ numerical methods for the forward modelling of galvanic corrosion problem. In order to obtain or verify results, these techniques are based mainly on boundary value problems (BVP) formulations, such as the finite element method (FEM), finite difference method (FDM) or boundary element method (BEM). These methods are successfully used and shown to be very accurate to solve BVPs. Some of them are also implemented in commercial software.

Situations may arise, where the boundary conditions are unknown and only some experimental data in certain locations are known. In this case, the problem is defined as an inverse one. This situation often occurs in many branches of science and mathematics where only the values of some model parameters can be obtained from observed data or measured data. Data on electric potential can be obtained in galvanic corrosion as a set of discrete data with one free parameter due to measuring potential differences. This situation, where measurements on electric potential can be provided as a set of discrete data within simply connected domain Ω imposes the problem to be inverse.

The theory of complex variables [1, 2] is used in order to connect current density and electric potential with one holomorphic function. The real part of it represents electric potential and the derivative is related to current density. The domain is divided into smaller subdomains where the holomorphic function is approximated with a quadratic function and its derivative with a piecewise linear function. These approximating functions obey continuity across subdomain interfaces.

A similar method was used in Irša and Galybin [3] with heat flow and in Galybin and Mukhamediev [4] and Mukhamediev et al. [5] with stress trajectories. The purpose of this work is to develop an efficient method for identification of current densities within a domain where some data on electric potential are provided.

2 Galvanic current density distribution in electrolyte

Current density is a measure of the density of flow of a conserved charge. The equation governing the distribution of potential and current flow in an electrolyte can be derived from the continuity equation governing charge conservation. The divergence of the current density **I** is equal to the negative rate of change of the charge density q [6]:

$$\nabla \cdot \mathbf{I} = -\frac{\partial q}{\partial t} \tag{1}$$

For steady state current distribution, $\partial q/\partial t = 0$ in uniform isotropic electrolyte, where the conductivity of electrolyte k is constant and the electric potential ϕ satisfies Laplace's equation:

$$\Delta\phi = 0 \tag{2}$$

The current density can be defined at any point as:

$$I_j = -k\frac{\partial\phi}{\partial x_j} \tag{3}$$

The fact that the electric potential $\phi(x, y)$ is harmonic, as a consequence of eqn (2), allows the use of the theory of complex variables for 2D potential problems. Introducing the harmonic conjugate function $\psi(x, y)$, where both ϕ and ψ satisfy Cauchy-Riemann equations (CR), one has the holomorphic function:

$$W(z) = \phi(x, y) + \mathrm{i}\psi(x, y) \tag{4}$$

where the real part is the electric potential and the imaginary part is a function harmonic conjugated, related to fluxes. The surfaces $\phi(x, y)$ = constant are equipotential and $\psi(x, y)$ = constant represent flux lines. Using CR and transformation rules in calculus, one can see that the derivative of holomorphic function W is related to the current density as follows:

$$H(z) \equiv \frac{\partial W}{\partial z} = \frac{\partial\phi}{\partial x} - \mathrm{i}\frac{\partial\phi}{\partial y} \tag{5}$$

Comparison of eqns (3) and (5) shows that derivative of holomorphic function, which has a real part equal to the electric potential is related to the conjugate current density $\tilde{I}$ by

$$H(z) = \tilde{I}\big/(-k).$$

3 Problem description

The problem is to recover the current density, based on experimental data for the electric potential. Let us consider a domain Ω, to be a simply connected domain or a subdomain of a bigger domain which is not necessarily simply connected and bounded. Let potential measurements be known at some discrete points (j = 1, …, N), located within Ω. The general problem is formulated as follows: *given the discrete data on electric potential* ϕ_j = 1 ... N, *find the field of current density within the domain* Ω. There are no other restrictions, such as data type or boundary conditions. The data can be redistributed uniformly or randomly.

3.1 On the uniqueness of the reconstructed solution

The problem deals with reconstruction of a holomorphic function, where the real and imaginary parts are harmonic and satisfy CR. Therefore having one function known, the second one can be easily obtained by integration of CR. Indeed, assume, for instance, that the real part of the holomorphic function is known analytically $\phi(x, y)$. CR allows finding the imaginary part by integration:

$$\psi(x,y) = \int \frac{\partial \varphi(x,y)}{\partial x} \mathrm{d}y + p \tag{6}$$

This formula indicates that the reconstruction can be achieved but it will always have one undetermined parameter as a free parameter. In this particular case, the free parameter is an additive constant in the imaginary part $\psi(x, y)$ of holomorphic function $W(z)$. After performing the differentiation of $W(z)$, the free parameter vanishes. It is therefore obvious that from the electric potential, one can recover the current density without any free parameter. In the case when data are known at discrete points, reconstruction, in general, is non-unique and can have any finite number of free parameters, which is evident from the following. Let an exact solution $f(z) = u(x, y) + iv(x, y)$ of the problem be found by any method. It is obvious, that the modulus must be real; therefore it satisfies the following:

$$\mathrm{Im}\left[f(z_j)e^{-i\omega(z_j)} \right] = 0, j = 1 ... N \tag{7}$$

where ω is the argument of $f(z)$. Let us consider a function $P(z)$ in the form:

$$P(z) = \prod_{j=1}^{m} C_j (z - z_j)^{m_j} \tag{8}$$

where C_j are arbitrary complex parameters, m_j are arbitrary positive integers and z_j are the data points. It is evident from eqn (8) that the sum, $f(z) + P(z)$, satisfies condition (7) and therefore it is also the solution of the problem.

In order to be meaningful, the solution has to be sought within a certain class of functions. Hereafter the solution is sought as a set of piecewise linear holomorphic functions. It will be shown that this set of functions allows reconstructing the current density with high accuracy.

4 Mathematical model

A similar mathematical formulation has been described in Irša and Galybin [3], where the required function was related to the modulus of holomorphic function. In this case the required current density is related directly to the derivative of the holomorphic function $W(z)$.

The domain Ω is divided into n smaller subdomains (elements) of an arbitrary shape, as shown in Fig. 1. In order to achieve a piecewise linear function for the current density, the function $W(z)$ must be approximated by a quadratic function. Both these approximations obey continuity along the interfaces (at certain collocation points located on the interfaces).

For function $H(z)$, related to current density, the approximating function is required to be the piece-wise linear function:

$$\chi_H^{(m)}(z) = a^{(m)} + b^{(m)} z, \quad m = 1 \dots n \tag{9}$$

For function $W(z)$, which has real part the electric potential, the approximating function is therefore the integral of $H(z)$:

$$\chi_W^{(m)}(z) = c^{(m)} + a^{(m)}z + \frac{1}{2}b^{(m)}z^2, \quad m = 1 \ldots n \tag{10}$$

Here $a^{(m)} = a_0^{(m)} + ia_1^{(m)}$, $b^{(m)} = b_0^{(m)} + ib_1^{(m)}$, $c^{(m)} = c_0^{(m)} + ic_1^{(m)}$ are 3 unknown complex constants; therefore 6 real constants are unknown in every element. The method is not restricted to these two particular approximating functions, however it will be demonstrated that the use of these two particular functions provides results with very high accuracy.

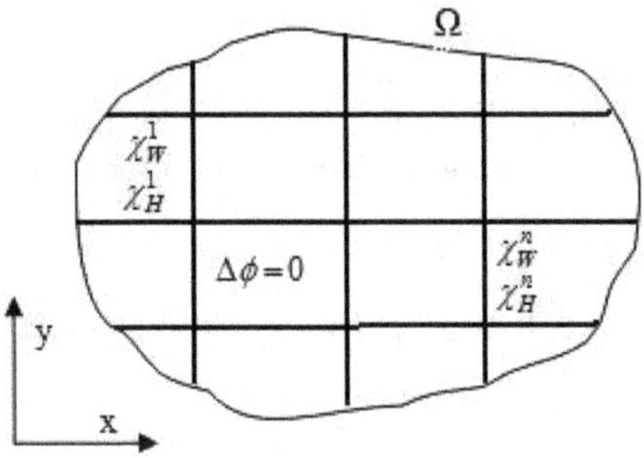

Figure 1: Discretization of the domain with approximating functions.

4.1 Discretization of the domain

In order to obey the continuity of the above mentioned functions, a grid was used consisting of square elements with collocation points (CP) on their interfaces. The CPs are distributed regularly.

The minimum number of nodes strictly depends on the number of unknowns in the approximating functions as well as the element type and number of known data. Should the system consist only from the continuity equations and no data imposed on electric potential, one can find the number of required collocation points as follows. Given that, the square elements are placed in the regular grid with J elements along x-axis and K elements along y-axis and 6 unknown real coefficients in every element, one finds the total number of unknowns N_{unkn} as:

$$N_{unkn} = 6N_{elem} = 6JK \tag{11}$$

The required number of collocation points depends on the number of element interfaces, which is $N_{\text{int}} = (J - 1)K + (K - 1)J$. Using the same number of collocation points per each interface, keeping in mind that separation of complex equations into real and imaginary parts presents two real equations at every collocation point, the required number of collocation points per interface is:

$$N_{cp} \geq \frac{N_{unkn}}{2_{\text{int}}} = \frac{6JK}{2[(J-1)K + (K-1)J]} = \frac{3JK}{2JK - J - K} \tag{12}$$

For a domain, where J or K is equal to 1, the number of collocation points per interface is $N_{cp} \geq 3$. In the general case, the number of collocation points per interface $N_{cp} > 1$, therefore $N_{cp} = 2$ is used further on. Given dense data on the electric potential, one could reduce the number of CP with respect to the unknowns and the total number of equations.

Other types of mesh could be used, such as triangular or polygonal with straight or curvilinear sides. The square grid is convenient from computational point of view because the distances between collocation points are regular, which has good influence on the structure of the matrix of the system.

4.2 Equations of continuity and condition equations

The problem consists of continuity equations and two types of condition equations. For the *k-th* collocation point, lying on the interface between the elements numbered *m* and *m+1*, the equation of continuity for *W*(*z*) is the following:

$$\chi_W^{(m)}(z_k) - \chi_W^{(m+l)}(z_k) = 0, \quad m = 1 \ldots n; \quad k = 1 \ldots N_{CP} \tag{13}$$

The second equation of continuity for *H*(*z*) is expressed following:

$$\chi_H^{(m)}(z_k) - \chi_H^{(m+l)}(z_k) = 0, \quad m = 1 \ldots n; \quad k = 1 \ldots N_{CP} \tag{14}$$

The condition equation containing data on electric potential ϕ comes from *W*(*z*), where the real part of it is equal to the electric potential, therefore:

$$\mathrm{Re}\left(\chi_W^{(m)}(z_j)\right) = \varphi_j, \quad j = 1 \ldots N \tag{15}$$

This condition equation is satisfied in elements, where the data occurs. It is obvious that from the collocation method, one can find the differences between interfering elements coefficients only. In the condition equation (15) all coefficients are presented except the imaginary part of *c,* which stays undetermined and cannot be obtained uniquely from the continuity equations. This coefficient is the free additive parameter discussed in section 3.1. This would decrease the rank of matrix and would require regularisation procedures. In order to avoid these procedures which contain additional parameters as regularization parameters among others, a second condition equation is used. The second condition equation is due to the free parameter c_1; it requires the average of all constants c_1 across whole domain to be zero. The condition is expressed as follows:

$$\sum_{i=1}^{N} \mathrm{Im}\left(c^{(i)}\right) = 0 \tag{16}$$

A system of linear algebraic equations is formed from eqns (13)–(16).

4.3 Solution of the linear system

By extracting the real and imaginary parts of all equations, that is, equations of continuity and condition equations, the real system of linear algebraic equations can be obtained and rewritten in a matrix form:

$$\mathbf{Ax} = \mathbf{b} \tag{17}$$

where $\mathbf{A} \in R^{m \times n}$, $\mathbf{b} \in R^m$ and $m > n$; $\mathbf{x}$ is the vector of the unknown real coefficients, of length n. The vector $\mathbf{x}$ consists of real and imaginary parts of unknown complex coefficients. Vector $\mathbf{b}$ is known exactly, while $\mathbf{A}$, the matrix of the system of linear algebraic equations, depends on CPs as well as the element type and size. Matrix $\mathbf{A}$ is not a square, the system is over-determined and therefore the left-hand side $\mathbf{Ax}$ does not exactly equal to $\mathbf{b}$; thus the system is inconsistent. However an approximate solution, $\mathbf{x}^*$, can be found by means of the least squares method [5] that minimise the residual:

$$\left\| \mathbf{Ax}^* - \mathbf{b} \right\|_2 \leq \varepsilon \tag{18}$$

where $\|...\|_2$ stands for the L_2 norm. If the arising system is well-posed and not large, then the inversion of the matrix does not meet any difficulties and the solution takes the form [7]:

$$\mathbf{x}^* = (\mathbf{A}^{\mathrm{T}}\mathbf{A})^{-1}\mathbf{A}^{\mathrm{T}}\mathbf{b} \tag{19}$$

The condition number *CN* is used in the numerical examples to control well-posedness of the system of linear algebraic equations. In the case of high *CN*, the methods for ill-posed problems should be used [8].

5 Numerical analysis

For the purpose of testing the methodology, an artificial function was chosen in the form:

$$W(z) = 54\left(z - \frac{\mathrm{i}}{3}\right) - 30\left(z - \frac{\mathrm{i}}{3}\right)^3 \tag{20}$$

Taking the real part and derivative of eqn (20), one has the electric potential and current density for the ideal function, which are used as given electric potential data and for comparison with the numerically obtained results for current density. The conductivity of the electrolyte is a multiplicative parameter to the current density, which was chosen as $k = 1$. Different types and structure of data were tested (ideal data and data subjected to errors).

5.1 Ideal data

This example used data obtained from eqn (20). It used 81 data uniformly redistributed within the domain and 25 elements as shown in Fig. 2.

The reconstructed electric potential and current are shown in Figs. 3 and 4. This configuration revealed accuracy with average error 4.7% by using 25 elements only. From Fig. 5, the improvement with refinement can be seen in residual (*r*/Equations), maximum and average errors as well as the behaviour of condition number *CN*.

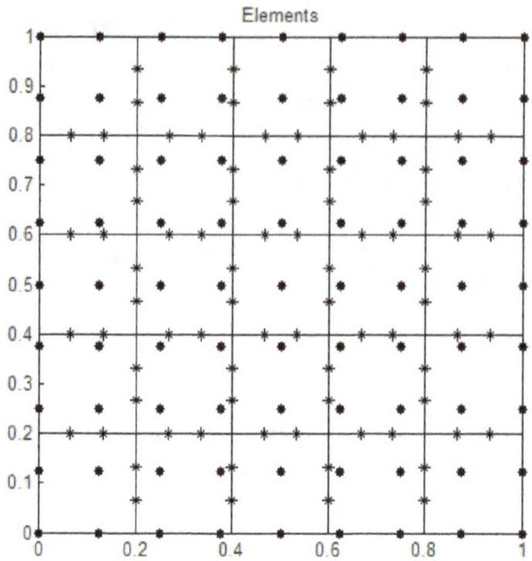

Figure 2: Discretized domain with given data on electric potential.

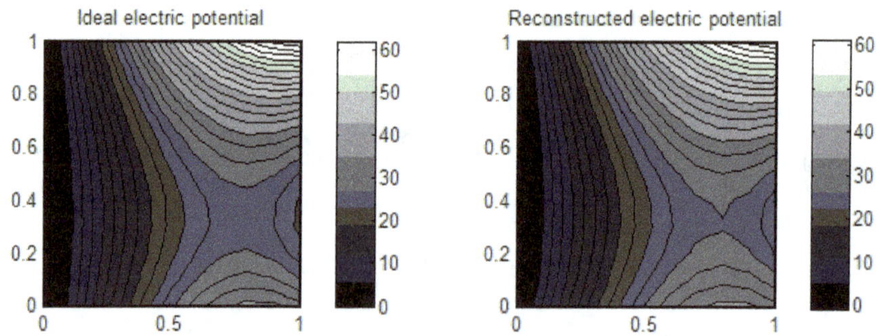

Figure 3: Ideal electric potential (left) and reconstructed electric potential (right).

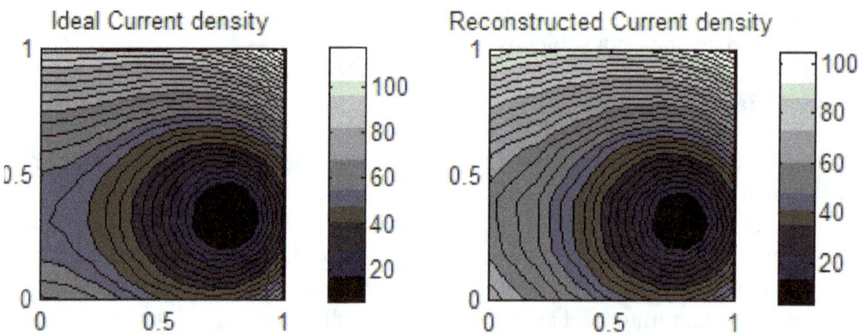

Figure 4: Ideal current density (left) and recovered current density (right).

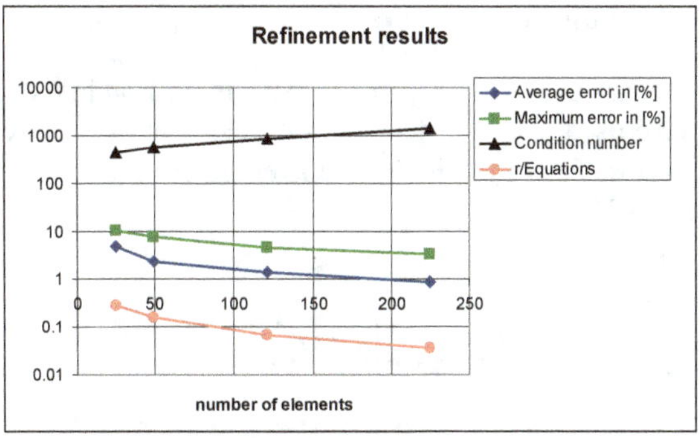

Figure 5: Refinement results for 81 ideal data points.

5.2 Data subjected to errors

Random error was introduced to the potential data to simulate experimental errors. The error used was ±10% on 81 uniformly redistributed data items and 49 elements. The results for electric potential and current density are shown in Figs. 6 and 7. The average error was 2.9%; the condition number was 571 and *r*/Equations was 0.2943.

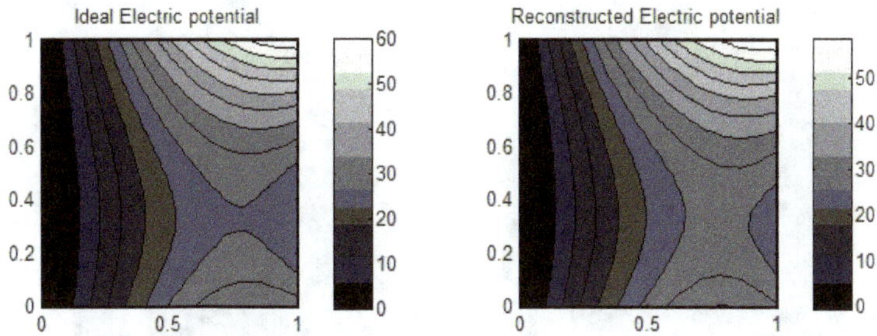

Figure 6: Ideal electric potential (left) and reconstructed electric potential from data subjected to errors (right).

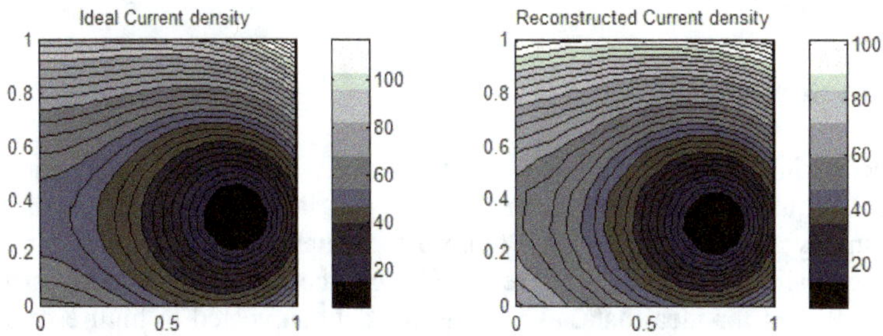

Figure 7: Ideal current density (left) and recovered current density from data subjected to errors (right).

5.3 Random data subjected to errors

The measurements are usually randomly redistributed. Random data subjected to errors ±10% were used; 81 data items and 49 elements were used as shown in Fig. 8. The results are shown in Fig. 9. The average error was 3.8%; *CN* was 754 and *r*/Equations = 0.2672.

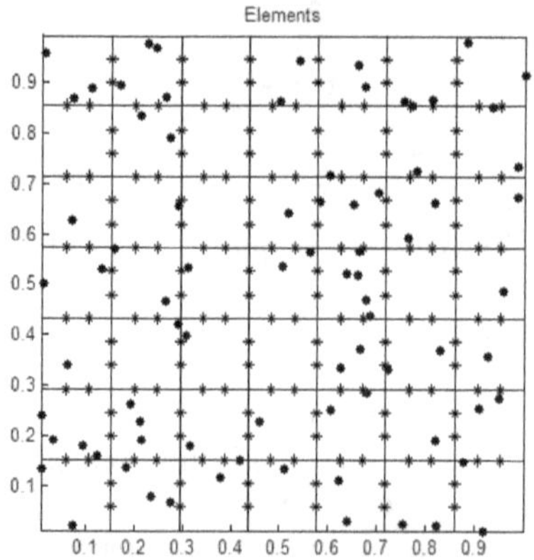

Figure 8: 81 random data subjected to errors and discretization.

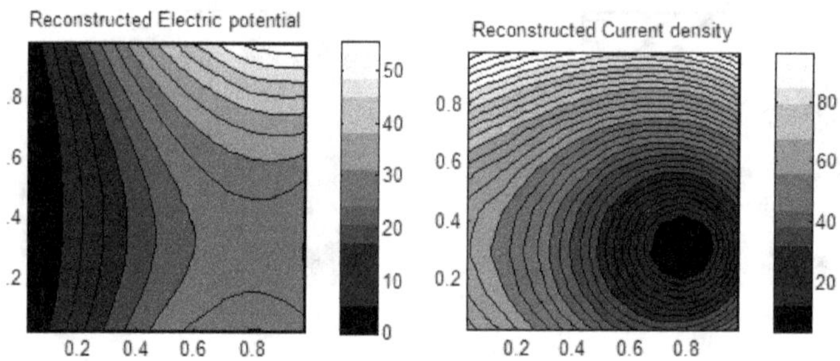

Figure 9: Reconstructed electric potential (left) and recovered current density (right).

6 Conclusion

The recovering of current density from data on electric potential, satisfying Laplace's equation was studied. In experiments, it is difficult or expensive to obtain many measurements and therefore numerical integration cannot be performed. The recovered results revealed high accuracy with an ideal artificial function *W* and for ideal data, as well as for data subjected to high errors. The method uses complex variable theory where one can obtain an holomorphic function *W* related to the electric potential and its derivative related to the current density.

Acknowledgement

The authors are grateful to EPSRC for the financial support of this work through the Research Grant EP/E032494/1.

References

[1] Rahman, M., *Complex Variables and Transform Calculus*, Computational Mechanics Publications: Southampton, 1997.

[2] Kwok, Y.K., *Applied Complex Variables*, Cambridge University Press: Cambridge, 2002.

[3] Irša, J. & Galybin, A.N., Heat flux reconstruction in grinding process from temperature data. *WIT Transactions on Modelling and Simulation*, **51**, pp. 413–423, 2009.

[4] Galybin, A.N. & Mukhamediev, Sh.A., Determination of elastic stresses from discrete data on stress orientations. *International Journal of Solids and Structures*, **41(18–19)**, pp. 5125–5142, 2004.

[5] Mukhamediev, Sh.A., Galybin, A.N. & Brady, B.H.G., Determination of stress fields in elastic lithosphere by methods based on stress orientations. *International Journal of Rock Mechanics and Mining Sciences*, **43(1)**, pp. 66–88, 2006.

[6] Jia, J.X., Song, G., Atrens, A., St John, D., Baynham, J. & Chandler, G., Evaluation of the BEASY program using linear and piecewise linear approaches for the boundary conditions. *Materials and Corrosion*, **55(11)**, pp. 845–852, 2004.

[7] Golub, G.H. & Van Loan, C.F., *Matrix Computations*, The John Hopkins Press Ltd: London, 1996.

[8] Tikhonov, A.N. & Arsenin, V.Y., *Solution of Ill-posed Problems*, Winston, Wiley: New York, 1977.

Author index

Abd El-Moneim A. 81
Abd-Elrhman Y. M. 81
Abosrra L. 107
Abud I. 61
Adey R. A. 245, 259
Amaya K. 211
Ashour A. F. 107
Ayllón E. S. 27, 69

Barbosa C. 61
Barnard N. C. 161
Bauer M. 15
Baynham J. M. W. 245, 259
Beloufa M. A. 47
Bigdeli Karimi M. 173
Brown S. G. R. 161

Caminha I. 61
Carcea A. G. 15

Galybin A. N. 289
Garland N. 3
Gepreel M. A. 81

Hadfield M. 39
Hirano M. 149
Hou B. 221
Hu J. 183
Hussein R. O. 133

Irša J. 289

Keide H. 61
Khan Z. 3, 39
Kimura Y. 91, 149
Kobayashi S. 81
Kolev H. 183
Koleva D. A. 183
Kunz M. 15

Lan Z. 221
Leiro M. 27, 69
Li B. 197

McIntyre N. S. 15
Mitchell S. C. 107
Muharemović Ad. 231
Muharemović Ai. 231
Mujezinović A. 231

Nakayama A. 211
Nascimento J. L. 61
Newman R. C. 15
Nie X. 133
Northwood D. O. 119, 133, 173

Peratta A. B. 245, 259
Peratta A. 289

Qin J. 15

Roesler C. R. M. 61

Saeed A. 3, 39
Sherry N. 15
Simpson T. 15
Smith R. 3
Stoilov V. 119, 173

Tamura N. 15
Toloei A. S. 119
Turković I. 231

Ulaganathan J. 15

Van Breugel K. 183

Wang X. 221
Wilton-Smith K. 39
Wu J. 271, 281

Xing S. 271, 281

Yamamoto N. 211
Yan Y. 271
Youseffi M. 107
Yun F. 281

Zhang G. 197

www.ingramcontent.com/pod-product-compliance
Lightning Source LLC
LaVergne TN
LVHW011848200926
848484LV00002B/419
9781784662493